Advancing Positioning, Navigation, and Timing (PNT) Service Using Satellite Navigation Technology

Advancing Positioning, Navigation, and Timing (PNT) Service Using Satellite Navigation Technology

Guest Editors

Ke Su
Liangliang Yuan
Yulong Ge
Amir Allahvirdi-Zadeh
Guo Chen

MDPI

Basel • Beijing • Wuhan • Barcelona • Belgrade • Novi Sad • Cluj • Manchester

Guest Editors

Ke Su
School of Space Information
Space Engineering University
Beijing
China

Liangliang Yuan
Shanghai Astronomical
Observatory
Chinese Academy of Sciences
Shanghai
China

Yulong Ge
School of Marine Science and
Engineering
Nanjing Normal University
Nanjing
China

Amir Allahvirdi-Zadeh
School of Earth and Planetary
Sciences
Curtin University
Perth
Australia

Guo Chen
GNSS Research Center
Wuhan University
Wuhan
China

Editorial Office
MDPI AG
Grosspeteranlage 5
4052 Basel, Switzerland

This is a reprint of the Special Issue, published open access by the journal *Remote Sensing* (ISSN 2072-4292), freely accessible at: https://www.mdpi.com/journal/remotesensing/special_issues/KK817ZF9W5.

For citation purposes, cite each article independently as indicated on the article page online and as indicated below:

Lastname, A.A.; Lastname, B.B. Article Title. *Journal Name* **Year**, *Volume Number*, Page Range.

ISBN 978-3-7258-6390-7 (Hbk)
ISBN 978-3-7258-6391-4 (PDF)
https://doi.org/10.3390/books978-3-7258-6391-4

Cover image courtesy of Ke Su

Contents

About the Editors

Ke Su

Ke Su is currently working at the Space Engineering University, China. He has been researching BDS/GNSS satellite navigation technologies and applications. He received a PhD degree from Shanghai Astronomical Observatory, Chinese Academy of Sciences, China. He is a member of the IAG prediction of ionospheric state and dynamics workgroup. He has published over 40 papers, among which more than 20 are SCI papers as the first or corresponding author, with a total citation count exceeding 600 and an H-index of 15. He has been appointed as a Young Editorial Board Member for *Satellite Navigation* and the *Journal of Projectiles, Rockets, Missiles and Guidance*, as well as a Guest Editorial Board Member of *Remote Sensing*. He has been a reviewer of more than 20 international journals, including *IEEE Transactions on Geoscience* and *Remote Sensing, Journal of Geodesy*, and *GPS solutions*, and so on. He has been awarded the national scholarship for doctoral students and master's degree candidate, the scholarship for the president of the CAS, the young excellent paper award at the international summit on BDS mass applications, and the young excellent paper award and the excellent oral presentation award at the China Satellite Navigation Conference.

Liangliang Yuan

Liangliang Yuan is an Associate Researcher at the Shanghai Astronomical Observatory, Chinese Academy of Sciences, China. His research focuses on ionospheric data assimilation and upper atmospheric coupling. He has led and contributed to over 10 research projects funded by prominent institutions, including the European Space Agency (ESA), the German Aerospace Center, the National Natural Science Foundation of China, and the Shanghai Science and Technology Commission. He has published more than 20 journal and conference papers. Additionally, he has been selected for prestigious talent programs such as the Shanghai Baiyulan Young Talent Program and the Pujiang Talent Program.

Yulong Ge

Yulong Ge is an Associate Professor at the School of Marine Science and Engineering, Nanjing Normal University, China. He holds several notable appointments, including Deputy Director of Science and Technology for Jiangsu Province, High-level Innovative and Entrepreneurial Talent (Double Innovation Doctor) of Jiangsu Province, and Committee Member of the Jiangsu BeiDou Satellite Navigation Standardization Technical Committee. He also serves as a Youth Editorial Board Member for *Satellite Navigation*. His research focuses on "BeiDou+" intelligent high-precision positioning and timing, marine surveying and mapping, and the development of intelligent equipment. Over the past five years, he has led or participated in more than 10 national, provincial, and local projects. As a first author or corresponding author, he has published over 20 SCI papers in journals including *Satellite Navigation, GPS Solutions*, and *Measurement*, and has co-authored more than 30 additional SCI papers.

Amir Allahvirdi-Zadeh

Amir Allahvirdi-Zadeh is a lecturer at the School of Earth and Planetary Sciences at Curtin University, Australia. His research focuses on the precise orbit determination of small satellites for space missions (Earth and beyond) and emerging LEO-PNT systems.

Guo Chen

Guo Chen is an Associate Researcher at the GNSS Research Center of Wuhan University. He is a key member of the International GNSS Service (IGS) Wuhan Combination Center (WCC) and has long been dedicated to research on BeiDou/GNSS product combination, performance evaluation, and precise positioning.

remote sensing

Editorial

Advancing Positioning, Navigation, and Timing (PNT) Service Using Satellite Navigation Technology

Ke Su [1,2,*], Liangliang Yuan [3], Yulong Ge [4], Amir Allahvirdi-Zadeh [5] and Guo Chen [6]

1 School of Space Information, Space Engineering University, Beijing 101416, China
2 Key Laboratory of Smart Earth, Beijing 100094, China
3 Shanghai Astronomical Observatory, Chinese Academy of Sciences, Shanghai 200030, China; llyuan@shao.ac.cn
4 School of Marine Science and Engineering, Nanjing Normal University, Nanjing 210023, China; geyulong@njnu.edu.cn
5 School of Earth and Planetary Sciences, Curtin University, Perth, WA 6845, Australia; amir.allahvirdizadeh@curtin.edu.au
6 GNSS Research Center, Wuhan University, Wuhan 430079, China; guo_chen@whu.edu.cn
* Correspondence: suke17@mails.ucas.ac.cn

1. Introduction

As a pivotal spatiotemporal infrastructure in the modern information society, satellite navigation provides global users with high-precision, all-weather, and round-the-clock Positioning, Navigation, and Timing (PNT) services. It constitutes an indispensable spatial infrastructure for today's national economic development and defense construction. Satellite navigation systems have found widespread applications across various sectors, including transportation, surveying and mapping, geographical information, meteorology, land and resources surveys, precision agriculture, and seismology, permeating every facet of the national economy.

Currently, four Global Navigation Satellite Systems (GNSSs) have fully completed their satellite constellation deployments, while two regional navigation satellite systems (RNSSs) and several satellite-based augmentation systems have also been largely established, with other satellite navigation systems are under planning. The discipline of satellite navigation emerged from the interdisciplinary convergence of geometry, electromagnetism, mechanics, and geodesy. With the expansion of its application domains and an increase in its significance, it has evolved into a relatively independent discipline. Looking ahead, the emergence of perception methods based on new physical principles is expected to give rise to new disciplines.

PNT serves as crucial foundational information for economic development and national defense advancement, as well as vital supporting information for the development of an intelligent society. PNT has permeated every aspect of our lives, from daily commuting to national security, and from scientific research to engineering applications, underscoring its indispensable status and role. With the increasing complexity of spatiotemporal measurement methods, PNT technology has gradually demonstrated a trend of multidisciplinary integration. In particular, with the emergence of satellite navigation being a colossal and complex system, the interdisciplinary characteristics of the PNT field have become even more pronounced. However, relying solely on any single PNT technology cannot meet the PNT needs of arbitrary users in any scenario. It is imperative to construct a PNT infrastructure network that provides full-domain coverage and integrates multiple physical principle-based information sources, forming a PNT application and service system characterized by multi-source data fusion and complementary advantages

 https://doi.org/10.3390/rs17233909

of multiple technologies. This not only represents an inevitable trend in the development of the international PNT system but also serves as the necessary path to promote PNT applications and enhance PNT service performance.

Currently, integrated PNT addresses the issue of PNT information sources, resilient PNT focuses on the optimal combination of PNT information sources, and micro-PNT tackles the challenges of miniaturization and low power consumption in PNT terminals. For the vast number of PNT users, having PNT infrastructure and resilient guidelines is not sufficient. It is also essential to address the issue of intelligent applications.

The Special Issue is dedicated to exploring the wide-ranging PNT technologies based on satellite navigation. We invite original research articles and reviews that delve into various aspects including, but not limited to, constellations, signals, orbit determination, PNT theory, algorithms, models, and their applications in engineering and Earth sciences, as well as multi-sensor integration. Topics of interest for this Special Issue include, but are not limited to, integrated PNT, resilient PNT, micro PNT, secure PNT, GNSS Precise Point Positioning (PPP) and PPP-Real-Time Kinematic (PPP-RTK), GNSS timing, GNSS orbiting determination and modeling, GNSS atmospheric sensing, GNSS ionosphere and space weather, integrated navigation and smart applications, satellite navigation countermeasure, future Low Earth Orbit (LEO) PNT, and broadband PNT constellations using signal of opportunity.

The Special Issue received a total of 38 submissions and ultimately published 9 research papers, all of which have undergone a rigorous review process. The objective of this Editorial is to provide an overview of the contributions from the studies in this Special Issue. Section 2 summarizes the individual articles hosted in the Special Issue in alphabetical order based on the publication time and article type, and Section 3 outlines some concluding remarks.

2. Overview of Contributions

Jiao et al. (Contribution 1) introduced a novel undifferenced model integrated with satellite clock offsets, which further transforms the inter-frequency clock bias (IFCB) into time-variant mixed observable-specific signal biases (OSBs) for both code and phase measurements. This innovation addresses the limitations inherent in traditional approaches while streamlining the bias correction process for multi-frequency PPP. The proposed model not only enhances the accuracy of mixed OSBs but also eliminates the adverse effects of receiver-dependent time-varying biases on satellite-based mixed OSB estimation. Quantitative evaluations demonstrate significant improvements, where the standard deviation (STD) and root mean square (RMS) of original OSBs are reduced by 7.5–60.9% and 9.4–66.1%, respectively, while those of epoch-differenced (ED) OSBs are improved by 50.0–87.5% and 60.0–88.9% for STD and RMS. Furthermore, PPP solutions utilizing the new mixed OSBs outperform those relying on conventional IFCB products. The findings validate the reliability, practicality, and efficacy of the proposed time-variant mixed OSB framework and the undifferenced model with integrated satellite clock offsets for multi-frequency PPP applications.

Tao et al. (Contribution 2) proposed a modified Multipath Hemispherical Map (MHM) approach to mitigate the multipath for single-frequency multi-GNSS tightly combined positioning. The method partitions the hemispherical space into a 36×9 grid with $10^\circ \times 10^\circ$ angular resolution, then utilizes elevation and azimuth angles as search parameters to determine the multipath correction value from the nearest grid point. To enhance computational efficiency without compromising accuracy, a k-dimensional tree structure is implemented for rapid nearest-neighbor searches. Experimental results demonstrate that the modified MHM improves mean positioning accuracy by 10.20%, 10.77%, and

9.29% for Global Positioning System (GPS), BeiDou Navigation Satellite System (BDS), and Galileo single-difference residuals, respectively, compared to conventional advanced sidereal filtering-based corrections. Precision enhancements reach 32.82%, 40.65%, and 31.97% in the east, north, and up components, respectively. The modified MHM exhibits superior performance and more consistent error mitigation across different satellite systems.

Zhao and Wei (Contribution 3) conducted a performance assessment of short-duration PPP using PPP-B2b correction products and introduced a backward smoothing technique to improve positioning accuracy during the initial convergence phase. The analysis revealed that the orbit and clock precision of BDS PPP-B2b products outperform those of GPS, with BDS achieving radial, along-track, and cross-track orbit accuracies of 0.059 m, 0.178 m, and 0.186 m, respectively, and clock synchronization accuracy within 0.13 ns. For hourly static PPP solutions, 50% of test sessions reached horizontal and vertical accuracy of 0.5 m and 0.1 m after convergence times of 4.5 min and 25 min, respectively. However, 7.07% to 23.79% of sessions failed to achieve 0.1 m accuracy due to insufficient GPS and BDS correction availability at specific stations. Simulated kinematic PPP required an additional 1–4 min to attain comparable accuracy to static PPP. The implementation of backward smoothing significantly enhanced positioning precision, achieving north, east, and up direction accuracies of 0.024 m, 0.046 m, and 0.053 m, respectively. In vehicle-based applications, forward PPP attained horizontal accuracy better than 0.5 m within 4 min, though errors could exceed 1.5 m and 3.0 m for east and up components during convergence. With smoothing applied, horizontal accuracy improved to better than 0.2 m, while vertical accuracy reached sub-0.3 m levels.

Chen et al. (Contribution 4) introduced an adaptive robust filtering algorithm featuring an optimized fading factor to address the divergence issues of traditional Kalman filters under severe heave motion and anomalous observation conditions. A multi-source information fusion framework was developed by integrating four key indicators, including satellite Positioning Dilution Of Precision (PDOP), solution quality metric (Q-value), effective satellite observation count (Satnum), and residual vector analysis. A dynamic weight adjustment mechanism was designed to enable real-time optimization of the fading factor, while a dual robust mechanism was constructed through the synergistic integration of robust estimation theory and adaptive filtering, employing a sequential update strategy. During the measurement update phase, observation weights were dynamically adjusted based on innovation covariance, and a fading memory factor was introduced in the time update phase to mitigate model error accumulation. Experimental evaluations demonstrated significant accuracy improvements compared to conventional Extended Kalman Filter (EKF), Sage-Husa adaptive filtering, and basic robust filtering approaches. In high-maneuver vehicle scenarios, three-dimensional positioning accuracy increased by 47.12%, 35.26%, and 9.58%, respectively, while shipborne heave motion scenarios showed corresponding improvements of 19.44%, 10.47%, and 8.28%. These findings provide an effective anti-interference solution for navigation systems operating in highly dynamic and complex environments.

Deng et al. (Contribution 5) proposed an orthogonal frequency division multiplexing (OFDM)-based LEO navigation system and conducted a detailed performance analysis. Leveraging 5G New Radio (NR) as the satellite transmission signal, the study identified NR signal components suitable for navigation services while introducing a novel zero-padding correlation (ZPC) receiver design. This ZPC technique effectively eliminates cyclic prefix (CP) and inter-carrier interference (ICI), significantly enhancing signal tracking precision. The power spectral density (PSD) of the NR navigation signal was mathematically derived, followed by a comprehensive evaluation of tracking accuracy under varying NR configurations, including bandwidth, spectral allocation, and signal composition parameters. An extended Kalman filter (EKF) was implemented to integrate pseudorange and pseudorange rate measurements for real-time positioning solutions. Simulation results revealed substantial improvements over conventional receivers, with ranging precision enhanced by 80% and positioning accuracy improved by 88.3%. The proposed ZPC receiver demonstrated centimeter-level navigation capabilities. This research provides a systematic analysis of LEO NR system navigation performance and offers valuable insights for the design of future LEO-based PNT architectures.

Wu et al. (Contribution 6) introduced a robust approach that leverages narrow-lane ambiguity resolution to compensate for receiver clock discontinuities, coupled with an adaptive sliding-window weighting strategy that optimally utilizes multi-epoch observational data. The methodology effectively addresses day-boundary jumps by implementing dynamic thresholding for error detection and minimizing the impact of incorrect ambiguity fixing. Experimental evaluations demonstrated that at an averaging interval of 76,800 s, the frequency stability for GPS, Galileo, and BDS integer precise point positioning (IPPP) solutions reached 4.838×10^{-16}, 4.707×10^{-16}, and 5.403×10^{-16}, respectively. For zero-baseline time transfer applications, GPS IPPP achieved stability at the 10^{-17} level, outperforming optical fiber time transfer in long-term differential comparisons. Across both short- and long-baseline scenarios, IPPP consistently surpassed conventional PPP float solutions and International GNSS Service (IGS) final products. Specifically, at a 307,200 s averaging interval, IPPP improved average frequency stability by approximately 29.3% relative to PPP and 32.6% compared to IGS final products.

Zheng et al. (Contribution 7) conducted a thorough evaluation of the Adaptive Stochastic Model (ASM) and its effect on PPP. The study first detailed the implementation of ASM using Variance Component Estimation (VCE) techniques. Experimental results revealed that ASM effectively captures varying observation conditions through dynamically estimated variance components, thereby enhancing both PPP float and fixed solutions particularly when predefined stochastic models prove insufficient. The approach also improved cycle-slip detection performance by tightening the stochastic constraints, reducing the missed detection rate from 19% to 8%. Furthermore, ASM accelerated both direct reconvergence and re-initialization processes following data interruptions, with reconvergence times shortened by 18% and re-initialization times by 55%, respectively.

Fan et al. (Contribution 8) introduced a tightly coupled integration model combining Pseudolite System (PLS) and GNSS at the observation level, where ambiguity resolution strategy plays a critical role in augmentation performance. The study proposes a novel ambiguity resolution approach that leverages the rapid convergence characteristics of PLS by independently fixing PLS ambiguities first, followed by cascading resolution of GNSS wide-lane and L1 ambiguities. The fixed ambiguities are subsequently incorporated as constraints in the filtering process to enhance solution robustness. Experimental evaluations demonstrated significant improvements in ambiguity fixing rates, particularly for short-duration augmentation scenarios, where the proposed method outperformed conventional approaches.

Lv et al. (Contribution 9) developed comprehensive modeling frameworks for both traditional and Artificial Intelligence (AI) approaches in GNSS PNT. The study systematically reviewed classical mathematical models, including polynomial, gray, Kalman filter, and time series methods, alongside AI-based techniques such as machine learning (ML), multilayer perceptron (MLP), recurrent neural networks (RNNs), and Transformer architectures. The technical attributes, practical applications, and inherent limitations of each model type were analyzed in depth. While AI-based models exhibited superior adaptability and performance in complex environments compared to classical methods, they demanded larger training datasets and significant computational resources. The research concluded with a comparative summary of strengths, weaknesses, and future development trajectories, providing actionable insights for advancing real-time, high-precision GNSS PNT solutions.

3. Concluding Remarks

The papers included in this Special Issue collectively demonstrate that field of PNT technology encompasses multiple disciplines and involves complex technical challenges. The advancement of PNT technology necessitates proactive integration with cutting-edge technologies such as big data and AI. It also requires research into multi-technology fusion development models and active exploration of PNT sensing technologies based on novel physical principles. Some studies in this Special Issue have proposed improvements to PNT technologies from the perspective of technical algorithms, while others have focused on enhancements through the adoption of novel technologies.

Based on the insights from this Special Issue, multiple promising avenues for future research become apparent as follows:

1. Research on challenges in spatiotemporal reference framework construction under strong interference conditions. The research primarily focuses on the anti-jamming capabilities of navigation systems, addressing adaptability issues in satellite constellations, signal architectures, and receiver performance, as well as the applicability of navigation systems under complex operational conditions [1,2].
2. Research on integrated PNT system technologies. Key research areas include deep-space PNT technologies, design and deployment of Lagrangian constellations, LEO-augmented PNT solutions, integrated navigation combining network communication technologies with GNSS, hybrid navigation approaches integrating inertial, celestial, gravimetric and magnetic, and terrain-matching navigation, along with underwater PNT systems [3–5].
3. Research on resilient PNT theories and methodologies. The research encompasses resilient integration techniques for multi-source PNT information, development of resilient functional and stochastic models for PNT data under complex environments, and establishment of fusion criteria, models, and algorithms for multi-source PNT information resilience [6–8].
4. Key theories and technologies for intelligent PNT service systems. Research priorities include PNT intelligent perception theories and sensor technologies, intelligent fusion of multi-source PNT information, and service architecture development for intelligent PNT systems adapted to diverse environmental conditions [9,10].

It is evident that PNT technology research remains in its nascent stage, with substantial efforts still required in satellite navigation-based PNT studies to enhance the continuity, availability, robustness, and reliability of PNT services.

Author Contributions: Writing—original draft preparation, K.S.; writing—review and editing, L.Y., Y.G., A.A.-Z. and G.C. All authors have read and agreed to the published version of the manuscript.

Funding: This research was funded by Scientific Research Key Laboratory Fund (Grant No. SYS-ZX02-2024-01), Engineering Project Fund (Grant No. 145AXL250004000X), National Natural Science Foundation of China (Grant No. 42505002), Natural Science Foundation of Shanghai (Grant No. 24YF2754300, 24PJA152) and the Faculty of Science and Engineering Research Program at Curtin University (Grant No. 2025).

Acknowledgments: The editors thank all contributing researchers for submitting their original work to this Special Issue and look forward to continued advances in this rapidly evolving field.

Conflicts of Interest: The authors declare no conflicts of interest.

List of Contributions

1. Jiao, G.; Su, K.; Fan, M.; Yang, Y.; Hu, H. BDS-3/GNSS Undifferenced Pseudorange and Phase Time-Variant Mixed OSB Considering the Receiver Time-Variant Biases and Its Benefit on Multi-Frequency PPP. *Remote Sens.* **2024**, *16*, 4433. https://doi.org/10.3390/rs16234433.
2. Tao, Y.; Liu, C.; Tong, R.; Zhao, X.; Feng, Y.; Wang, J. Multipath Mitigation in Single-Frequency Multi-GNSS Tightly Combined Positioning via a Modified Multipath Hemispherical Map Method. *Remote Sens.* **2024**, *16*, 4679. https://doi.org/10.3390/rs16244679.
3. Zhao, L.; Zhai, W. Assessment of PPP Using BDS PPP-B2b Products with Short-Time-Span Observations and Backward Smoothing Method. *Remote Sens.* **2025**, *17*, 25. https://doi.org/10.3390/rs17010025.
4. Chen, Z.; Liu, Y.; Liu, S.; Wang, S.; Yang, L. An Improved Fading Factor-Based Adaptive Robust Filtering Algorithm for SINS/GNSS Integration with Dynamic Disturbance Suppression. *Remote Sens.* **2025**, *17*, 1449. https://doi.org/10.3390/rs17081449.
5. Deng, L.; Yang, Y.; Ma, J.; Wu, T.; Qian, X.; Li, H. Breaking the Cyclic Prefix Barrier: Zero-Padding Correlation Enables Centimeter-Accurate LEO Navigation via 5G NR Signals. *Remote Sens.* **2025**, *17*, 2116. https://doi.org/10.3390/rs17132116.
6. Wu, K.; Qin, W.; Lv, D.; Wu, W.; Wei, P.; Yang, X. Robust and Adaptive Ambiguity Resolution Strategy in Continuous Time and Frequency Transfer. *Remote Sens.* **2025**, *17*, 2878. https://doi.org/10.3390/rs17162878.
7. Zheng, Y.; Sun, Y.; Zhou, Y.; Wang, S.; Liu, Y. Variance Component Estimation (VCE)-Based Adaptive Stochastic Modeling for Enhanced Convergence and Robustness in GNSS Precise Point Positioning (PPP). *Remote Sens.* **2025**, *17*, 3071. https://doi.org/10.3390/rs17173071.
8. Fan, C.; Yao, Z.; Wang, J.; Lu, M. Cascaded Ambiguity Resolution for Pseudolite System-Augmented GNSS PPP. *Remote Sens.* **2025**, *17*, 3149. https://doi.org/10.3390/rs17183149.
9. Lv, Y.; Meng, Z.; Wang, G.; Liu, M.; Yan, E. Review of Research on Satellite Clock Bias Prediction Models in GNSS. *Remote Sens.* **2025**, *17*, 3177. https://doi.org/10.3390/rs17183177.

References

1. Defraigne, P. GNSS Time and Frequency Transfer. In *Springer Handbook of Global Navigation Satellite Systems*; Teunissen, P.J.G., Montenbruck, O., Eds.; Springer International Publishing: Cham, Switzerland, 2017; pp. 1187–1206.
2. Han, C.; Liu, L.; Cai, Z.; Lin, Y. The space–time references of BeiDou Navigation Satellite System. *Satell. Navig.* **2021**, *2*, 18. [CrossRef]
3. Yang, Y.; Mao, Y.; Sun, B. Basic performance and future developments of BeiDou global navigation satellite system. *Satell. Navig.* **2020**, *1*, 1. [CrossRef]
4. Ge, Y.; Liu, Y.; Lyu, D.; Wu, M.; Wang, Y.; Cao, X.; Shen, F. LEO satellite clock modeling and its benefits for real-time LEO PPP timing. *Measurement* **2025**, *242*, 116035. [CrossRef]
5. Zhang, Z.; Tang, Y.; Zheng, Z.; Qiao, D.; Chen, H.; Zhang, T. Research on artificial intelligence technology for Solar system boundary exploration missions. *J. Deep Space Explor.* **2024**, *11*, 523–537. [CrossRef]
6. Su, K.; Jiao, G. Estimation of BDS pseudorange biases with high temporal resolution: Feasibility, affecting factors, and necessity. *Satell. Navig.* **2023**, *4*, 17. [CrossRef]
7. Critchley-Marrows, J.J.R.; Verspieren, Q. Ensuring PNT resilience in a time of navigation uncertainty. *Space Policy* **2025**, *72*, 101665. [CrossRef]
8. Yang, Y. Resilient PNT concept frame. *Acta Geod. Cartogr. Sin.* **2018**, *47*, 893.

9. Yang, Y.; Yang, C.; Ren, X. PNT intelligent services. *Acta Geod. Cartogr. Sin.* **2021**, *50*, 1006.
10. Wang, X.; Liu, J.; Liu, Z.; Yuan, Z. Measurement and Evaluation of Marine Intelligent Transportation PNT Data Based on BDS and DGNSS. *IOP Conf. Ser. Mater. Sci. Eng.* **2020**, *719*, 012069. [CrossRef]

remote sensing

MDPI

Article

BDS-3/GNSS Undifferenced Pseudorange and Phase Time-Variant Mixed OSB Considering the Receiver Time-Variant Biases and Its Benefit on Multi-Frequency PPP

Guoqiang Jiao [1], Ke Su [2,3,4,*], Min Fan [1], Yuze Yang [5] and Huaquan Hu [2,3]

1 Beijing Institute of Tracking and Telecommunications Technology, Beijing 100094, China; jiaoguoqiang18@mails.ucas.ac.cn (G.J.); fanmin@bittt.cn (M.F.)
2 Department of Space Information, Space Engineering University, Beijing 101416, China; hqhu@hgd.edu.cn
3 Key Laboratory of Smart Earth, Beijing 100094, China
4 National Key Laboratory of Space Target Awareness, Beijing 101416, China
5 Shanghai Astronomical Observatory, Chinese Academy of Sciences, Shanghai 200030, China; yzyang@shao.ac.cn
* Correspondence: suke17@mails.ucas.ac.cn; Tel.: +86-176-2101-9466

Abstract: The legacy Global Navigation Satellite System (GNSS) satellite clock offsets obtained by the dual-frequency undifferenced (UD) ionospheric-free (IF) model absorb the code and phase time-variant hardware delays, which leads to the inconsistency of the precise satellite clock estimated by different frequencies. The dissimilarity of the satellite clock offsets generated by different frequencies is called the inter-frequency clock bias (IFCB). Estimates of the IFCB typically employ epoch-differenced (ED) geometry-free ionosphere-free (GFIF) observations from global networks. However, this method has certain theoretical flaws by ignoring the receiver time-variant biases. We proposed a new undifferenced model coupled with satellite clock offsets, and further converted the IFCB into the code and phase time-variant mixed observable-specific signal bias (OSB) to overcome the defects of the traditional model and simplify the bias correction process of multi-frequency precise point positioning (PPP). The new model not only improves the mixed OSB performance, but also avoids the negative impact of the receiver time-variant biases on the satellite mixed OSB estimation. The STD and RMS of the original OSB can be improved by 7.5–60.9% and 9.4–66.1%, and that of ED OSB (it can reflect noise levels) can be improved by 50.0–87.5% and 60.0–88.9%, respectively. Similarly, the corresponding PPP performance for using new mixed OSB is better than that of using the traditional IFCB products. Thus, the proposed pseudorange and phase time-variant mixed OSB concept and the new undifferenced model coupled with satellite clock offsets are reliable, applicable, and effective in multi-frequency PPP.

Keywords: inter-frequency clock bias (IFCB); observable-specific signal bias (OSB); epoch-differenced (ED); undifferenced; multi-frequency precise point positioning

1. Introduction

With GNSS modernization and development, the traditional GPS-only dual-frequency model is gradually transitioning to multi-GNSS (GPS, BDS, GLONASS, Galileo, and regional system) and multi-frequency tendencies. The GPS Block IIF satellites can transmit L1, L2, and L5 signals [1]. The global BeiDou navigation satellite system (BDS-3) satellites can broadcast the B1C, B2a, B2b, and B2 (B2a + B2b) signals as well as the legacy B1I and B3I signals [2]. Similarly, the new GLONASS satellites have begun to transmit code division multiple access (CDMA) G1a, G2a, and G3 signal and frequency division multiple access (FDMA) G1 and G2 signals [3]. Galileo can transmit E1, E5a, E5b, E5, and E6 signals [4]. The multi-frequency observations can be beneficial in precise point positioning (PPP), cycle slip detection, precise clock offset estimation (PCE), and so on [5–7].

Normally, the legacy precise satellite clock offsets released by GNSS analysis centers (ACs) are generated by a dual-frequency ionosphere-free (IF) combination model [8]. For example, the recommended basic frequency pair for GPS, Galileo, and BDS are L1/L2, E1/E5a, and B1I/B3I signals [1,9,10]. Due to the characteristics of multi-frequency multi-channel GNSS signals, there is an inter-frequency bias between different frequencies, and there is an intra-frequency deviation between different channels of the same frequency. The code and phase time-variant hardware delay will be absorbed into satellite clock offsets, which leads to the inconsistency of the precise satellite clock offsets estimated by different frequencies [11]. This inconsistency is known as the inter-frequency clock bias (IFCB) [1]. The IFCB can be divided into code IFCB (CIFCB) and phase IFCB (PIFCB) [12]. The CIFCB is a linear combination of code hardware delays, and it is also called the differential code bias (DCB) or pseudorange observable-specific signal bias (OSB) [13]. The DCB is usually determined by the carrier-to-code leveling (CCL) method [14,15]. Most International GNSS Service (IGS) and International GNSS Monitoring and Assessment System (iGMAS) ACs such as Centre for Orbit Determination in Europe (CODE), Deutsches Zentrum für Luft- und Raumfahrt (DLR), Chinses Academy of Sciences (CAS), WhuHan University (WHU), and Shanghai Astronomical Observatory (SHA) use the CCL method to obtain DCB [16–19]. The PIFCB is a linear combination of phase time-variant hardware delays, which is usually generated by epoch-differenced (ED) geometry-free (GF) ionosphere-free (IF) linear combination (GFIF) [12]. At present, IGS and iGMAS ACs have not yet officially released IFCB, and the OSB files obtained from GNSS AC mainly release pseudorange and phase OSB [3,20]. Nevertheless, many references have conducted research on it. Pan, et al. [21] used the ED GFIF model to obtain GPS L5 PIFCB, and researched characteristics to model and predict it [22]. Similarly, Gong, et al. [23] used the ED GFIF model to generate the PIFCB and analyzed its periodic characteristics to develop the corresponding prediction model. Su, et al. [24] converted the PIFCB obtained by the above method into carrier-phase OSB, which greatly simplified the complexity of the client algorithm. Although this method has high computational efficiency, it has certain theoretical flaws. However, the correct method assumes that the PIFCB of the same satellite at different receivers has good consistency and the receiver PIFCB can be combined into satellite PIFCB. Actually, receiver hardware delays are greatly affected by the ground environment variation [25]. If the receiver hardware delays change, it will have a negative impact on the satellite PIFCB estimation. The IFCB can also be calculated by estimating the satellite clock offset for all frequencies. Guo and Geng [26] provided the GPS L1/L2 and L5-only types of satellite clock products for the triple-frequency PPP. Jiao, et al. [27] established a multi-frequency precise satellite clock offset estimation model to obtain clock offsets and the corresponding IFCB. Fan, et al. [28] first modeled the IFCB and then brought the model coefficients into the undifferenced model to obtain the corresponding IFCB model coefficients, which can improve IFCB estimation accuracy. Geng, et al. [29] merged IFCB into phase OSB at the third frequency, which simplifies the PPPAR process. But, it is not very convenient for multi-frequency float PPP. However, these methods require the estimation of the ambiguity parameters for the basic frequency as well as the third frequency. The ambiguity parameters will increase with the increase in observations, and its corresponding computational burden is enlarged [30]. Therefore, it is very meaningful to develop an IFCB estimation method with high computational efficiency and accuracy.

With this background, we developed the new undifferenced IFCB estimation model considering the receiver time-variant biases, and further converted it into code and phase time-variant mixed OSB to improve multi-frequency PPP (MFPPP) performance. The new model avoids the negative impact of receiver time-variant biases on the satellite mixed OSB, and it has good accuracy and low noise. First, the mathematical methods of the traditional ED GFIF-phase time-variant OSB model and new undifferenced mixed OSB model coupled with satellite clock offsets and an analysis of their characteristics are presented in detail. Then, the processing datasets and the corresponding processing strategies are presented. Naturally, the experimental analysis about the advantages of the new undifferenced mixed

OSB model and its benefits on MFPPP are introduced in the following section. Finally, the conclusions are summarized.

2. Methods

This section begins with the GNSS pseudorange and phase observation models. The traditional epoch-differenced phase-only GFIF OSB estimation model and the new undifferenced models coupled with satellite clock offsets are developed in detail. The IFCB is converted to OSB type for simplifying the client calculation.

2.1. General Observation Models and OSB Concept

The GNSS original observation equations with the time-variant parts of pseudorange and phase hardware delays can be written as follows [31]:

$$\begin{cases} p^s_{r,j}(t) = \rho^s_r(t) + dt_r(t) - dt^s(t) + b_{r,j} - b^s_j + b_{r,j}(t) - b^s_j(t) + T^s_r(t) + I^s_{r,j}(t) + \varepsilon_{p_j}(t) \\ \phi^s_{r,j}(t) = \rho^s_r(t) + dt_r(t) - dt^s(t) + B_{r,j} - B^s_j + B_{r,j}(t) - B^s_j(t) + T^s_r(t) - I^s_{r,j}(t) + N^s_{r,j} + \varepsilon_{\phi_j}(t) \end{cases} \tag{1}$$

where $p^s_{r,j}(t)$ and $\phi^s_{r,j}(t)$ are the pseudorange and phase observations on frequency j in units of meter, respectively; $\rho^s_r(t)$ is the geometrical range from satellite s to receiver r at epoch t; $dt_r(t)$ and $dt^s(t)$ denote the original receiver and satellite clock offsets; and $b_{r,j}$ and b^s_j represent the receiver and satellite time-invariant uncalibrated code delays (UCDs), while $b_{r,j}(t)$ and $b^s_j(t)$ represent the corresponding time-variant parts; similarly, $B_{r,j}$ and B^s_j depict the time-invariant uncalibrated phase delays (UPDs) and $B_{r,j}(t)$ and $B^s_j(t)$ are the time-dependent parts; $T^s_r(t)$ depicts the slant tropospheric delay; $I^s_{r,j}(t)$ is the slant ionospheric delay; $N^s_{r,j}$ denotes the integer ambiguity; and $\varepsilon_{p_j}(t)$ and $\varepsilon_{\phi_j}(t)$ are the pseudorange and phase observation noises containing multipath and unmodeled errors.

The IGS and iGMAS ACs typically employ dual-frequency undifferenced IF combination observations to generate satellite clock offsets [31,32]:

$$\begin{cases} p^s_{r,IF_{ij}}(t) = \rho^s_r(t) + dt_{r,IF_{ij}}(t) - dt^s_{IF_{ij}}(t) + T^s_r(t) + \bar{\varepsilon}_{pIF_{ij}}(t) \\ \phi^s_{r,IF_{ij}}(t) = \rho^s_r(t) + dt_{r,IF_{ij}}(t) - dt^s_{IF_{ij}}(t) + T^s_r(t) + \overline{N}^s_{r,IF_{ij}} + \varepsilon_{\phi IF_{ij}}(t) \end{cases} \tag{2}$$

with

$$\begin{cases} dt_{r,IF_{ij}}(t) = dt_r(t) + b_{r,IF_{ij}} + B_{r,IF_{ij}}(t) \\ dt^s_{IF_{ij}}(t) = dt^s(t) + b^s_{IF_{ij}} + B^s_{IF_{ij}}(t) \\ \bar{\varepsilon}_{pIF_{ij}}(t) = \varepsilon_{pIF_{ij}}(t) + b_{r,IF_{ij}}(t) - b^s_{IF_{ij}}(t) - B_{r,IF_{ij}}(t) + B^s_{IF_{ij}}(t) \\ \overline{N}^s_{r,IF_{ij}} = N^s_{r,IF_{ij}} + B_{r,IF_{ij}} - B^s_{IF_{ij}} - b_{r,IF_{ij}} + b^s_{IF_{ij}} \end{cases} \tag{3}$$

where $(\cdot)_{IF_{ij}} = \alpha_{ij} \cdot (\cdot)_i + \beta_{ij} \cdot (\cdot)_j$, $\alpha_{ij} = \frac{f_i^2}{f_i^2 - f_j^2}$, $\beta_{ij} = \frac{-f_j^2}{f_i^2 - f_j^2}$, IF_{ij} denotes the dual-frequency IF-combined operation; i and j are the IGS/iGMAS recommended frequency, where GPS is L1/L2, BDS is B1I/B3I, and Galileo is E1/E5a; $dt_{r,IF_{ij}}(t)$ and $dt^s_{IF_{ij}}(t)$ denote receiver and satellite IF-combined clock offsets; and $\bar{\varepsilon}_{pIF_{ij}}(t)$ and $\overline{N}^s_{r,IF_{ij}}$ are the recombined pseudorange observation noises and IF ambiguity, respectively. The satellite clock offsets published by IGS and iGMAS ACs will absorb the UCD, time-variant UPD, and reference clock, which can be written as follows [11]:

$$d\bar{t}^s_{IF_{ij}}(t) = dt^s(t) + b^s_{IF_{ij}} + B^s_{IF_{ij}}(t) + dt_{ref} = dt^s_{IF_{ij}}(t) + dt_{ref} \tag{4}$$

where $d\bar{t}^s_{IF_{ij}}(t)$ is the actual estimated clock offsets and dt_{ref} is the reference clock.

The satellite clock offsets absorbed with the hardware delays make the measurement inconsistent at the specific frequency. The satellite clock offsets for the specific frequency can be expressed as follows:

$$\begin{cases} d\overline{t}_i^s(t) = dt_{IF_{ij}}^s(t) + \delta_{p_i} + \delta_{\phi_i}(t) + dt_{ref} = dt_{IF_{ij}}^s(t) + \delta_{M_i}^s(t) + dt_{ref} \\ d\overline{t}_j^s(t) = dt_{IF_{ij}}^s(t) + \delta_{p_j} + \delta_{\phi_j}(t) + dt_{ref} = dt_{IF_{ij}}^s(t) + \delta_{M_j}^s(t) + dt_{ref} \\ d\overline{t}_k^s(t) = dt_{IF_{ij}}^s(t) + \delta_{p_k} + \delta_{\phi_k}(t) + dt_{ref} = dt_{IF_{ij}}^s(t) + \delta_{M_k}^s(t) + dt_{ref} \end{cases} \tag{5}$$

where $d\overline{t}_i^s(t)$, $d\overline{t}_j^s(t)$, and $d\overline{t}_k^s(t)$ denote the satellite clock offsets for the basic ith, jth, and the third kth frequency signals. δ_{p_i} and δ_{p_j} denote the pseudorange OSB for the baisc frequency signals; δ_{p_k} denotes the pseudorange OSB for the third frequency signal; and, in analogy, $\delta_{\phi_i}(t)$, $\delta_{\phi_j}(t)$, and $\delta_{\phi_k}(t)$ are carrier-phase time-variant OSB for the basic and the third frequency. Normally, the definition of satellite IFCB is the bias between the satellite clock offsets for the different frequency, which usually contains CIFCB (also known as DCB or pseudorange OSB) and PIFCB (also known as the phase time-variant OSB). The δ_{p_k} and $\delta_{\phi_k}(t)$ can be combined into the mixed IFCB with OSB type $\delta_{M_k}^s(t)$.

Generally, the pseudorange OSB is usually determined by the CCL method. However, the phase time-variant OSB for the basic frequency cannot be calculated by the methods of obtaining pseudorange OSB. As shown in (2), it will be absorbed into the other parameters such as clock offsets and residuals. As for the dual-frequency PPP, the phase time-variant OSB for the basic frequency will not have any negative impact. Because this bias is recombined into other parameters, the server algorithms for satellite clock offsets and the client PPP algorithms are strictly self-consistent. Hence, the carrier-phase time-variant OSB values at the basic frequency signals can be set to zero. The carrier-phase time-variant OSB values at the basic and the third frequency signals can be written as follows.

$$\begin{cases} \left.\begin{aligned} \delta_{\phi_i}^s(t) = 0 \\ \delta_{\phi_j}^s(t) = 0 \end{aligned}\right\} baseline\ frequency \\ \delta_{\phi_k}^s(t) = \alpha_{ij} \cdot (1 - \mu_k/\mu_j) \cdot B_i^s(t) - \beta_{ij} \cdot (\mu_k - 1) \cdot B_j^s(t) - B_k^s(t) \end{cases} \tag{6}$$

Equation (6) indicates that the carrier-phase time-variant OSB at different frequencies, and μ_i and μ_k denote the frequency-dependent multiplier factor.

2.2. Traditional GFIF Phase Time-Variant OSB Model

The phase time-variant OSB is commonly generated by the ED GFIF model. The GFIF combination between the third and the basic frequency can be written as follows [33]:

$$\phi_{r,GFIF_{ijk}}^s(t) = \phi_{r,IF_{ik}}^s(t) - \phi_{r,IF_{ij}}^s(t) = \delta_{r,\phi IF_{ik}}(t) - \delta_{\phi IF_{ik}}^s(t) + \overline{N}_{r,GFIF_{ijk}}^s \tag{7}$$

where $\overline{N}_{r,GFIF_{ijk}}^s$ denotes the differential IF-phase ambiguity. If there is no cycle slip, and the epoch difference strategy can be used to eliminate ambiguity, which can be written as follows:

$$\Delta\delta_{r,\phi IF_{ik}}(t, t-1) - \Delta\delta_{\phi IF_{ik}}^s(t, t-1) = \phi_{r,GFIF_{ijk}}^s(t) - \phi_{r,GFIF_{ijk}}^s(t-1) \tag{8}$$

where $\Delta\delta_{r,\phi IF_{ik}}(t, t-1)$ and $\Delta\delta_{\phi IF_{ik}}^s(t, t-1)$ denote the differential IF PIFCB.

Because the PIFCB of the same satellite at different stations has good consistency, the receiver PIFCB can be combined into satellite PIFCB [33], which reads as follows:

$$\Delta\delta_{r,\phi IF_{ik}}^s(t, t-1) = \phi_{r,GFIF_{ijk}}^s(t) - \phi_{r,GFIF_{ijk}}^s(t-1) \tag{9}$$

where $\Delta\delta_{r,\phi IF_{ik}}^s(t, t-1)$ denotes the corresponding IFCB for the paired receiver and satellite.

Assuming that satellite s can be tracked by m common viewing stations in the observational network at epoch t and $t-1$, the epoch-differenced PIFCB is expressed as follows:

$$\begin{cases} \Delta\delta^s_{\phi IF_{ik}}(t,t-1) = \dfrac{\sum\limits_{r=1}^{m}\left[\Delta\delta^s_{\phi IF_{ik}}(t,t-1)\cdot\omega^s_r(t,t-1)\right]}{\sum\limits_{r=1}^{m}\left[\omega^s_r(t,t-1)\right]} \\ \omega^s_r(t,t-1) = \begin{cases} \sin(Ele^s_r(t,t-1)), Ele^s_r(t,t-1) < 40^\circ \\ 1, Ele^s_r(t,t-1) \geq 40^\circ \end{cases} \end{cases} \tag{10}$$

where $Ele^s_r(t,t-1)$ represents the average elevation at epoch t and $t-1$. The PIFCB can be accumulated through epoch-differenced values, as shown below:

$$\delta^s_{\phi IF_{ik}}(t) = \delta^s_{\phi IF_{ik}}(t_0) + \sum_{t=t_0+1}^{t}\Delta\delta^s_{\phi IF_{ik}}(t,t-1) \tag{11}$$

with

$$\delta^s_{\phi IF_{ik}}(t) = B^s_{IF_{ik}}(t) - B^s_{IF_{ij}}(t) \tag{12}$$

where t_0 is the initial epoch time and the initial undifferenced PIFCB, $\delta^s_{\phi IF_{ik}}(t_0)$, is usually set to zero. $\Delta\delta^s_{\phi IF_{ik}}(t,t-1)$ is the ED IFCB between epoch t and $t-1$. $\delta^s_{\phi IF_{ik}}(t)$ can be accumulated by the $\Delta\delta^s_{\phi IF_{ik}}(t,t-1)$ from the initial epoch time. The $\delta^s_{\phi IF_{ik}}(t)$ is the IF-combined PIFCB, which is difficult to be corrected in the complex PPP models such as non-basic frequency PPP, undifferenced PPP, MFPPP, and so on. According to the SINEX_BIAS Version 1.0 [34], the bias in the form of OSB type can better adapt to multi-frequency observations and the complex PPP algorithms. Combining (6) and (12), the IF-combined PIFCB can be transformed into the phase time-variant OSB, which can be expressed as follows.

$$\delta^s_{\phi_k}(t) = \beta_{ik}^{-1}\cdot\delta^s_{\phi IF_{ik}}(t) \tag{13}$$

The phase time-variant OSB for basic and third frequency can generated by using Equations (6) and (13).

2.3. Undifferenced Mixed OSB Model Coupled with Satellite Clock Offsets

Through the comprehensive analysis of the above calculation process, it can be found that the ED GFIF model has three major flaws. The first is that it neglects the receiver PIFCB. It assumes that the PIFCB of the same satellite at different receivers has good consistency and the receiver PIFCB can be combined into satellite PIFCB. The second is that it neglects the relationship between two adjacent epochs because it obtains PIFCB by averaging and accumulating. The third and most important is the MFPPP need to correct both pseudorange OSB and PIFCB, which makes the client algorithms more complex. For traditional methods, two bias corrections are required to achieve MFPPP, which makes the client algorithms more complex. To avoid the above defects, the undifferenced pseudorange and phase time-variant mixed OSB model coupled with satellite clock offsets can be expressed as follows:

$$\begin{cases} p^s_{r,IF_{ik}}(t) = \rho^s_r(t) + dt_{r,IF_{ik}}(t) - dt^s_{IF_{ij}}(t) - \delta^s_{IF_{ik}}(t) + T^s_r(t) + \bar{\varepsilon}_{pIF_{ik}}(t) \\ \phi^s_{r,IF_{ik}}(t) = \rho^s_r(t) + dt_{r,IF_{ik}}(t) - dt^s_{IF_{ij}}(t) - \delta^s_{IF_{ik}}(t) + T^s_r(t) + \overline{N}^s_{r,IF_{ik}} + \varepsilon_{\phi IF_{ik}}(t) \end{cases} \tag{14}$$

where $\delta^s_{IF_{ik}}(t)$ is the satellite pseudorange and phase time-variant mixed IFCB.

The IGS legacy clocks are fixed throughout the undifferenced mixed OSB model; hence, we further rewrite (14) by introducing (4) as follows:

$$\begin{cases} p^s_{r,IF_{ik}}(t) + d\bar{t}^s_{IF_{ij}}(t) = \rho^s_r(t) + d\bar{t}_{r,IF_{ik}}(t) - \delta^s_{IF_{ik}}(t) + T^s_r(t) + \bar{\varepsilon}_{pIF_{ik}}(t) \\ \phi^s_{r,IF_{ik}}(t) + d\bar{t}^s_{IF_{ij}}(t) = \rho^s_r(t) + d\bar{t}_{r,IF_{ik}}(t) - \delta^s_{IF_{ik}}(t) + T^s_r(t) + \overline{N}^s_{r,IF_{ik}} + \varepsilon_{\phi IF_{ik}}(t) \end{cases} \tag{15}$$

where $d\bar{t}^{s}_{IF_{ij}}(t)$ is IGS legacy clock offsets and $d\bar{t}_{r,IF_{ik}}(t)$ is the recombined receiver clock offsets, which absorb the reference clock dt_{ref}. Obviously, there is a rank-deficient problem between $d\bar{t}_{r,IF_{ik}}(t)$ and $\delta^{s}_{IF_{ik}}(t)$. To avoid the negative impact of receiver hardware delay on the satellite IFCB estimation, the zero-mean condition (ZMC) for a constellation is used to separate the receiver and satellite IFCB.

Naturally, there are two technical routes to obtain satellite IFCB according to whether it is with or without pseudorange OSB correction for basic frequency. If the basic frequency observations are corrected by pseudorange OSB, the third satellite IFCB will not contain satellite pseudorange OSB, which can be expressed as follows.

$$\delta^{s}_{M_{IF_{ik}}}(t) = \beta_{ik} \cdot (b^{s}_{k} - b^{s}_{i}) + B^{s}_{IF_{ik}}(t) - B^{s}_{IF_{ij}}(t) = \delta^{s}_{p_{IF_{ik}}} + \delta^{s}_{\phi_{IF_{ik}}}(t) \tag{16}$$

The IF-combined mixed IFCB $\delta^{s}_{M_{IF_{ik}}}(t)$ can be further converted into OSB type, which is expressed as follows.

$$\begin{cases} \left.\begin{array}{l} \delta^{s}_{M_i}(t) = \delta_{p_i} = \beta_{ij} \cdot \left(b^{s}_{i} - b^{s}_{j}\right) \\ \delta^{s}_{M_j}(t) = \delta_{p_j} = -\alpha_{ij} \cdot \left(b^{s}_{i} - b^{s}_{j}\right) \end{array}\right\} \text{basic frequency pair} \\ \delta^{s}_{M_k}(t) = b^{s}_{k} - b^{s}_{i} \\ \qquad +\alpha_{ij} \cdot (1 - \mu_k/\mu_i) \cdot B^{s}_{i}(t) - \beta_{ij} \cdot (\mu_k - 1) \cdot B^{s}_{j}(t) - B^{s}_{k}(t) \\ \qquad = \delta_{p_k} + \delta_{\phi_k}(t) = \beta^{-1}_{ik} \cdot \delta^{s}_{IF_{ik}}(t) \end{cases} \tag{17}$$

If the basic frequency observations are not corrected by pseudorange OSB, the estimated mixed IFCB will absorb satellite pseudorange OSB. Hence, the mixed IFCB for the third frequency can be given as follows.

$$\delta^{s}_{M_{IF_{ik}}}(t) = b^{s}_{IF_{ik}} - b^{s}_{IF_{ij}} + B^{s}_{IF_{ik}}(t) - B^{s}_{IF_{ij}}(t) = \delta^{s}_{p_{IF_{ik}}} + \delta^{s}_{\phi_{IF_{ik}}}(t) \tag{18}$$

The mixed IFCB $\delta^{s}_{M_{IF_{ik}}}(t)$ can be further converted into OSB type, which is expressed as follows.

$$\begin{cases} \left.\begin{array}{l} \delta^{s}_{M_i}(t) = 0 \\ \delta^{s}_{M_j}(t) = 0 \end{array}\right\} \text{basic frequency pair} \\ \delta^{s}_{M_k}(t) = \alpha_{ij} \cdot (1 - \mu_k/\mu_i) \cdot b^{s}_{i} - \beta_{ij} \cdot (\mu_k - 1) \cdot b^{s}_{j} - b^{s}_{k} \\ \qquad +\alpha_{ij} \cdot (1 - \mu_k/\mu_i) \cdot B^{s}_{i}(t) - \beta_{ij} \cdot (\mu_k - 1) \cdot B^{s}_{j}(t) - B^{s}_{k}(t) \\ \qquad = \delta_{p_k} + \delta_{\phi_k}(t) = \beta^{-1}_{ik} \cdot \delta^{s}_{IF_{ik}}(t) \end{cases} \tag{19}$$

Equation (19) indicates that the OSB for the third frequency is composed of the time-invariant UCD and time-variant UPD, which differs from Equation (17) in including and excluding pseudorange OSB for basic frequency.

In summary, the traditional IFCB model is based on three assumptions: (1) the PIFCB of the same satellite at different stations has good consistency and the receiver PIFCB can be combined into satellite PIFCB; (2) pseudorange and phase IFCB are determined by CCL and ED GFIF models, respectively; (3) the PIFCB is determined by averaging multiple GNSS tracking stations. These assumptions also reflect some defects. If the time-varying phase delays of different GNSS receivers are inconsistent, it will have a negative impact on the satellite PIFCB estimation. It is undeniable that traditional methods are more efficient than undifferenced models in terms of computational efficiency because they do not require the calculation of ambiguity. However, the new method only has one set of ambiguity parameters, so the computational efficiency is still very high. In addition, the pseudorange OSB is required for basic frequency observations, and both pseudorange OSB and phase time-variant OSB corrections are required for the third frequency observation in MFPPP. Obviously, this way is too complicated. In addition, the ACs such as CODE, DLR, CAS, WHU, and SHA basically use CCL to generate satellite pseudorange OSB products.

This will lead to the poor self-consistency of those biases. The new method can avoid complex bias correction. We do not need to make pseudorange OSB corrections for the basic frequency observations. We only need to make the pseudorange and phase time-variant mixed OSB corrections for the third frequency observation, which significantly optimizes the terminal positioning algorithm. The method of (19) is used to obtain the mixed OSB.

3. Experiment Setup

Approximately, the 130 stations collected from IGS MGEX for the period day of year (DOY) from 001 to 030, 2021, are used to obtain GPS, BDS-3, and Galileo satellite mixed OSB. The distribution of the selected GNSS stations for OSB estimation can be found in Figure 1.

The traditional phase time-variant ED OSB model is relatively simple. It can be calculated according to Equations (8)–(14), and its description will not be repeated again [33]. The undifferenced mixed OSB estimation processing strategies are illustrated in Table 1. The GPS L1/L2, BDS-3 B1I/B3I, and Galileo E1/E5a are treated as the basic frequency, while GPS L5, BDS-3 B1C and B2a, and Galileo E5, E5b, and E6 are used to estimate satellite IFCB. The satellite orbits and clock offsets are fixed to the WHU final products, and the station coordinates are fixed to IGS weekly SINEX-file coordinates [3,35]. The ZMC method for a constellation is used to eliminate rank deficiency between receiver and satellite IFCB. As for tropospheric delay, the dry parts are corrected by the modified Hopfield model based on the Global Pressure and Temperature 3 (GPT3) and Vienna Mapping Functions 3 (VMF3) models, while the wet parts are estimated by setting the parameter based on the wet part of VMF3 [36,37]. The Phase Center Offset (PCO) and Phase Center Variations (PCVs) for GPS, BDS-3, and Galileo multi-frequency observations are corrected using the IGS antenna file (igs14.atx). In the bidirectional Kalman filter (forward and backward) of IFCB processing, the ambiguities are estimated as float constants, the ZWD is estimated as a random walk process, and the receiver clock is estimated as white noise. To avoid gross errors, and retain the IFCB's original characteristics, the satellite IFCB is estimated as a random walk process and the receiver IFCB is estimated as white noise. In the forward Kalman Filter of PPP processing, the processing strategy is basically the same as that of mixed OSB estimation. For more details for PPP processing strategies, please refer to [5,38].

Table 1. Data processing strategy for undifferenced mixed OSB estimation.

Items	Strategies
Basic frequency pair	GPS L1/L2, BDS-3 B1I/B3I, Galileo E1/E5a
Estimated frequency	GPS L5, BDS-3 B1C/B2a, Galileo E5b/E5/E6
Elevation cutoff	7 degrees
Weighting	Elevation weight [sin(elevation)]
Filter type	Bidirectional Kalman filter (forward + backward)
Satellite orbit and clock offsets	Fixed to WHU MGEX final precise products
Receiver coordinate	Fixed to IGS SINEX-file coordinates
Satellite IFCB	Estimated as random walk (10^4 m^2/s) [5]
Reference IFCB	Zero-mean condition
Receiver clock offsets	Estimated as white noise
Tropospheric delay	Dry part: modified Hopfield model Wet part: estimated as random walk (10^{-9} m^2/s)
Ionospheric delay	Eliminated first order by IF observations
Satellite and receiver antenna	IGS MGEX values (igs14.atx)
Phase windup effect	Corrected [39]
Relativistic effect	Corrected [40]
Earth rotation	Corrected [40]
Tide effect	Solid earth, pole, and ocean tide [40]
Ambiguity	Estimated as float
DCB	Parameter recombination

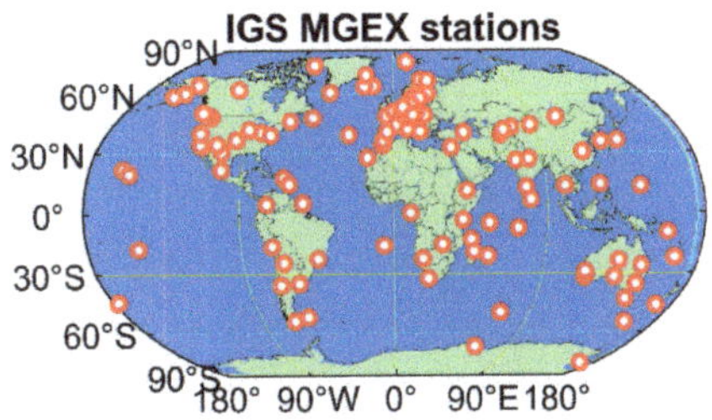

Figure 1. Distribution of the selected GNSS tracking stations for satellite mixed OSB estimation.

4. Results

This section presents the result of the OSB obtained from the ED GFIF model and the new undifferenced model coupled with satellite clock offsets. First, the mixed OSB service system framework is introduced based on the above theoretical analysis. Then, two OSB solving methods are compared in terms of their characteristics and MFPPP performances.

4.1. Pseudorange and Phase Time-Variant Mixed OSB Service System

We established the multi-GNSS pseudorange and phase time-variant mixed OSB service system based on the above theoretical discussions. Figure 2 illustrates the flowchart of the mixed OSB service system. First, the data download module automatically downloads GNSS observations, precise orbits, and clock offsets from IGS MGEX and iGMAS. The IGS MGEX observation values and the precise orbit clock products of the WHU IGS and iGMAS analysis center are downloaded to generate the mixed OSB in this manuscript. If the precise orbit and clock offsets are missing, the ED GFIF methods are used as a backup algorithm for obtaining the phase-only time-variant IFCB. The phase-only time-variant IFCB can be transformed into OSB type. Typically, MFPPP requires both phase-only time-variant IFCB and DCB corrections [24,41]. We synthesize the additional DCB products and phase-only time-variant OSB into the mixed OSB on the server side, which can simultaneously correct both pseudorange and phase observations. This way, it can reduce the complexity of client algorithms. If the precise orbit and clock offsets are not missing, the new undifferenced IFCB model coupled with satellite clock offsets methods is used to generate the mixed IFCB. Furthermore, this IFCB can be transformed into OSB. The client only needs to correct this bias to achieve MFPPP. Note that, unless otherwise specified, the mixed OSB in the following text refers to the pseudorange and phase time-variant mixed OSB.

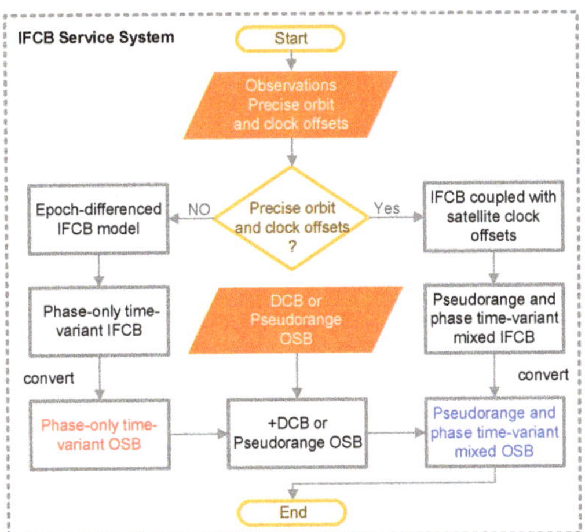

Figure 2. Multi-GNSS pseudorange and phase time-variant mixed OSB service system.

4.2. Characteristics of the Mixed OSB

The mixed OSB obtained from Figure 2 contains pseudorange and phase time-variant OSB. The zero-mean constraint of satellite mixed OSB obtained from the two schemes may not be consistent, which introduces systematic errors in comparing the results. In order to better illustrate the satellite mixed OSB situation for two schemes, it is necessary to convert all the mixed OSB to the same reference datum. To simplify the description, Scheme 1 and Scheme 2 represent the mixed OSB obtained from the ED GFIF model and undifferenced model, respectively.

To analyze the characteristics at each frequency, Figures 3–5 show the time series of the mixed OSB for GPS L5, BDS-3 B1C, B2a, and Galileo E5b, E5, and E6 signals on DOY 011 in 2021. The time resolution is 30 s. Figure 6 shows the corresponding mixed OSB amplitudes, in which x-axes depict the PRN of each satellite (for clarity and simplicity, specific numbers are not labeled). It is not difficult to find that the OSB for GPS BLOCK-IIF and BDS-3 signals have obvious amplitudes, while the GPS BLOCK-IIA, BLOCK-IIR, and Galileo OSB amplitudes are relatively small. The amplitude of GPS BLOCK-IIF L5 OSB is in the decimeter range, while the amplitude of BDS-2 B2I and BDS-3 B1C and B2a OSB is in the centimeter range. The OSB for Galileo E5b, Galileo E5, and Galileo E6 is around 1~2 cm. The OSB for BDS-3 B1C and BDS-3 B2a is around 1~3 cm. We all know that the accuracy of phase observation values is about 0.3 cm, so the IFCB correction should be considered in PPP. By comparing the OSB of Scheme 1 and Scheme 2, it can be seen that the trend of their changes is basically the same. The results further proved that the new undifferenced method is reasonable. But, undeniably, there are some differences in some satellites between the two schemes, especially G18, G23, C44, C45, and C46. There are four reasons for this phenomenon: (1) Scheme 1 assumes that the phase time-variant OSB of the same satellite at different stations has good consistency and the receiver phase time-variant OSB can be combined into satellite phase time-variant OSB. As for Scheme 2, the zero-mean condition for all satellites is introduced to separate the receiver and satellite OSB, which avoids the negative of the receiver on satellite OSB estimation. (2) Due to the use of ZMC in Scheme 2, all OSBs obtained from Scheme 2 will absorb the datum for a satellite constellation, while Scheme 1 will not. In other words, the OSBs obtained from Scheme 2 are similar to estimating satellite clock offsets, which will absorb the OSB for the reference satellite. Therefore, there are differences between the two schemes. (3) The number of BDS-3 C44/C45/C46 observations is relatively small, which affects the estimation accuracy [5]. (4) From a theoretical analysis of the formula, some parameters of the undifferenced model have certain correlations. The undifferenced model requires satellite orbit, satellite clock offsets, PCO, and PCV to achieve the estimation of OSB. Inevitably, the accuracy of these products will affect the estimation of undifferenced OSB. Therefore, the results with the new method suffer from irrational systematic errors. However, the undifferenced OSB model has better coupling with orbit and clock errors. The corresponding self-consistency between the satellite products is stronger, which will be beneficial for PPP.

Those systematic errors introduced by precise products can be eliminated through the ED method. Therefore, the ED method can reflect the noise level. To further compare the advantages and disadvantages of the two schemes, Figures 7–9 depict the corresponding mixed ED OSB for GPS L5, BDS-3 B1C, B2a, and Galileo E5b, E5, and E6 signals. It is not hard to observe that the ED OSB of Scheme 2 is better than that of Scheme 1 in terms of stability and accuracy. The noise of Scheme 1 is significantly higher than that of Scheme 2. To more objectively illustrate the performance of Schemes 1 and 2, Table 2 illustrates the statistical results and improvement of Scheme 2 in January 2021. The STD improvement of the undifferenced OSB is 7.5, 68.4, 38.6, 13.3, 60.9, and 22.2%, and the RMS can be improved by 9.4, 66.1, 34.8, 11.8, 56.5, and 20.7% with respect to GPS L5, BDS-3 B1C, B2a, and Galileo E5b, E5, and E6 signals. Moreover, the STD improvement of the ED OSB is 60.0, 80.0, 87.5, 75.0, 50.0, and 75.0%, and the RMS can be improved by 60.0, 83.3, 88.9, 75.0, 66.7, and 80.0%C. There are significant differences in improvement rates among different satellite systems or frequencies. There are two main reasons: (1) the undifferenced mixed OSB for different satellite systems are frequencies that have different scales, and (2) the number of observations for different signals is different, so there are differences in the improvement.

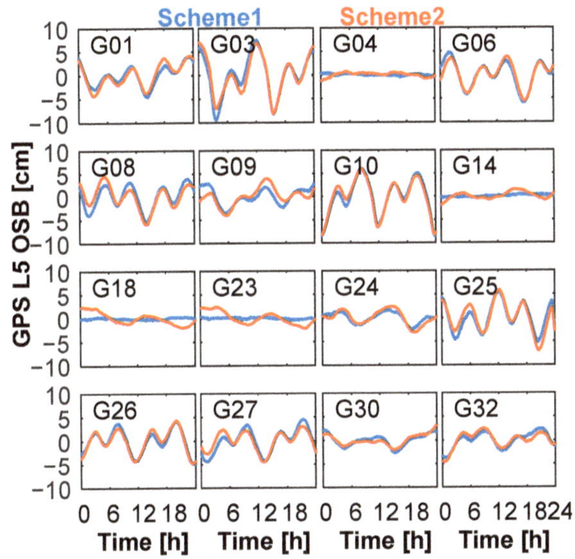

Figure 3. GPS L5 mixed OSB on DOY 011, 2021.

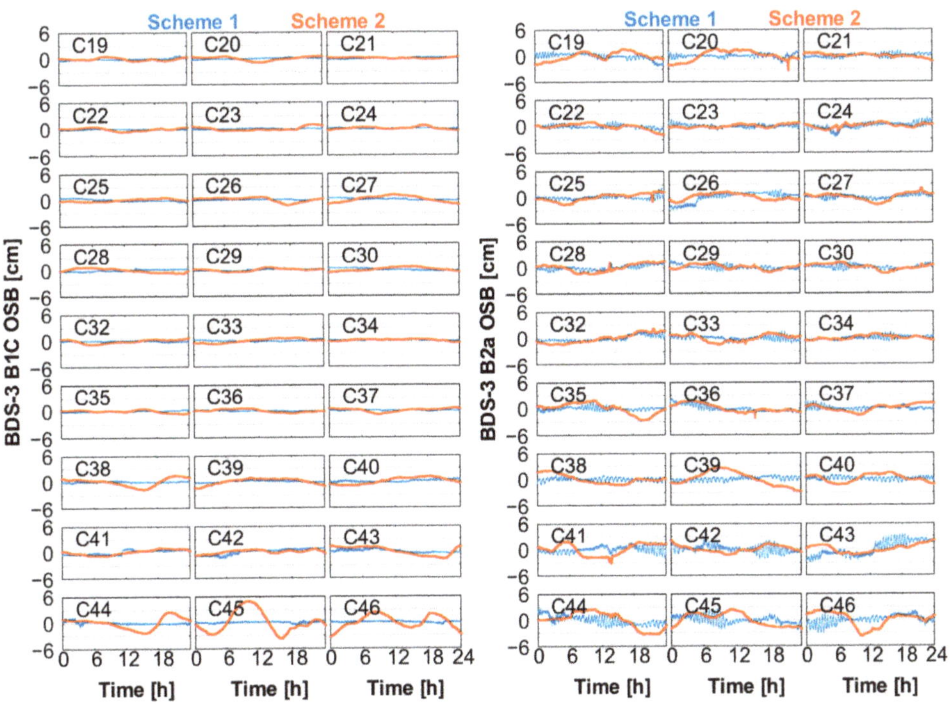

Figure 4. The BDS-3 B1C and B2a mixed OSB on DOY 011, 2021.

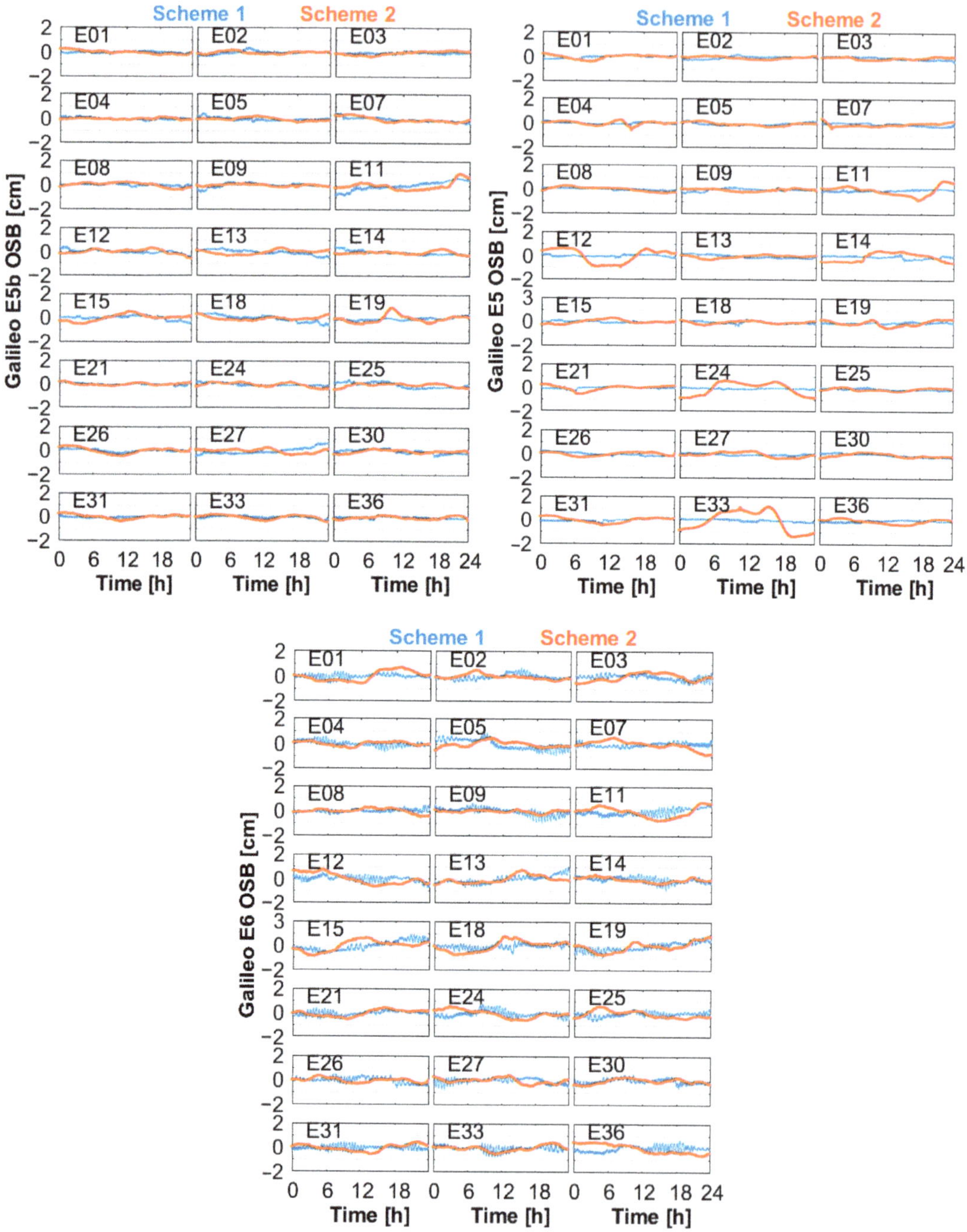

Figure 5. The Galileo E5b, E5, and E6 mixed OSB on DOY 011, 2021.

Overall, these discoveries illustrate that the stability and accuracy of the new undifferenced OSB estimation algorithm compared to the ED IFCB algorithm.

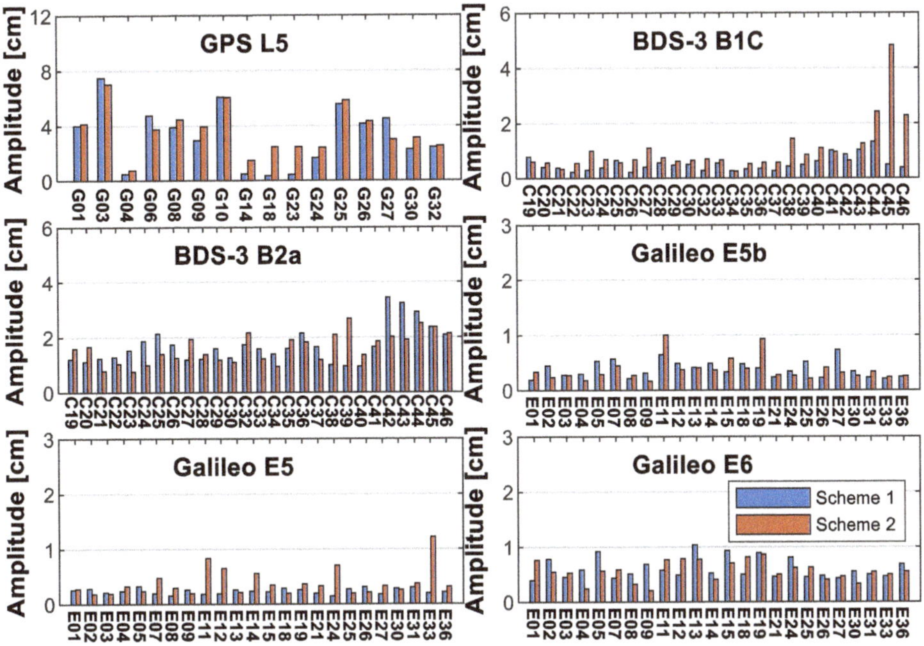

Figure 6. Amplitudes of the mixed OSB of GPS L5, BDS-3 B1C, B2a, and Galileo E5b, E5, and E6 signals on DOY 011, 2021.

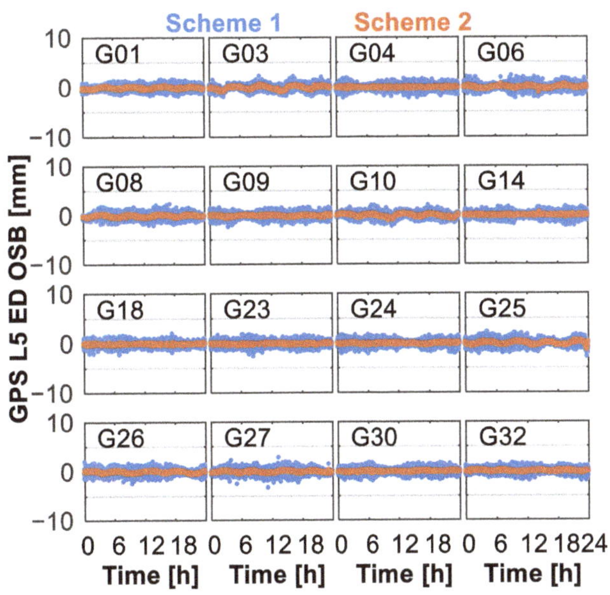

Figure 7. Pseudorange and phase time-variant mixed epoch-differenced (ED) OSB for GPS L5 signal on DOY 011, 2021.

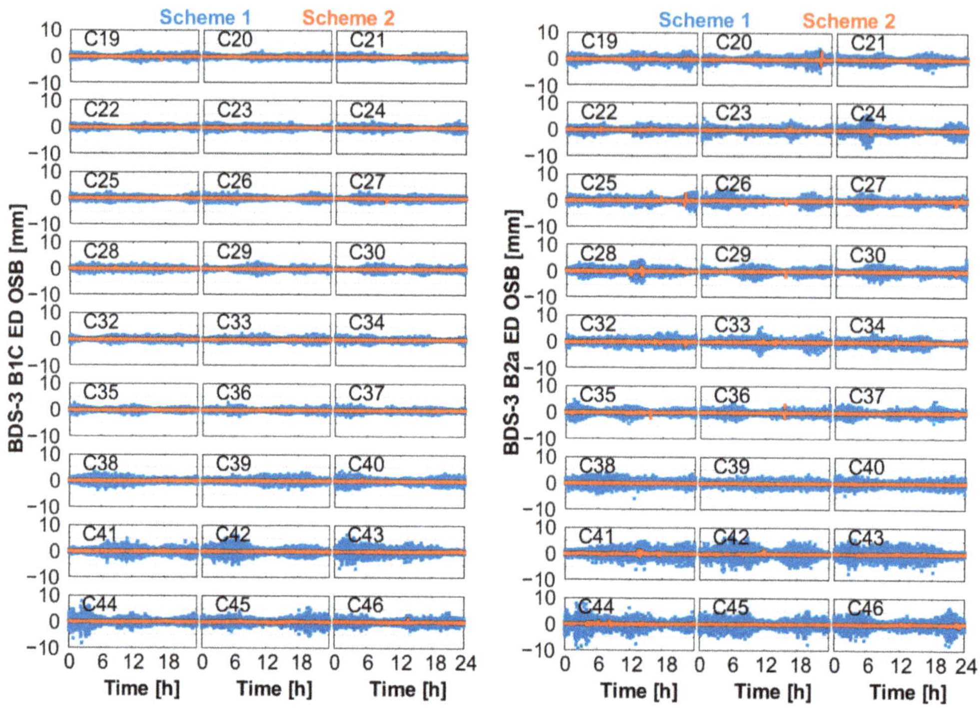

Figure 8. Pseudorange and phase time-variant mixed ED OSB for BDS-3 B1C and B2a signals on DOY 011, 2021.

Notably, it can be found that there is a systematic periodic bias in the OSB obtained from Scheme 1 at BDS B2a and Galileo E6 signals from Figures 4 and 5. Some of the literature suggests that this systematic periodic bias is related to the number of available observations at the GNSS tracking station. Due to the small number of MGEX stations capable of tracking BDS-3 B2a and Galileo E6 signals, the accuracy and reliability of the estimated OSB are poor [24]. The OSB estimation is affected by the receiver antenna environment and receiver and antenna quality, which results in periodic errors [1]. Unfortunately, these explanations cannot explain this systematic error at the BDS B2a and Galileo E6 signals. However, it can be found that the mixed OSB value obtained from Scheme 2 does not have significant periodic bias. By comparing the two algorithms, it can be seen that the epoch-differenced algorithm is based on the assumption that the IFCB of the same satellite at different stations has good consistency and the receiver phase time-variant OSB can be combined into satellite phase time-variant OSB. However, it is necessary to separate the satellite and receiver phase time-variant OSB in the new algorithm. To confirm that this small periodic bias is caused by some receiver phase time-variant OSB, Figure 10 shows the receiver OSB for all receiver types in the observation network. It can be seen that the JAVAD TRE-3 receiver OSB for BDS-3 B2a and Galileo E6 exhibits significant periodic variations; the OSB of other receivers is relatively stable and does not have significant deviations. Figure 11 further depicts the receiver OSB for multiple stations with the JAVAD TRE-3 receiver. This is just more evidence for the theory that JAVAD TRE-3 receiver hardware delay exhibits periodic variations at the BDS-3 B2a and Galileo E6 signals. Scheme 1 ignores the receiver phase time-variant OSB, which negatively impacts the estimation of the satellite OSB. On the contrary, the new undifferenced method uses zero-mean constraints to separate the satellite and receiver OSB, which effectively solves this problem.

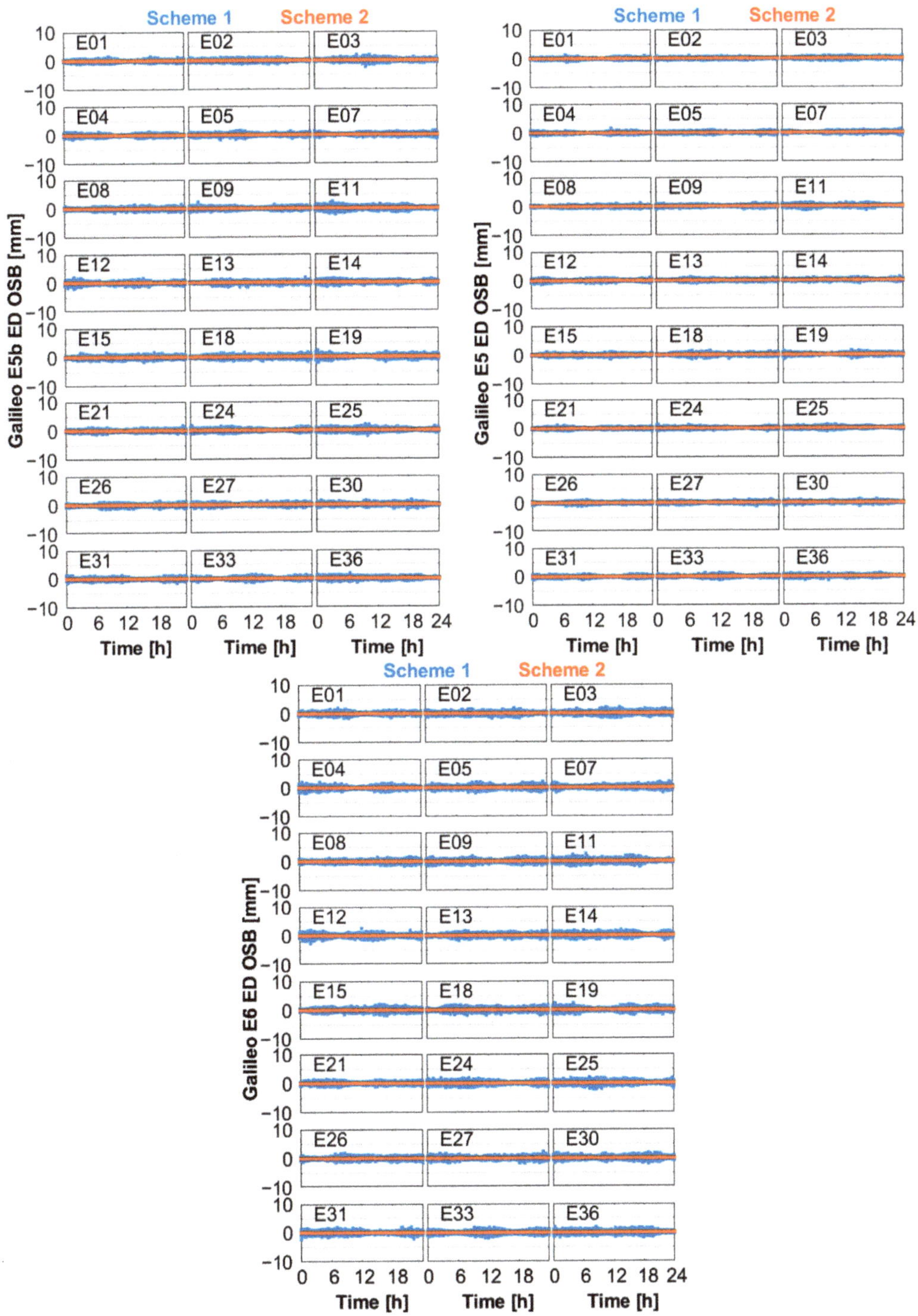

Figure 9. Daily satellite pseudorange and phase time-variant mixed OSB for Galileo E5b, E5, and E6 signals on DOY 011, 2021.

Table 2. Statistical results and improvement of epoch-differenced and undifferenced OSB for GPS L5, BDS-3 B1C, B2a, and Galileo E5b, E5, and E6 signals in January 2021.

Type (Unit: mm)		GPS	BDS-3		Galileo		
		L5	B1C	B2a	E5b	E5	E6
STD UD OSB	Scheme 1	18.6	5.7	8.8	1.5	2.3	2.7
	Scheme 2	17.2	1.8	5.4	1.3	0.9	2.1
	Improvement	7.5%	68.4%	38.6%	13.3%	60.9%	22.2%
RMS UD OSB	Scheme 1	20.3	5.9	9.2	1.7	2.3	2.9
	Scheme 2	18.4	2.0	6.0	1.5	1.0	2.3
	Improvement	9.4%	66.1%	34.8%	11.8%	56.5%	20.7%
STD ED OSB	Scheme 1	0.5	0.5	0.8	0.4	0.2	0.4
	Scheme 2	0.2	0.1	0.1	0.1	0.1	0.1
	Improvement	60.0%	80.0%	87.5%	75.0%	50.0%	75.0%
RMS ED OSB	Scheme 1	0.5	0.6	0.9	0.4	0.3	0.5
	Scheme 2	0.2	0.1	0.1	0.1	0.1	0.1
	Improvement	60.0%	83.3%	88.9%	75.0%	66.7%	80.0%

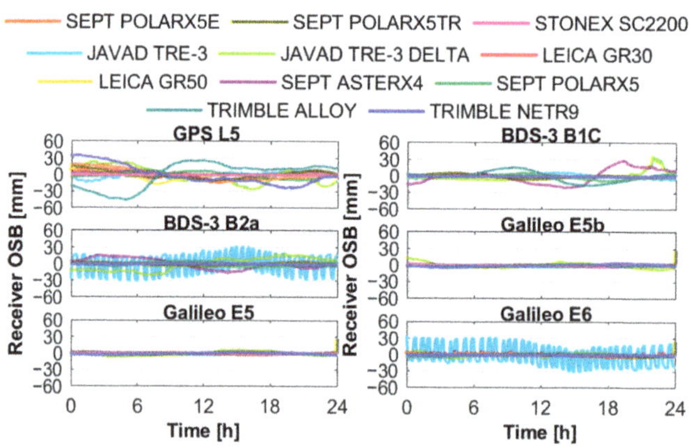

Figure 10. Daily receiver pseudorange and phase time-variant mixed OSB for GPS L5, BDS-3 B1C and B2a, and Galileo E5b, E5, and E6 signals on DOY 011, 2021.

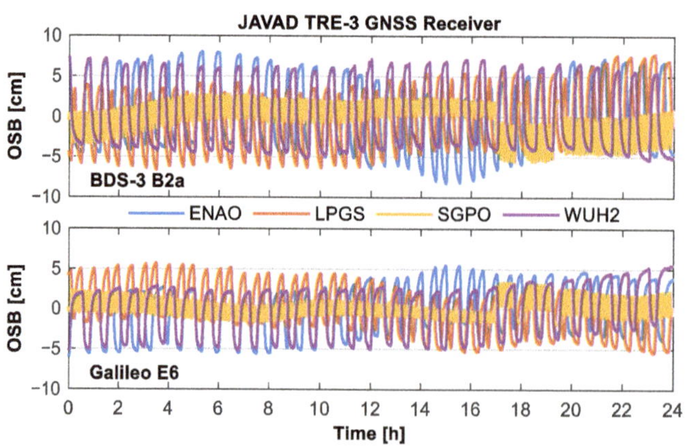

Figure 11. Daily JAVAD TRE-3 receiver pseudorange and phase time-variant mixed OSB at BDS-3 B2a and Galileo E6 signals on DOY 011, 2021.

4.3. The Advantages of the Mixed OSB in MFPPP

To evaluate the benefits of the mixed OSB on MFPPP, we carried out GPS L1 + L2 + L5, BDS-3 B1I + B3I + B2a, BDS-3 B1I + B3I + B1C + B2a, and Galileo E1 + E5a + E5b + E5 + E6 multi-frequency PPP. First, the PPP results and phase residuals were given. Then, the statistical PPP results of all GNSS stations were statistically analyzed in positioning accuracy and initial convergence time.

The observation data for 100 selected stations in a month are used to evaluate GPS, BDS-3, and Galileo MFPPP performance. It is worth noting that the PPP solutions without OSB are the result of only correcting the DCB. Figure 12 depicts the positioning error of the GPS, BDS-3, and Galileo MFPPP. It can be clearly found that GPS, BDS-3, and Galileo MFPPP can be improved in positioning accuracy and convergence time by correcting the satellite pseudorange and phase time-variant mixed OSB. Furthermore, the OSB estimated by Scheme 2 is superior to that of Scheme 1 in stability, accuracy, and noise level, and its PPP performance is relatively more stable and better than that of Scheme 1 in positioning accuracy and initial convergence.

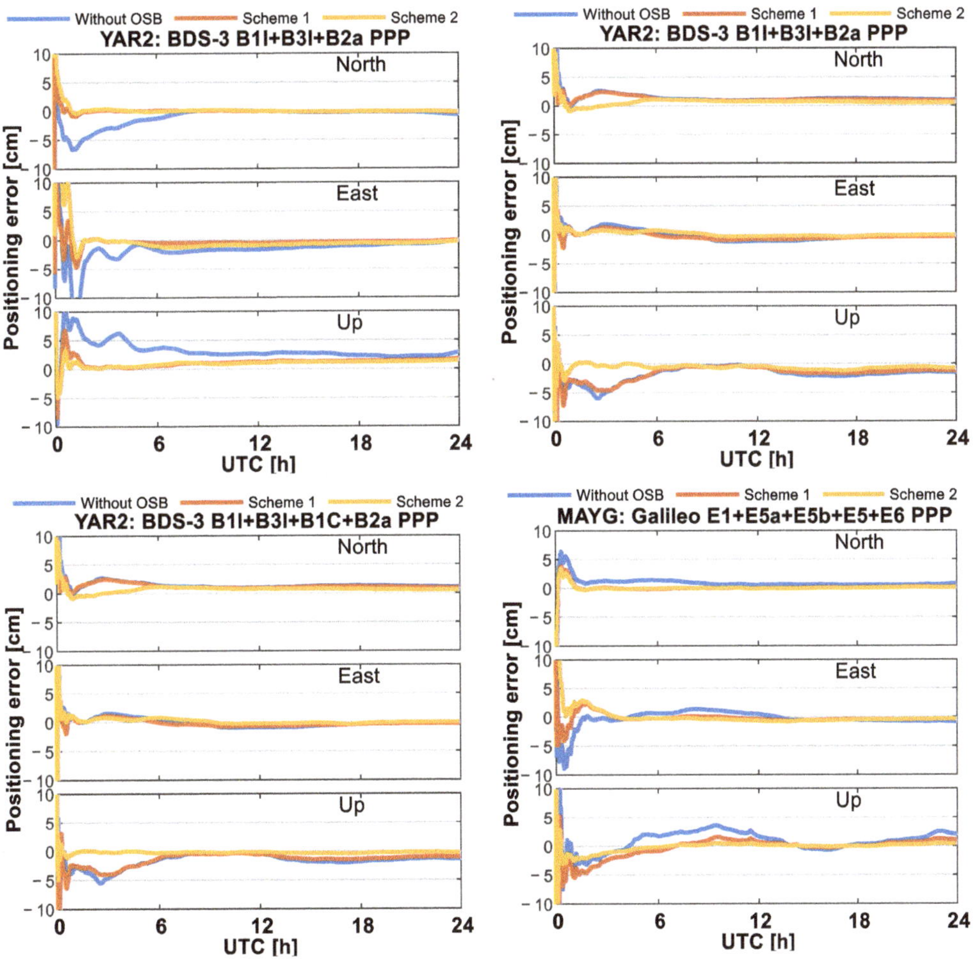

Figure 12. Positioning error of the GPS L1 + L2 + L5, BDS-3 B1I + B3I + B2a, BDS-3 B1I + B3I + B1C + B2a, and Galileo E1 + E5a + E5b + E5 + E6 multi-frequency PPP models on DOY 011, 2021.

Figure 13 illustrates the GPS, BDS-3, and Galileo MFPPP phase residuals, where different colors identify different satellites. The carrier-phase residuals are analyzed for the 24 h result. Significant systematic errors exist in the GPS, BDS-3, and Galileo MFPPP results without OSB correction. The RMS values of the carrier-phase residuals are obviously reduced with OSB corrections.

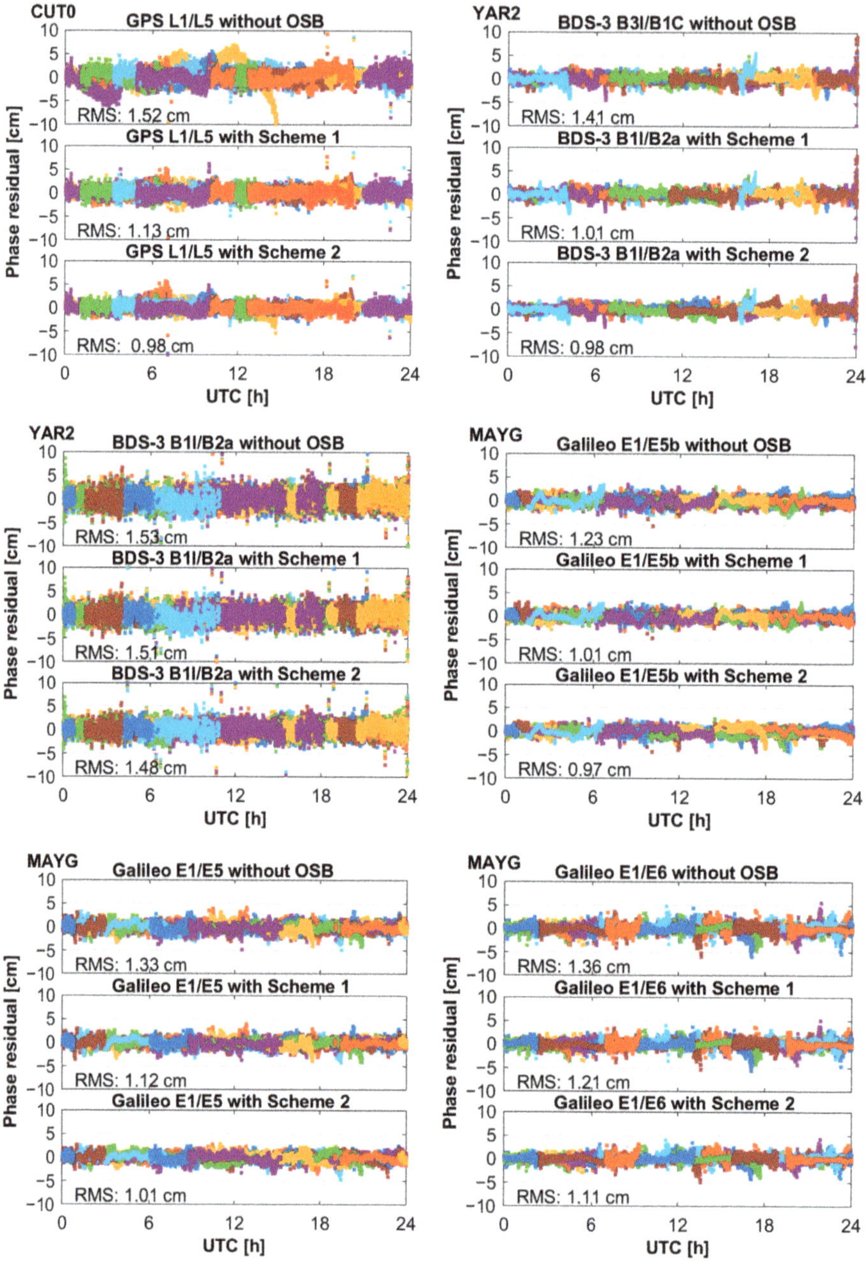

Figure 13. Phase residuals of the GPS L1 + L2 + L5, BDS-3 B1I + B3I + B2a, BDS-3 B1I + B3I + B1C + B2a, and Galileo E1 + E5a + E5b + E5 + E6 multi-frequency PPP on DOY 011, 2021.

Figure 14 depicts the boxplot of the convergence time for GPS L1 + L2 + L5, BDS-3 B1I + B3I + B2a, BDS-3 B1I + B3I + B1C + B2a, and Galileo E1 + E5a + E5b + E5 + E6 multi-frequency PPP. Moreover, Figure 15 indicates the corresponding boxplot of the positioning accuracy in the north, east, and up components. The corresponding median and mean values are also depicted in the figures. The convergence epoch is defined as the 3D positioning errors kept within 10 cm from the current epoch to the next 20 epochs. The multi-frequency PPP positioning performance in the convergence time and positioning accuracy is significantly improved with the mixed OSB correction. Specifically, the PPP performance of Scheme 2 is superior to Scheme 1 due to the low noise and high accuracy of OSB obtained from Scheme 2. For instance, the GPS L1 + L2 + L5 PPP mean convergence time is reduced from 29.40 min to 24.20 min and then to 20.30 min, the BDS-3 B1I + B3I + B2a PPP mean convergence time is reduced from 34.7 min to 30.50 min and then to 27.30 min, the BDS-3 B1I + B3I + B1C + B2a PPP mean convergence time is reduced from 29.5 min to 25.50 min and then to 21.40 min, and the Galileo E1 + E5a + E5b + E5 + E6 PPP mean convergence time is reduced from 20.90 min to 17.50 min and then to 12.30 min, respectively. The GPS L1 + L2 + L5 PPP positioning accuracy is improved from (0.78, 0.95, 1.53) cm to (0.61, 0.75, 1.30) cm and then to (0.54, 0.60, 1.11) cm in the north, east, and up components. The BDS-3 B1I + B3I + B2a triple-frequency PPP positioning accuracy is improved from (0.83, 1.03, 1.93) cm to (0.75, 0.88, 1.55) cm and then to (0.61, 0.70, 1.31) cm in the north, east, and up components. The BDS-3 B1I + B3I + B1C + B2a PPP positioning accuracy is improved from (0.68 0.94, 1.33) cm to (0.51, 0.70, 1.20) cm and then to (0.44, 0.60, 1.10) cm in the north, east, and up components. The Galileo E1 + E5a + E5b + E5 + E6 PPP positioning accuracy is improved from (0.53 0.93, 1.21) cm to (0.43, 0.68, 1.11) cm and then to (0.38, 0.60, 1.01) cm in the north, east, and up components.

In summary, the new undifferenced mixed OSB model coupled with satellite clock offsets avoids the negative impact of the receiver time-variant biases on the satellite OSB estimation. At the same time, the noise of the mixed OSB obtained from the new method is lower, and the new method combines the pseudorange and phase time-varying OSB, which greatly simplifies the client bias correction and can be better applied to multi-frequency PPP applications.

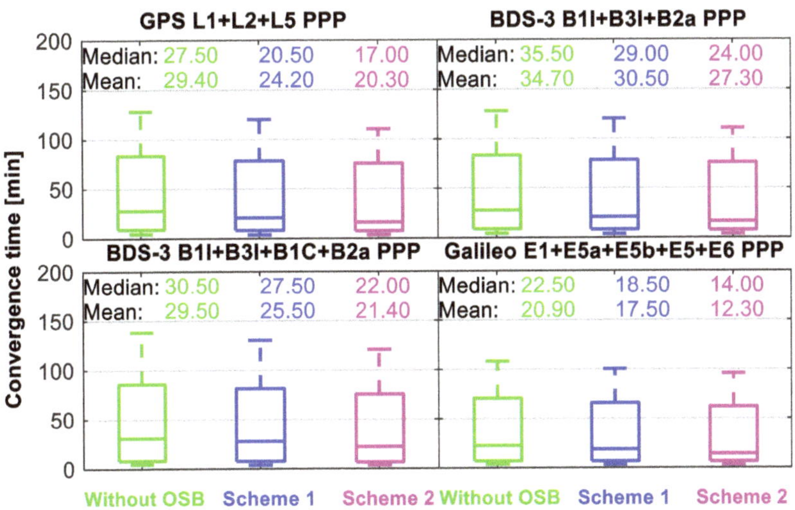

Figure 14. Boxplot of the convergence time for GPS L1 + L2 + L5, BDS-3 B1I + B3I + B2a, BDS-3 B1I + B3I + B1C + B2a, and Galileo E1 + E5a + E5b + E5 + E6 multi-frequency PPP models.

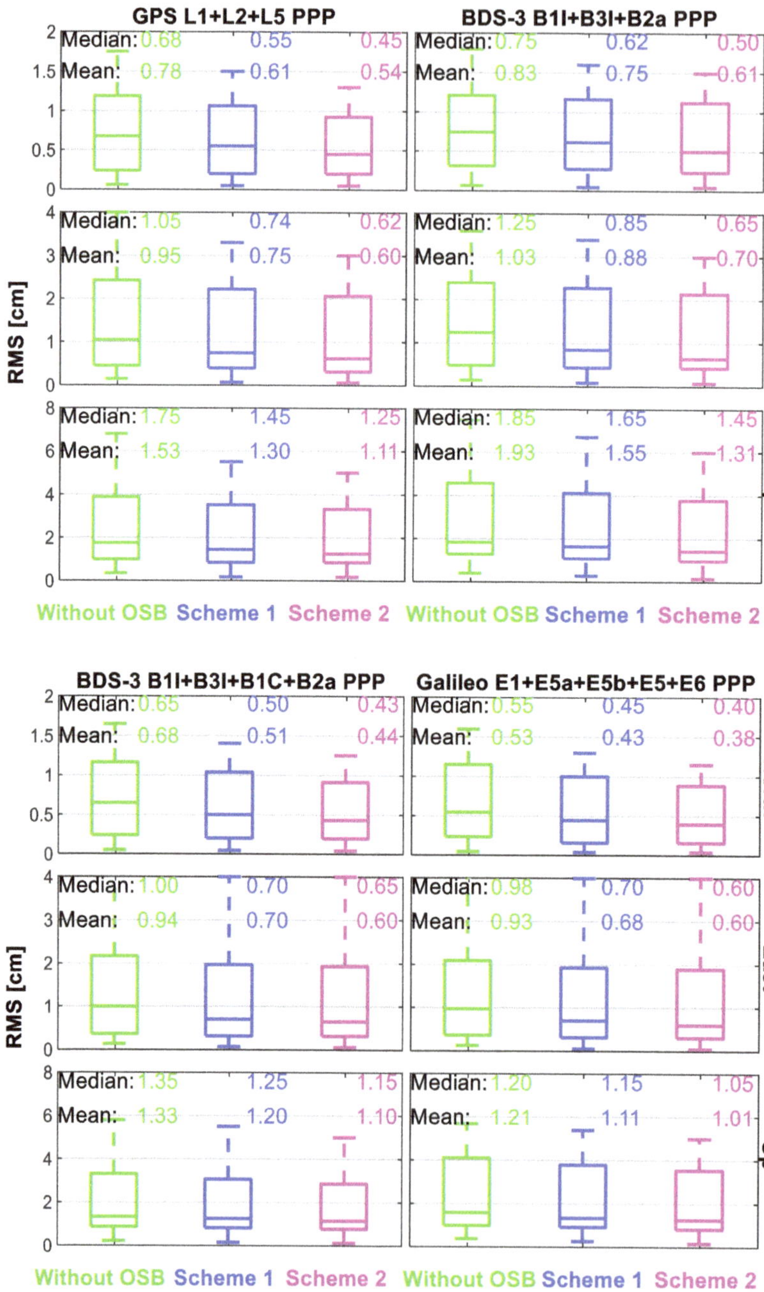

Figure 15. Boxplot of the positioning accuracy for GPS L1 + L2 + L5, BDS-3 B1I + B3I + B2a, BDS-3 B1I + B3I + B1C + B2a, and Galileo E1 + E5a + E5b + E5 + E6 multi-frequency PPP models.

5. Conclusions

The inconsistency of the precise satellite clock estimated by different frequencies is defined as the IFCB. Unfortunately, the traditional phase time-variant OSB model has high noise and low accuracy and is affected by the receiver time-variant biases. The multi-frequency PPP needs to correct both pseudorange OSB and phase time-variant mixed OSB, which makes the client algorithms more complex. For traditional methods, two bias corrections are required to achieve multi-frequency PPP, which makes the client algorithms more complex. To overcome those flaws and simplify the bias correction process of multi-frequency PPP, the definition of pseudorange and carrier-phase time-variant mixed OSB and a new undifferenced model coupled with satellite clock offsets are presented.

The new undifferenced mixed OSB model coupled with satellite clock offsets avoids the negative impact of the receiver time-variant biases on the satellite mixed OSB estimation. For instance, the JAVAD TRE-3 receiver OSB for BDS-3 B2a and Galileo E6 exhibits significant periodic variations, which lead to periodic bias in satellite mixed OSB for the ED GFIF model. The STD improvement of the undifferenced OSB is 7.5–60.9%, and the RMS can be improved by 9.4–66.1%. Similarly, the STD improvement of the ED OSB can be improved by 50.0–87.5%, and the RMS can be improved by 60.0–88.9%. Hence, the new undifferenced mixed OSB with characteristics of low noise level and high accuracy is more suitable for MFPPP.

With the mixed OSB correction, the GPS and BDS-3 MFPPP performance is obviously improved. With the undifferenced mixed OSB correction, the mean convergence time of GPS, BDS-3, and Galileo PPP can be reduced by several minutes, and the positioning accuracy can be improved by 11.5–20.0% for GPS L1 + L2 + L5 from (0.61, 0.75, 1.30) cm to (0.54, 0.60, 1.11) cm, 15.5–20.5% for BDS-3 B1I + B3I + B2a from (0.75, 0.88, 1.55) cm to (0.61, 0.70, 1.31) cm, 8.3–14.2% for BDS-3 B1I + B3I + B1C + B2a from (0.51, 0.70, 1.20) cm to (0.44, 0.60, 1.10) cm, and 9.0–11.7% for Galileo E1 + E5a + E5b + E5 + E6 from (0.43, 0.68, 1.11) cm to (0.38, 0.60, 1.01) cm compared with the traditional IFCB, respectively.

The GNSS pseudorange and phase time-variant mixed OSB concept and the new undifferenced model coupled with satellite clock offsets demonstrate that it is reasonable and beneficial for the GNSS field.

Author Contributions: Conceptualization, G.J. and K.S.; methodology, G.J.; software, G.J.; validation, G.J., K.S. and M.F.; formal analysis, G.J.; investigation, G.J.; resources, K.S.; data curation, K.S.; writing—original draft preparation, G.J.; writing—review and editing, M.F.; visualization, Y.Y.; supervision, K.S.; project administration, G.J. and H.H; funding acquisition, K.S. and H.H. All authors have read and agreed to the published version of the manuscript.

Funding: The work is supported by Key Laboratory Fund (No 614201004012209).

Data Availability Statement: The original contributions presented in the study are included in the article, further inquiries can be directed to the corresponding author.

Acknowledgments: The authors gratefully acknowledge WHU for providing precise orbit and clock products (ftp://igs.gnsswhu.cn/pub/ (accessed on 1 October 2024)). Many thanks to the IGS MGEX for providing the observation data (ftp://igs.gnsswhu.cn/pub/ (accessed on 1 October 2024)).

Conflicts of Interest: The authors declare no conflicts of interest.

References

1. Montenbruck, O.; Hugentobler, U.; Dach, R.; Steigenberger, P.; Hauschild, A. Apparent clock variations of the Block IIF-1 (SVN62) GPS satellite. *GPS Solut.* **2011**, *16*, 303–313. [CrossRef]
2. Yang, Y.; Liu, L.; Li, J.; Yang, Y.; Zhang, T.; Mao, Y.; Sun, B.; Ren, X. Featured services and performance of BDS-3. *Sci. Bull.* **2021**, *66*, 2135–2143. [CrossRef] [PubMed]
3. Dach, R.; Bockmann, E. *International GNSS Service Technical Report 2022 (IGS Annual Report)*; IGS Central Bureau and University of Bern, Bern Open Publishing: Bern, Switzerland, 2023.
4. Falcone, M.; Hahn, J.; Burger, T. Galileo. In *Springer Handbook of Global Navigation Satellite Systems*; Teunissen, P.J.G., Montenbruck, O., Eds.; Springer International Publishing: Cham, Switzerland, 2017; pp. 247–272.
5. Jiao, G.; Song, S.; Jiao, W. Improving BDS-2 and BDS-3 joint precise point positioning with time delay bias estimation. *Meas. Sci. Technol.* **2020**, *31*, 025001. [CrossRef]
6. Li, J.; Yang, Y.; Xu, J.; He, H.; Guo, H. GNSS multi-carrier fast partial ambiguity resolution strategy tested with real BDS/GPS dual- and triple-frequency observations. *GPS Solut.* **2013**, *19*, 5–13. [CrossRef]

7. Li, X.; Liu, G.; Li, X.; Zhou, F.; Feng, G.; Yuan, Y.; Zhang, K. Galileo PPP rapid ambiguity resolution with five-frequency observations. *GPS Solut.* **2019**, *24*, 24. [CrossRef]
8. Montenbruck, O.; Steigenberger, P.; Prange, L.; Deng, Z.; Zhao, Q.; Perosanz, F.; Romero, I.; Noll, C.; Stürze, A.; Weber, G.; et al. The Multi-GNSS Experiment (MGEX) of the International GNSS Service (IGS)—Achievements, prospects and challenges. *Adv. Space Res.* **2017**, *59*, 1671–1697. [CrossRef]
9. Steigenberger, P.; Deng, Z.; Guo, J.; Prange, L.; Song, S.; Montenbruck, O. BeiDou-3 orbit and clock quality of the IGS Multi-GNSS Pilot Project. *Adv. Space Res.* **2023**, *71*, 355–368. [CrossRef]
10. Steigenberger, P.; Montenbruck, O. Galileo status: Orbits, clocks, and positioning. *GPS Solut.* **2016**, *21*, 319–331. [CrossRef]
11. Ye, S.; Zhao, L.; Song, J.; Chen, D.; Jiang, W. Analysis of estimated satellite clock biases and their effects on precise point positioning. *GPS Solut.* **2017**, *22*, 16. [CrossRef]
12. Li, H.; Li, B.; Lou, L.; Yang, L.; Wang, J. Impact of GPS differential code bias in dual- and triple-frequency positioning and satellite clock estimation. *GPS Solut.* **2016**, *21*, 897–903. [CrossRef]
13. Su, K.; Jiao, G. Estimation of BDS pseudorange biases with high temporal resolution: Feasibility, affecting factors, and necessity. *Satell. Navig.* **2023**, *4*, 17. [CrossRef]
14. Villiger, A.; Schaer, S.; Dach, R.; Prange, L.; Sušnik, A.; Jäggi, A. Determination of GNSS pseudo-absolute code biases and their long-term combination. *J. Geod.* **2019**, *93*, 1487–1500. [CrossRef]
15. Wang, N.; Yuan, Y.; Li, Z.; Montenbruck, O.; Tan, B. Determination of differential code biases with multi-GNSS observations. *J. Geod.* **2015**, *90*, 209–228. [CrossRef]
16. Håkansson, M.; Jensen, A.B.O.; Horemuz, M.; Hedling, G. Review of code and phase biases in multi-GNSS positioning. *GPS Solut.* **2016**, *21*, 849–860. [CrossRef]
17. Jin, X.; Song, S.; Zhou, W.; Cheng, N. Multi-GNSS global ionosphere modeling enhanced by virtual observation stations based on IRI-2016 model. *J. Geod.* **2022**, *96*, 81. [CrossRef]
18. Ren, X.; Zhang, X.; Xie, W.; Zhang, K.; Yuan, Y.; Li, X. Global Ionospheric Modelling using Multi-GNSS: BeiDou, Galileo, GLONASS and GPS. *Sci. Rep.* **2016**, *6*, 33499. [CrossRef]
19. Wang, N.; Li, Z.; Duan, B.; Hugentobler, U.; Wang, L. GPS and GLONASS observable-specific code bias estimation: Comparison of solutions from the IGS and MGEX networks. *J. Geod.* **2020**, *94*, 74. [CrossRef]
20. Guo, J.; Geng, J.; Zeng, J.; Song, X.; Defraigne, P. GPS/Galileo/BDS phase bias stream from Wuhan IGS analysis center for real-time PPP ambiguity resolution. *GPS Solut.* **2024**, *28*, 67. [CrossRef]
21. Pan, L.; Zhang, X.; Guo, F.; Liu, J. GPS inter-frequency clock bias estimation for both uncombined and ionospheric-free combined triple-frequency precise point positioning. *J. Geod.* **2018**, *93*, 473–487. [CrossRef]
22. Pan, L.; Zhang, X.; Li, X.; Liu, J.; Guo, F.; Yuan, Y. GPS inter-frequency clock bias modeling and prediction for real-time precise point positioning. *GPS Solut.* **2018**, *22*, 76. [CrossRef]
23. Gong, X.; Gu, S.; Lou, Y.; Zheng, F.; Yang, X.; Wang, Z.; Liu, J. Research on empirical correction models of GPS Block IIF and BDS satellite inter-frequency clock bias. *J. Geod.* **2020**, *94*, 36. [CrossRef]
24. Su, K.; Jin, S.; Jiao, G. GNSS carrier phase time-variant observable-specific signal bias (OSB) handling: An absolute bias perspective in multi-frequency PPP. *GPS Solut.* **2022**, *26*, 71. [CrossRef]
25. Zhang, B.; Teunissen, P.J.G.; Yuan, Y. On the short-term temporal variations of GNSS receiver differential phase biases. *J. Geod.* **2016**, *91*, 563–572. [CrossRef]
26. Guo, J.; Geng, J. GPS satellite clock determination in case of inter-frequency clock biases for triple-frequency precise point positioning. *J. Geod.* **2017**, *92*, 1133–1142. [CrossRef]
27. Jiao, G.; Song, S.; Su, K. Improving undifferenced precise satellite clock estimation with BDS-3 quad-frequency B1I/B3I/B1C/B2a observations for precise point positioning. *GPS Solut.* **2022**, *27*, 28. [CrossRef]
28. Fan, L.; Shi, C.; Li, M.; Wang, C.; Zheng, F.; Jing, G.; Zhang, J. GPS satellite inter-frequency clock bias estimation using triple-frequency raw observations. *J. Geod.* **2019**, *93*, 2465–2479. [CrossRef]
29. Geng, J.; Wen, Q.; Zhang, Q.; Li, G.; Zhang, K. GNSS observable-specific phase biases for all-frequency PPP ambiguity resolution. *J. Geod.* **2022**, *96*, 11. [CrossRef]
30. Jiao, G.; Song, S.; Su, K. Analysis of Galileo five-frequency precise satellite clock estimation models and its effect on multi-frequency PPP. *Measurement* **2023**, *206*, 112297. [CrossRef]
31. Leick, A.; Rapoport, L.; Tatarnikov, D. GNSS Positioning Approaches. In *GPS Satellite Surveying*; Wiley: Hoboken, NJ, USA, 2015; pp. 257–399.
32. Li, X.; Wang, Q.; Wu, J.; Yuan, Y.; Xiong, Y.; Gong, X.; Wu, Z. Multi-GNSS products and services at iGMAS Wuhan Innovation Application Center: Strategy and evaluation. *Satell. Navig.* **2022**, *3*, 20. [CrossRef]
33. Pan, L.; Zhang, X.; Li, X.; Liu, J.; Li, X. Characteristics of inter-frequency clock bias for Block IIF satellites and its effect on triple-frequency GPS precise point positioning. *GPS Solut.* **2016**, *21*, 811–822. [CrossRef]
34. Schaer, S. *SINEX BIAS-Solution (Software/Technique) INdependent EXchange Format for GNSS BIASes Version 1.00*; IGS Workshop on GNSS Biases: Bern, Switzerland, 2016.
35. Guo, J.; Xu, X.; Zhao, Q.; Liu, J. Precise orbit determination for quad-constellation satellites at Wuhan University: Strategy, result validation, and comparison. *J. Geod.* **2015**, *90*, 143–159. [CrossRef]

36. Hopfield, H.S. Two-quartic tropospheric refractivity profile for correcting satellite data. *J. Geophys. Res.* **1969**, *74*, 4487–4499. [CrossRef]
37. Landskron, D.; Bohm, J. VMF3/GPT3: Refined discrete and empirical troposphere mapping functions. *J. Geod.* **2018**, *92*, 349–360. [CrossRef] [PubMed]
38. Su, K.; Jin, S.; Jiao, G. Assessment of multi-frequency global navigation satellite system precise point positioning models using GPS, BeiDou, GLONASS, Galileo and QZSS. *Meas. Sci. Technol.* **2020**, *31*, 064008. [CrossRef]
39. Beyerle, G. Carrier phase wind-up in GPS reflectometry. *GPS Solut.* **2008**, *13*, 191–198. [CrossRef]
40. Petit, G.; Luzum, B. *IERS Conventions (2010)*; No. IERS-TN-36; Bureau International des Poids et Mesures Sevres (France): Saint-Cloud, France, 2010.
41. Xia, Y.; Pan, S.; Zhao, Q.; Wang, D.; Gao, W. Characteristics and modelling of BDS satellite inter-frequency clock bias for triple-frequency PPP. *Surv. Rev.* **2018**, *52*, 38–48. [CrossRef]

remote sensing

MDPI

Article

Multipath Mitigation in Single-Frequency Multi-GNSS Tightly Combined Positioning via a Modified Multipath Hemispherical Map Method

Yuan Tao [1,2], Chao Liu [1,3,*], Runfa Tong [1,2], Xingwang Zhao [1], Yong Feng [2] and Jian Wang [4]

1 School of Spatial Information and Geomatics Engineering, Anhui University of Science and Technology, Huainan 232001, China; yuantao@cumt.edu.cn (Y.T.); tb24160010a41@cumt.edu.cn (R.T.); xwzhao@aust.edu.cn (X.Z.)
2 School of Environment and Spatial Informatics, China University of Mining and Technology, Xuzhou 221116, China; tb24160017a41@cumt.edu.cn
3 School of Mining and Geomatics Engineering, Hebei University of Engineering, Handan 056038, China
4 School of Geomatics and Urban Spatial Informatics, Beijing University of Civil Engineering and Architecture, Beijing 102616, China; wangjian@bucea.edu.cn
* Correspondence: chliu1@aust.edu.cn; Tel.: +86-182-5542-0251

Abstract: Multipath is a source of error that limits the Global Navigation Satellite System (GNSS) positioning precision in short baselines. The tightly combined model between systems increases the number of observations and enhances the strength of the mathematical model owing to the continuous improvement in GNSS. Multipath mitigation of the multi-GNSS tightly combined model can improve the positioning precision in complex environments. Interoperability of the multipath hemispherical map (MHM) models of different systems can enhance the performance of the MHM model due to the small multipath differences in single overlapping frequencies. The adoption of advanced sidereal filtering (ASF) to model the multipath for each satellite brings computational challenges owing to the characteristics of the multi-constellation heterogeneity of different systems; the balance efficiency and precision become the key issues affecting the performance of the MHM model owing to the sparse characteristics of the satellite distribution. Therefore, we propose a modified MHM method to mitigate the multipath for single-frequency multi-GNSS tightly combined positioning. The method divides the hemispherical map into 36×9 grids at $10° \times 10°$ resolution and then searches with the elevation angle and azimuth angle as independent variables to obtain the multipath value of the nearest point. We used the k-d tree to improve the search efficiency without affecting precision. Experiments show that the proposed method improves the mean precision over ASF by 10.20%, 10.77%, and 9.29% for GPS, BDS, and Galileo satellite single-difference residuals, respectively. The precision improvements of the modified MHM in the E, N, and U directions were 32.82%, 40.65%, and 31.97%, respectively. The modified MHM exhibits greater performance and behaves more consistently.

Keywords: multi-GNSS; tightly combined model; multipath mitigation; MHM; ASF

1. Introduction

The Global Navigation Satellite System (GNSS) double-difference technique can quickly provide precise position information and achieve mm or even higher precision on a global level after the carrier phase ambiguities are fixed [1,2]. Therefore, the double-difference technique is widely applied in fields such as deformation monitoring [3–6]. Most of the errors can be mitigated in a short baseline using the double-difference technique, including the satellite clock error, receiver clock error, tropospheric delay, and ionospheric delay [7–9]. However, multipaths that depend on the station observation environment cannot be eliminated using existing methods, including differential techniques and parameterization [10]. Therefore, several researchers have proposed site selection, and hardware- and software-based methods for mitigating multipaths.

Firstly, site selection reduces the impact of multipath effects, mainly by choosing open environments, but it cannot avoid the limitations of objective conditions [11]. Secondly, hardware-based multipath suppression techniques are mainly performed by improving the "narrow correlator" [12] of the receiver or using a multipath suppression antenna [13]; however, these methods cannot eliminate multipaths and the instruments are costly [14]. Finally, for software-based methods that exploit the spatio-temporal repeatability of multipath effects in static environments, sidereal filtering (SF) and a multipath hemispherical map (MHM) can be implemented to generate multipath mitigation models for real-time GNSS data processing [15].

Genrich and Bock (2006) proposed SF to mitigate a multipath based on the constant environment of the station and the repeatability of GPS satellites [13]. The wavelet decomposition technique was applied to extract a carrier-phase multipath from GPS double-difference observation residuals. Advanced sidereal filtering (ASF) uses each satellite repeat period instead of the mean repeat period to effectively mitigate multipaths [16]. The orbital repeat period shift time method was used to calculate the repeat period of satellites, which has the advantages of simple calculation and better orbital real-time performance [17]. Zhong et al. (2010) and Ye et al. (2015) used the "zero mean" assumption and the ASF method to effectively mitigate a multipath [9,18]. However, calculating the repeat period for each satellite undoubtedly increases the complexity of multipath mitigation in the case of large differences in repeat periods. In addition, the operational period of each satellite changes slowly over time, reducing the real-time effect of mitigating multipaths [19].

Therefore, lookup table methods, including MPS (multipath stacking) and MHM, effectively mitigate the multipath by calculating the mean of the multipath in a higher-resolution discrete grid based on the fact that the multipath depends only on the orbital positions in the sky when the environment of the main multipath is kept constant [15,20]. However, an excessively high resolution causes insufficient multipath mitigation in part of the grid, and an excessively low resolution causes a high-frequency multipath in the grid to be difficult to mitigate because of the sparse distribution characteristics of satellite orbits. The multipath hemispherical model based on trend surface analysis (T-MHM) fits the multipath in the grid through a linear regression model, which was effectively mitigated for both high-frequency and low-frequency multipaths [21,22], but there was a pathological problem with the regression equations solving in some of the grids [11]. Tian et al. (2023) used the least squares method to mitigate the multipath and improve the positioning precision by utilizing the spatial autocorrelation characteristic of the multipath [4]. The resolution of the grid limits the mitigation effect of the multipath with different frequencies using MHM and its extensions [23], or it is difficult to fit a complex nonlinear multipath model using a certain established function model; the multipath also consumes a large amount of computer memory during the modeling process, which leads to insufficient processing efficiency. Ren et al. (2023) proposed an MHM model with geographic cut-off elevation constraints that mask unsatisfactory signals while reducing low-frequency multipaths, which helps GNSS deformation monitoring in complex environments [24]. Pan et al. (2024) used multipath modeling as a regression task to fit multipaths in the spatial domain using a machine learning approach, demonstrating that machine learning-based multipath mitigation methods are effective and easy to use [25].

With the rapid development of multi-frequency multi-constellation GNSS, more than 130 GNSS satellites are available. The latest generation of GNSS satellites has three or more frequencies. It is expected that receivers developed by manufacturers will primarily support two signal frequencies: 1575.42 MHz and 1176.45 MHz since the major systems have overlapping frequency signals (except GLONASS), increasing the compatibility of GNSS positioning. The overlapping frequencies between BDS-3/GPS/Galileo increase the probability of a tight combination of the three systems [26]. The tightly combined model adds additional observations and improves the strength of the model by utilizing a single-reference satellite to process observations from all the system satellites involved [27,28]. In addition, the increasing number and frequency of satellites not only increases the number of GNSS observations but also reduces the effect of the multipath on positioning precision [14]. Therefore, in dense vegetation and tall building environments where observation satellites are obscured, the application of the tightly combined model [29] and multipath mitigation have important significance. Geng et al. (2024) established an interoperable MHM mitigating multipath using the overlap frequency signals of GPS, Galileo, and BDS-3, showing that the interoperable GPS/Galileo/BDs-3 MHM is able to improve the spatial resolution, modeling efficiency, and correction performance [30]. The carrier phase residuals based on GPS, BDS-3, and Galileo overlapping frequency show a strong correlation at the same spatial position. Lu et al. (2023) proposed that the multipath joint modeling and correction method can improve the positioning performance more than the independent modeling and correction [31].

Taking into account the above multiple factors, we extracted the multipath from the single-frequency multi-system tightly combined double-difference observation equations and established an MHM model with the station as the center and elevation and azimuth as the axes. We then divided the hemispherical map into 36×9 grids at $10^\circ \times 10^\circ$ resolution. The elevation and azimuth angles of the satellites were input into the model and the K-nearest neighbor (KNN) algorithm was utilized to search for the nearest points and obtain the values of the corresponding multipaths. Previous KNN search algorithms have low search efficiency, and in this study, we used the k-d tree to improve search efficiency. The k-d tree was proposed by Jon Louis Bentley in 1975 [32]. It is a tree data structure (binary tree) used in multidimensional space to solve the problem of data organization and search in high-dimensional space, which can significantly speed up the search operations in multidimensional space, especially when the data volume is large, such as nearest neighbor search, range search, and point queries.

The rest of this paper is organized as follows. Section 2 describes the principles of the tightly combined model and the orbital repeat period of multi-GNSS satellites. The proposed method for multipath mitigation is described in Section 3. Section 4 verifies the performance of the proposed method through experiments. Finally, Section 5 summarizes the study.

2. Principle

2.1. GPS/Galileo/BDS Tightly Combined Model and Multipath Extraction

The selection method of the reference satellite is used as the basic condition to distinguish between loosely and tightly combined models in the process of establishing multi-GNSS double-difference observation equations. Selecting reference satellites separately in each system is a loosely combined model, and selecting a common reference satellite in all systems is a tightly combined model.

The difference between the reference satellites and the other satellites in the system eliminates most of the systematic errors, but each double difference in the system consumes the observations of the three reference satellites in the GPS/Galileo/BDS-3 loosely combined model.

Early experiments on GPS/Galileo have demonstrated that there is differential inter-system bias (DISB) when different types of receivers observe L1/E2 and L5/E5a signals at both ends of the baseline [28], and the existence of DISB is mainly caused by hardware delays associated with different receivers and satellite signals of different systems. In this regard, the DISB must be estimated and corrected for GPS/Galileo/BDS tightly combined observations using different receiver types. The GPS/Galileo/BDS tightly combined double-difference model that considers DISB can be obtained by choosing a GPS satellite as a reference satellite between the GPS/Galileo/BDS systems. Table 1 lists the GPS/Galileo/BDS system signal frequencies and wavelengths.

Table 1. Frequency and wavelength of GPS/Galileo/BDS signals.

GNSS Systems	Frequency Band	Frequency (MHz)	Wavelength (cm)
GPS/Galileo/BDS-3	L1/E1/B1C	1575.42	19.03
GPS	L2	1227.60	24.42
GPS/Galileo/BDS-3	L5/E5a/B2a	1176.45	25.48
Galileo/BDS-3	E5b/B2b	1207.14	24.83

When the receiver type is different, the DISB of the GPS/Galileo/BDS-3 tightly combined mode is more stable and can be corrected in advance using estimation. When using overlapping frequency data from GPS/Galileo/BDS-3 of the same type of receiver for the tightly combined mode, the DISB is close to 0 [26]. Therefore, the GPS/Galileo/BDS-3 tightly combined double-difference model without DISB can be obtained by selecting a GPS satellite as the common reference satellite in GPS/Galileo/BDS-3:

$$\begin{cases} P_{kl,i}^{1_G s_G} = \rho_{kl}^{1_G s_G} + e_{kl,i}^{1_G s_G} \\ \phi_{kl,i}^{1_G s_G} = \rho_{kl}^{1_G s_G} + \lambda_i N_{kl,i}^{1_G s_G} + \varepsilon_{kl,i}^{1_G s_G} \\ P_{kl,i}^{1_G q_E} = \rho_{kl}^{1_G q_E} + e_{kl,i}^{1_G q_E} \\ \phi_{kl,i}^{1_G q_E} = \rho_{kl}^{1_G q_E} + \lambda_i N_{kl,i}^{1,q_E} + \varepsilon_{kl,i}^{1_G q_E} \\ P_{kl,i}^{1_G b_C} = \rho_{kl}^{1_G b_C} + e_{kl,i}^{1_G b_C} \\ \phi_{kl,i}^{1_G b_C} = \rho_{kl}^{1_G b_C} + \lambda_i N_{kl,i}^{1,b_C} + \varepsilon_{kl,i}^{1_G b_C} \end{cases} \tag{1}$$

where ϕ is the carrier phase observation, P is the pseudo-range observation, ρ is the geometric distance between the satellite and GNSS receiver, λ is the frequency wavelength, N is ambiguity, e and ε are the observation noise of the pseudo-range and carrier phase, 1_G indicates that one GPS satellite is used as a common reference satellite, s_G, q_E, and b_C are GPS, Galileo, and BDS non-reference satellites. The GNSS receivers k and l receive the pseudo-range and carrier phase observations of the s satellites of the GPS, q satellites of the Galileo, and b satellites of the BDS, at the overlap frequency $i(i = 1, 2, \ldots, f)$.

The double-difference technique eliminates most of the systematic errors, including orbital error, receiver clock error, and satellite clock error, and the tropospheric and ionospheric delays are ignored in short baseline positioning. Therefore, the carrier phase double-difference residuals are mainly multipath and noise, and the double-difference residuals for each satellite need to be extracted from the observation equations to mitigate the multipath. The double-difference equation is linearized to obtain the following equation:

$$\boldsymbol{V} = \boldsymbol{A}x + \boldsymbol{B}y - \boldsymbol{L} \tag{2}$$

where x is the unknown baseline vector; y is the integer ambiguity; $\boldsymbol{A}$ and $\boldsymbol{B}$ are the design matrices of the unknown baseline vector and integer ambiguity; $\boldsymbol{L}$ is the carrier phase constant term; $\boldsymbol{V}$ is the double-difference residuals.

2.2. Orbital Repeat Period of Multi-GNSS Satellites

The motion of the satellite relative to the station is repeatability; therefore, the repeatability of the multipath effect depends on the orbital motion of the satellite. The repeat period of the multipath is essentially the same as that of the satellite relative to the station. Therefore, SF and its extension methods rely on the time delay of the satellite.

We analyze the orbital repeat period of multi-GNSS by calculating the amount of time delay for different satellite orbital repeats from the broadcast ephemeris in 2022. The angular velocity n of the satellite can be obtained at each ephemeris broadcasting moment based on the orbital radius and angular velocity correction provided in the ephemeris:

$$n = \sqrt{\mu}/a^{\frac{3}{2}} + n_c \tag{3}$$

where a is the orbital half-length axis of the different satellites, μ is the gravitational constant of Earth for the different satellites, and n_c is the mean angular velocity correction. The orbital repeat period of different system satellites was calculated as following:

$$t = R \times T_e - O \times 2\pi \cdot n \quad (4)$$

where T_e is the rotational period of the Earth, which is calculated to be 86,164.0996 s through the mean solar time, R and O are integers prime to each other, R is the rotation cycle of the Earth, and O is the number of orbits accomplished in one orbital repeat period of the satellite. R and O for different constellations of satellites are listed in Table 2.

Table 2. Values of R and O for different constellations of satellites. R means the rotation cycle of the Earth; O means the number of orbits accomplished in one orbital repeat period of the satellite.

Satellites	R (Cycle)	O (Cycle)
GPS	1	2
BDS GEO/IGSO	1	1
BDS MEO	7	13
Galileo	10	7

In a stable environment, the multipath is only related to the geometric relationship between the receiver, the reflection source, and the satellite position. Therefore, it is necessary to consider the satellite's operating period and calculate the satellite's sub-satellite point. Due to the heterogeneous characteristics of multi-GNSS constellations, the operating period of each type of satellite is considered separately. Therefore, it is necessary to calculate the time delay of the repeat period of each satellite's sub-satellite point. The satellite's time delay is calculated based on the orbital parameters given by the broadcast ephemeris, and the single-difference residuals are aligned according to the time delay of each satellite to implement the ASF method to reduce the subsequent multipath impact.

Figure 1 shows the time delays of orbital repeat periods of the GPS, BDS, and Galileo satellites for 2022. GPS satellites G10, G13, G14, and G27 change within 4 s in a year. The remaining satellites show variations of less than 15 s in a year and no more than 2 s in a month. This indicates that GPS satellites have precise orbit control. The time delays of the geostationary orbit (GEO) satellites of the BDS were in the range of 215–260 s, with a maximum of 12 s of daily variation. The medium earth orbit (MEO) satellite has a time delay of approximately 1700 s and operates more stably. The time delay of the inclined geosynchronous orbit (IGSO) satellite was in the range of 210–260 s, with a variation of no more than 20 s a month. The Galileo satellite time delays are above 2400 s and do not change for more than 30 s in a year. The time delays exhibit significant variations with satellite operation time, and ASF modeling of the multipath requires the calculation of time delays for each satellite. In addition, there are significant orbital maneuvers for GPS satellites and IGSO satellites of BDS, with maneuver adjustment periods of approximately 12 and 6 months, respectively. The variable time delays and satellite maneuvers will limit the application of ASF in real time.

The time delays of most of the satellites in the figure exhibit short-period perturbations and long-period linear properties. The time delays of satellites of different systems vary considerably, and those of satellites of the same system are still not the same. Some satellites are affected by satellite maneuvers. It is difficult to achieve multi-GNSS multipaths that are mitigated using repeatability modeling methods in the coordinate domain. This limits the advantages of ASF for multipath mitigation.

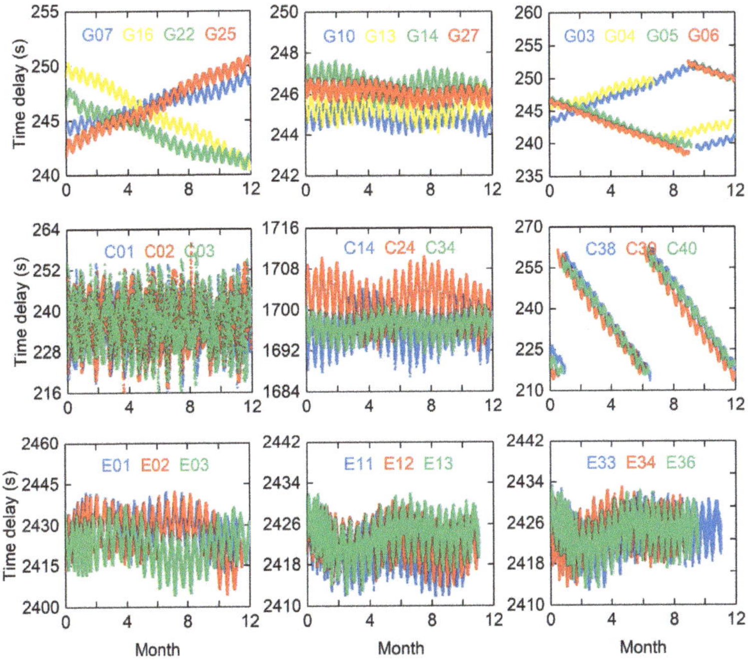

Figure 1. Time delays of orbital repeat period of GPS, BDS, and Galileo satellites in 2022.

3. Proposed Method

The established modified MHM model is based on the mechanism of spatial repeatability of multipaths, while the temporal repeatability of multipaths is derived from spatial repeatability. The modified MHM model was established with the station as the center and elevation and azimuth angles as the coordinate axes. We divided the hemispherical map into 3 × 3, 6 × 6, and 60 × 60 grids according to 20° × 20°, 10° × 10°, and 1° × 1° resolutions in the range of elevation angles of 30°–90° and azimuth angles of 120°–180°, respectively. After experimentation, we found that the precision difference between the three resolution grid models is very small (the improvements in the single-difference residual multipath mitigation of DOY92 corresponding to 20° × 20°, 10° × 10°, and 1° × 1° are 36.67%, 36.68%, and 36.68%, respectively), but the times to search for all points for the 20° × 20°, 10° × 10°, and 1° × 1° resolution grids are 3:45:23, 0:15:35, and 1:23:15, respectively. This indicates that the grid resolution has no effect on the precision of the multipath model, but affects the efficiency of the search, which is best using the 10° × 10° resolution. Therefore, we divided the hemispherical map into 36 × 9 grids at a 10° × 10° resolution, as shown in Figure 2. Satellite single-difference residuals were loaded into this model based on their positions. In the subsequent multipath mitigation process, the elevation and azimuth angles of the satellite were input into the model to search for the nearest point. The searched multipath values were introduced into the observation equation for multipath mitigation, realizing the purpose of real-time multipath mitigation.

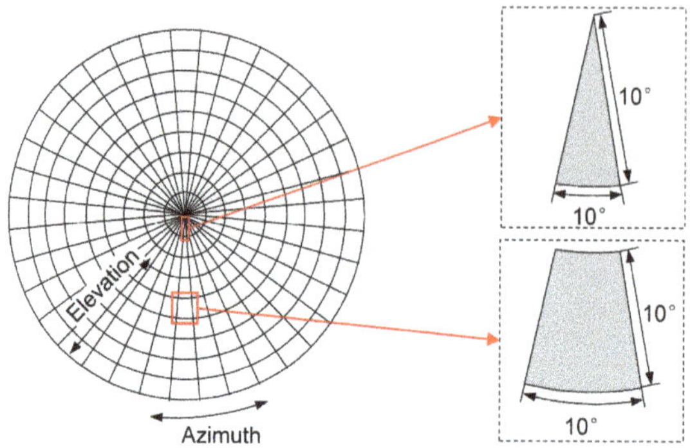

Figure 2. Hemispherical map with two typical grids.

Previously, the search method generally used the K-nearest neighbor (KNN) algorithm [33], which was implemented by traversing all points. Calculating the distance of query points from all points is very time consuming when the dataset is large. Therefore, to improve the efficiency of KNN, we can consider using special data structures to store data to reduce the number of distance calculations. In this study, a k-d tree was used to increase the KNN search efficiency. The basic principle of the KNN is as follows:

$$D = \{d_1, d_2, \cdots, d_c\}\ \mathrm{KNN}_{k_1} \Rightarrow \{m_1, m_2, \cdots, m_u, \cdots m_{k_1}\} \tag{5}$$

where D presents all c data points; the nearest k_1 points of query point d in D are found by KNN search. The composition of KNN_{k_1} must satisfy the following conditions.

$$\|d, d_m\| \leq \|d, d_j\| \tag{6}$$

where $\|d, d_m\|$ is the present distance between d and d_m; $j \in [1, c] - KNN_{k_1}$.

The k-d tree iteratively splits the space with hyperplanes and constructs binary trees, allowing the KNN to search for the logarithmic time complexity. At each level of the binary tree, the data were divided into two groups based on the dimension with the highest variance. A k-d tree must be constructed before performing a KNN search, as shown in Figure 3. The first step is to construct the root node in a hyperrectangle that contains all data points in space; the next step is to select a node on the coordinate axes, create a hyperplane perpendicular to the coordinate axes through the node, and divide the hyperrectangle into two subregions; the third step is to repeat the second step in the case of at least one data point in the subregion; otherwise, stop the implementation.

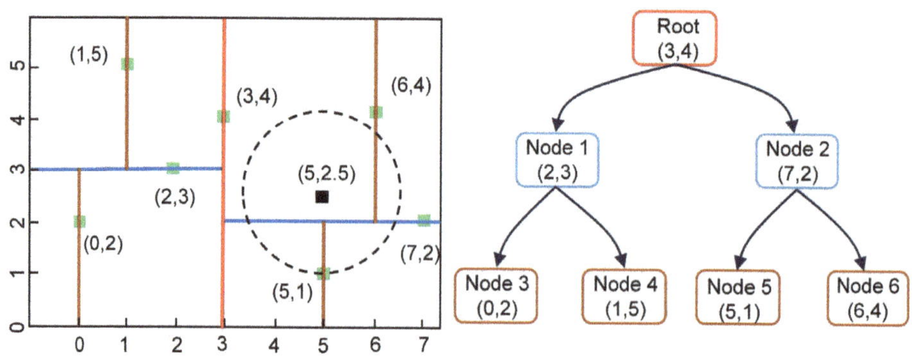

Figure 3. Construction process of the k-d tree.

The process of multipath mitigation in multi-GNSS tightly combined positioning based on the modified MHM model is shown in Figure 4.

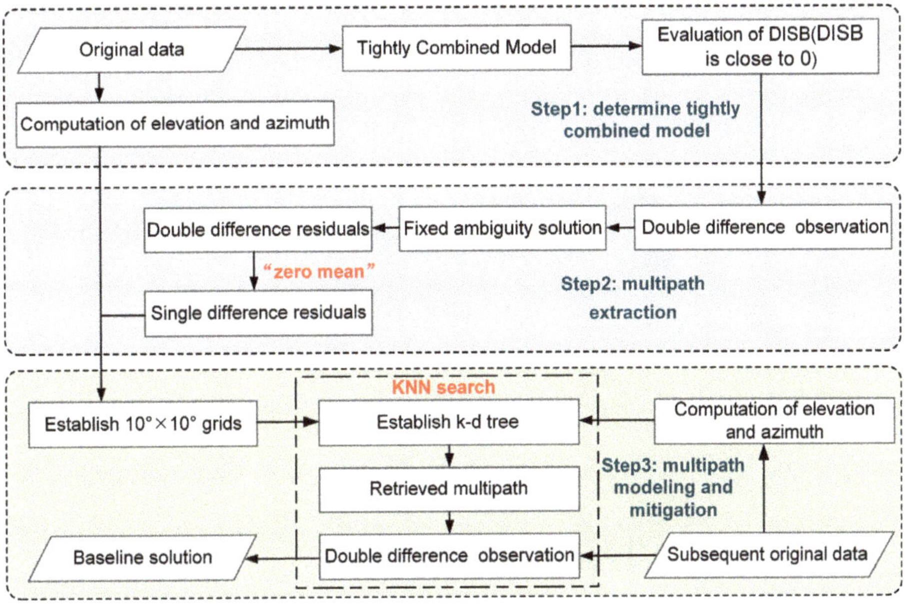

Figure 4. The processing flow of the modified MHM model.

Step 1: Determine the tightly combined model. After evaluating the size and stability of the DISB estimated from the GPS/BDS/Galileo tight combination, the DISB of the same type of receiver is close to zero, so we use the tightly combined multi-GNSS model.

Step 2: Multipath extraction. The double-difference observation equation is first composed, then the fixed ambiguity rate is enhanced by a priori baseline vectors to obtain the carrier phase double-difference residuals, and further the single-difference residuals are obtained by the "zero-mean" assumption.

Step 3: Multipath modeling and mitigation. The above single-difference residuals are formed into a $10^\circ \times 10^\circ$ grid according to the elevation and azimuth angles, and a k-d tree is built in each grid. The elevation and azimuth angles of the satellite on the subsequent original data were used as query points by using the KNN search in the k-d tree to obtain the retrieved multipath, which is brought into the double-difference observation equation for real-time multipath mitigation. Finally, baseline precision was evaluated.

4. Experiment Analysis

4.1. Data Collection

In order to verify the multipath mitigation performance of the modified MHM method for GPS/Galileo/BDS tightly combined positioning, the baseline length between the two stations was set to be 12.92 m during data collection, and both the base station and the rover station were erected on the roof of the School of Spatial Information and Geomatics Engineering, Anhui University of Science and Technology, China. The observation environment is illustrated in Figure 5. The multipath error mainly comes from the wall and the ground, the height of the base station antenna is about 2.3 m, and the height of the wall around the rover station is about 1.5 m. The north side of the rover station is at ground level, with the receiver at a height of about 17 m above ground level, and the south side is on the roof of the building, with a distance of about 13 m from the opposite wall. This multipath error source can meet the requirements we deal with in the experiment. We use two receivers of the same type for orientation processing to eliminate the antenna phase-center variation, and the DISB between the systems can be negligible in the short baseline. Ionospheric and tropospheric errors can be eliminated using the double-difference observation equation, and ambiguity is fixed using the LAMBDA algorithm [34]. The receivers we use can receive the GPS triple-frequency, Galileo triple-frequency, and BDS five-frequency observations. Data were collected from 23 March to 16 April 2022 (DOY82-DOY106) for a total of 25 days, which covers the minimum time required for Galileo satellites with a repeat period of ten days, with a data sampling rate of 1 s, a satellite cutoff elevation angle of 15°, and a PDOP threshold set to 8.

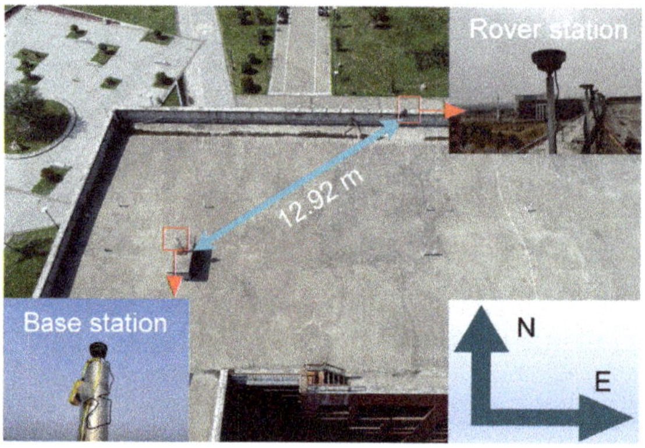

Figure 5. Observation environment around the station.

4.2. Single-Difference Residuals Analysis

Before conducting formal experiments, it is necessary to distinguish between factors due to poor satellite geometry and the multipath due to ground reflectors. Figure 6 shows the position dilution of precision (PDOP), horizontal dilution of precision (HDOP), vertical dilution of precision (VDOP), and the number of observable satellites for multi-GNSS. The number of satellites is maintained between 20 and 30 in the entire period, which is more observable than the single systems because of the tightly combined multi-GNSS single-frequency observation. The PDOP, VDOP, and HDOP describe the strength of the satellite geometric distribution on the positioning precision. To avoid poor satellite results in the spatial geometric distribution, an empirical value of eight was set as the PDOP threshold when the data were collected. As can be seen in Figure 6, both the PDOP and HDOP are below 1.5, while the VDOP is below 1, indicating that the positioning measurements are sufficiently accurate in both the horizontal and vertical directions. Therefore, it can be considered that the main factor affecting positioning precision is the multipath.

As the elevation angle of the satellite decreases, the GNSS signal is interfered with more ground reflectors during propagation; therefore, the multipath signal becomes more complex. It is necessary to evaluate multipaths from different elevation angles and satellite systems. The double-difference residuals are converted to single-difference residuals based on the "zero-mean" assumption, and the elevation angle corresponding to the single-difference residuals at each epoch is solved. The single-difference residuals are analyzed instead of the multipath because the single-difference residuals are mainly composed of the multipath and random noise, and the total sum of the random noise is 0. The elevation angles were grouped from low to high at 5° intervals, and the root mean square (RMS) of the single-difference residuals was calculated for each group. Figure 7 demonstrates the effect of single-difference residuals with increasing elevation angle. The RMS of the single-difference residuals of the three systems gradually decreased as the elevation angle increased, and eventually remained between 20 mm and 30 mm. Fluctuations were observed between 45° and 55° and between 85° and 90°. The differences in the single-difference residuals with the elevation angle among the three systems were small.

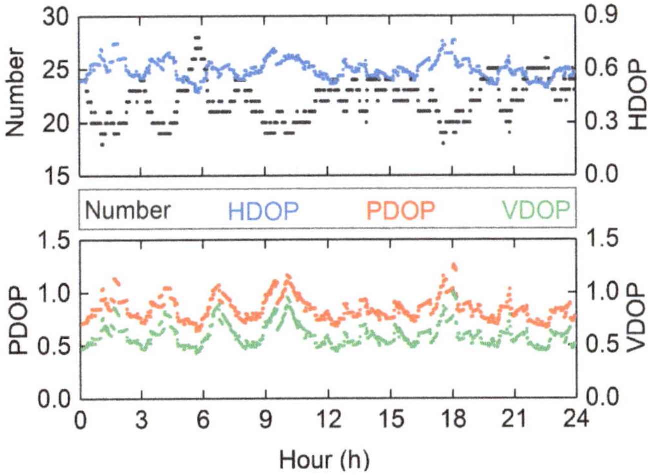

Figure 6. PDOP, HDOP, VDOP, and the number of observable satellites for multi-GNSS.

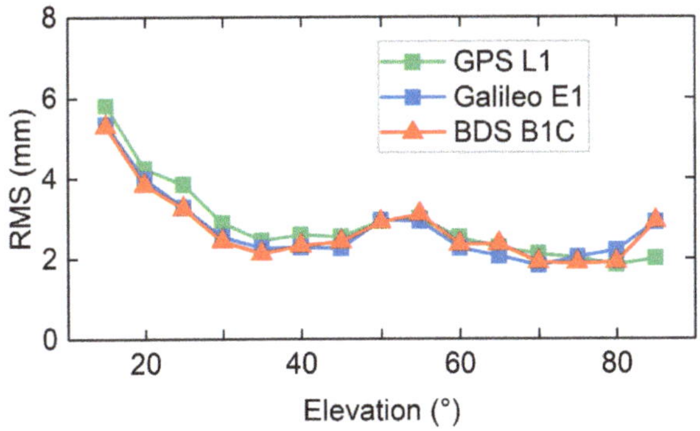

Figure 7. RMS of single-difference residuals for different elevation.

Figure 8a shows the single-difference residuals of the GPS satellite in DOY82-84, and it can be seen that the trend of single-difference residuals of this satellite for three consecutive days is highly correlated. The single-difference residuals of the previous day after removing this time delay are highly correlated with the current single-difference residuals because there is a time delay between the satellite single-difference residuals of the previous day and the current day. Time delays can be calculated using the broadcast ephemeris. Figure 8b shows the distribution of single-difference residuals of Galileo satellites in DOY82, DOY83, and DOY93, which have inconsistent visibility and lower correlation of their single-difference residuals in two adjacent days. Galileo satellites have a more significant correlation between DOY82 and DOY92, with a correlation of 0.6436, despite a time interval of 10 days, and a much longer time delay than GPS satellites. Figure 8c,d show single difference residual results for the MEO and IGSO satellites of the BDS, respectively. From the distribution of the single-difference residuals of DOY82, DOY83, and DOY89, it can be concluded that there is no obvious similarity between the single-difference residuals of DOY82 and DOY83, and there are differences in the visibility of the satellites. However, the single-difference residuals of DOY82 and DOY89 present a strong correlation, and it is verified that the orbital repeat period of the MEO satellites of the BDS is 7 days. The fluctuations of the single-difference residuals of IGSO satellites in DOY82, DOY83, and DOY84 shows the strongest correlation in two consecutive days, and their time delays are also closer to those of GPS, which show repeatability within a sidereal day. However, the correlation of the single-difference residuals of DOY82 and DOY84 was lower than that of the two adjacent days.

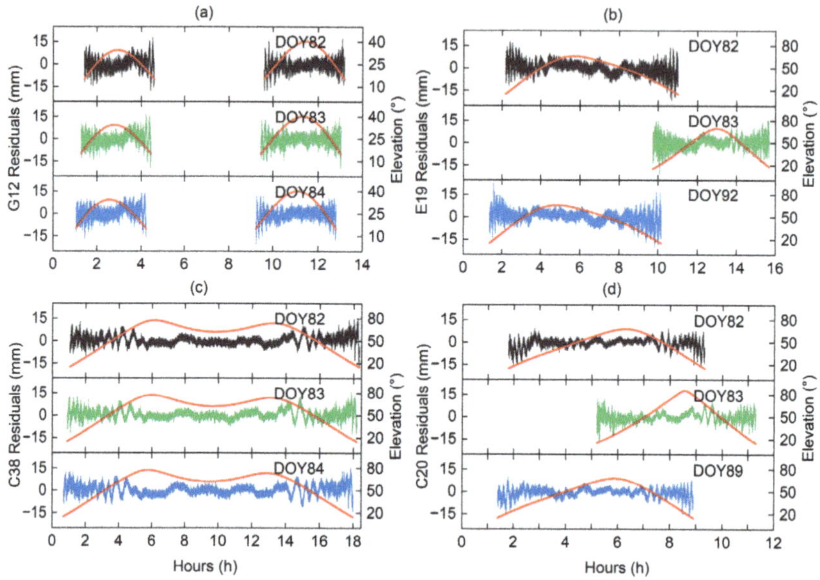

Figure 8. Single-difference residuals and elevation of GPS MEO G12 (**a**), Galileo MEO E19 (**b**), BDS-3 MEO C38 (**c**) and IGSO C20 (**d**).

4.3. Multipath Mitigation by Modified MHM

To analyze the multipath mitigation performance of the modified MHM method in multi-GNSS single frequency combined positioning, we first mitigated the multipath on the single-difference residuals, and second compared the differences between single and multiple systems in the coordinate series. Figure 9 shows the original single-difference residuals for the G12 satellite and the multipath obtained by the modified MHM and the residuals mitigated by the modified MHM. As shown in the figure and local figure, the single residuals obtained by the modified MHM were consistent with the trend of the original residuals. The noise at both ends of the mitigated residuals is relatively large, which is consistent with the previous analysis. In the low elevation angle region, the signal-to-noise ratio of the satellite is small, and the noise is larger.

Figure 10 shows the hemispherical maps of the original single difference residual and that mitigated using the modified MHM method. GPS, Galileo, and BDS satellites have severe multipath effects at elevation angles of 15–30°, and the single difference residual varies significantly with the elevation angle within the black rectangle in Figure 10d. After modified MHM mitigation of the multipath, as shown in Figure 10h, the color changes from red-blue to yellow-green in the region of 15–30° elevation angle, and the multipath was mitigated in the region above 30°.

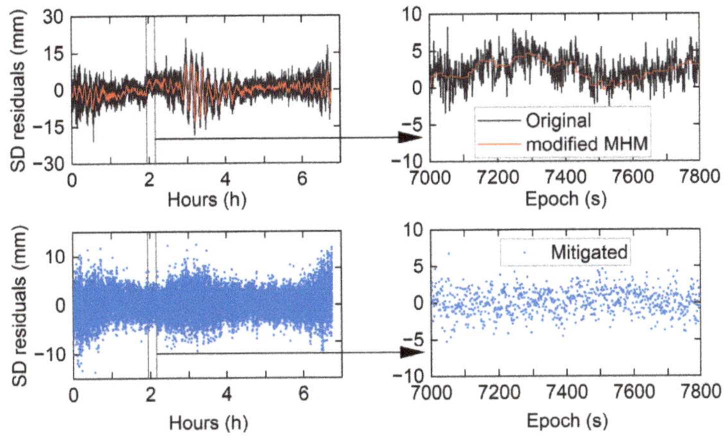

Figure 9. Original single-difference residuals for the G12 satellite and the multipath obtained by modified MHM and the residuals mitigated by modified MHM.

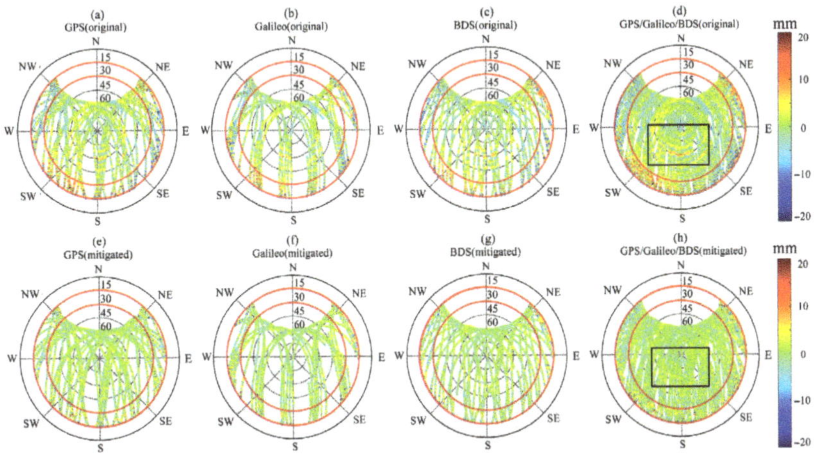

Figure 10. Hemispherical maps of the original single difference residual (**a**–**d**) and mitigated by using the modified MHM method (**e**–**h**) for GPS, Galileo, BDS, and GPS/Galileo/BDS combined observations, respectively. (The black rectangles in (**d**,**h**) are used to mark the areas in the sky map where multipath errors and improvements are evident).

We processed the GPS, BDS, Galileo, and GPS/BDS/Galileo combined observations to distinguish between single- and multiple-system multipath mitigation performances. Figure 11 shows the RMS of the original residuals and the residuals mitigated by the modified MHM of DOY92 in the E/N/U directions. G, C, E, and G/C/E represent GPS, BDS, Galileo, and GPS/BDS/Galileo, respectively. In the E direction, the RMS of the original residuals of GPS/BDS/Galileo is significantly lower than that of a single system, which is 1.79 mm; after multipath mitigation, the RMS of GPS/BDS/Galileo decreases by 0.41 mm, 0.36 mm, and 0.34 mm, respectively, compared to GPS, BDS, and Galileo. In the N direction, the RMS of the original residuals of Galileo is lower than that of the single systems and multiple systems, but the RMS of the residuals after multipath mitigation of BDS is the lowest. In the U direction, the original RMS of GPS/BDS/Galileo is 1.88 mm, 0.81 mm, and 2.50 mm lower than that of GPS, BDS, and Galileo, respectively; after mitigating the multipath, the GPS/BDS/Galileo is reduced to 3.00 mm, significantly lower than the other results. We mitigated the multipath of seven consecutive days using the modified MHM method and obtained the following results. Table 3 displays the mean precision improvements of GPS, BDS, Galileo, and GPS/BDS/Galileo. We found that the precision improvements of GPS/BDS/Galileo are generally higher than those of the single systems. However, it should be noted that in the U direction, the RMS of BDS is higher than that of GPS/BDS/Galileo, before and after multipath mitigation. This unexpected result explains that the increase in the number of satellites does not necessarily lead to an improvement in positioning precision, but because of the increase in the number of satellites, the impact of the multipath effect will also increase.

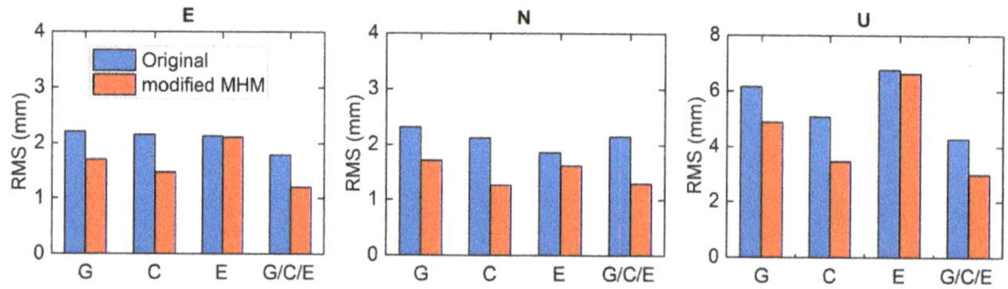

Figure 11. RMS of the original residuals and residuals mitigated by the modified MHM of DOY92 in the E/N/U directions.

Table 3. Mean precision improvement of GPS, BDS, Galileo, and GPS/BDS/Galileo.

Direction	Mean Precision Improvement (%)			
	GPS	BDS	Galileo	GPS/BDS/Galileo
E	18.41	25.56	5.60	29.19
N	18.22	29.76	12.80	31.42
U	17.61	29.20	7.39	24.73

4.4. Comparison with Other Methods in the Observation Domain and the Coordinate Domain

To compare the multipath mitigation performance of the ASF and modified MHM methods for multi-GNSS, experiments were conducted separately. The results of four representative satellites before and after multipath mitigation with single-difference residuals were selected from the three systems. Figure 12 shows the original single-difference residuals of G12, E19, C20, and C38 and the residuals mitigated by the ASF and modified MHM methods. The original single-difference residuals mainly included multipath and random noise. After multipath mitigation of the ASF and modified MHM methods, the residuals only contain random noise and the residual multipath, and the mitigation effect can be intuitively seen according to the fluctuation of the residual series. It can be intuitively found that the amplitude changes of the G12, C38, and E19 residuals are relatively small, indicating that the multipath has been eliminated compared with the original residuals and after multipath mitigation by ASF and modified MHM, respectively. After the modified MHM mitigates the multipath, the RMS values of G12, C20, and E19 are 2.29 mm,1.38 mm, and 2.28 mm, respectively. After ASF mitigates the multipath, the RMS values of G12, C20, and E19 are 2.37 mm, 1.32 mm, and 2.27 mm, respectively. We found that the difference between the two methods was not significant. As for the residual of C38 after multipath mitigation, the RMS of the modified MHM is 0.27 mm lower than that of the ASF, and the performance of the modified MHM is slightly better than that of the ASF. It is worth mentioning that the MEO repeat period of BDS causes a significant "end effect" after ASF processes multipath, while the MHM approach is not affected.

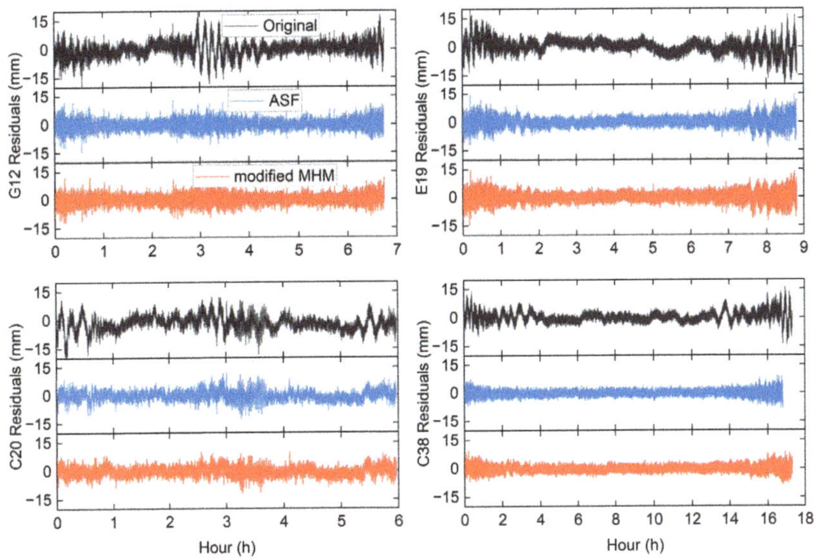

Figure 12. Original single-difference residuals of G12, E19, C20, and C38 and the residuals mitigated by ASF and modified MHM methods.

We calculated the precision improvement of all satellite single-difference residuals after multipath mitigation by ASF and modified MHM for 25 consecutive days to further verify the difference between the two methods. Figure 13 shows the precision improvement of the GPS, BDS, and Galileo single-difference residuals. The red and blue dashed lines represent the mean improvements in ASF and modified MHM, respectively. As can be seen from the figure, in terms of GPS single-difference residuals, although the precision improvement of both methods is declining, the precision improvement of the modified MHM is significantly greater than that of the ASF intuitively, and the mean precision improvement of the modified MHM is 10.17% higher than that of the ASF. For BDS single-difference residuals, the precision improvement of the modified MHM is greater than that of the ASF within 15 days, and the fluctuation of the ASF is significantly larger, with mean precision improvements of 32.93% and 22.16% for ASF and modified MHM, respectively. For Galileo, the fluctuation trends of the modified MHM and ASF were similar, but in DOY93-98, DOY102, and DOY103, the mean precision improvement of ASF was 15.12% lower than that of modified MHM. The mean precision improvement of the modified MHM over ASF was 10.20%, 10.77%, and 9.29% for the single-difference residuals for the GPS, BDS, and Galileo satellites, respectively.

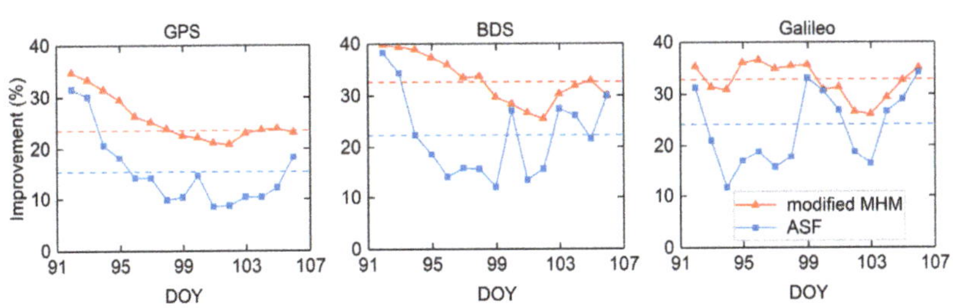

Figure 13. Precision improvement of GPS, BDS, and Galileo single-difference residuals, respectively.

The performance of ASF can be influenced by two main factors: (1) the precision of multipath extraction from carrier phase residuals; (2) the precision of each satellite's repeat period. Additionally, multipath effects are most similar in adjacent repeat periods and tend to diverge over time. Although the repeat period of GPS satellites is relatively more stable compared to BDS-3 and Galileo (Figure 1), the multipath differences continue to increase, resulting in a gradual decline in ASF performance. In contrast, the repeat periods of BDS and Galileo are not fully consistent with GPS and exhibit greater variability. This inconsistency leads to abrupt changes in ASF performance on DOY 100 and DOY 99. In summary, statistical analysis of the single-difference residuals before and after multipath mitigation shows that the MHM has greater performance and behaves more consistently.

The purpose of the multipath mitigation of double-difference observations is to improve the positioning precision, and the mitigated double-difference observations are used to fix the ambiguity and further estimate the baseline coordinates. The baseline coordinates of the station were determined based on the single epoch model, and the precise reference coordinates of the station were averaged over multiple consecutive days without mitigation for the multipath model. Uncorrected double-difference observations were also used to generate baseline coordinates for comparison purposes. Figure 14 shows the DOY92 original baseline series and the series after multipath mitigation using the ASF and modified MHM methods. The distribution of the baseline series can be seen in Figure 15. After mitigating the multipath, the series obeys the normal distribution, indicating that the multipath has been eliminated. The E, N, and U components are differenced with accurate reference station coordinates and then are fast Fourier transformed to obtain the power spectral density (PSD) to further compare the differences between the two methods in the frequency domain.

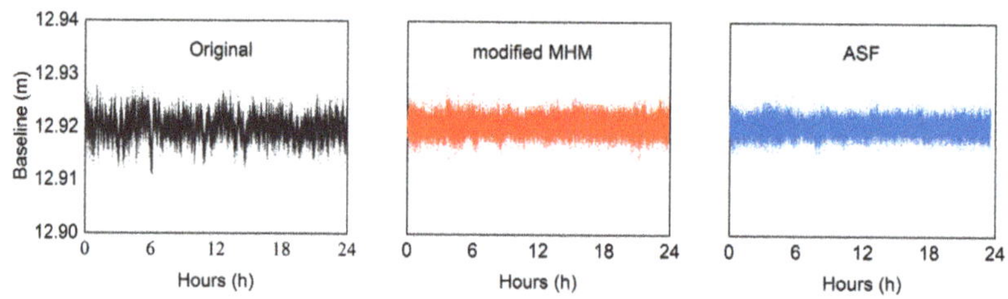

Figure 14. Original baseline series and the series after multipath mitigation by using ASF and modified MHM methods in DOY92.

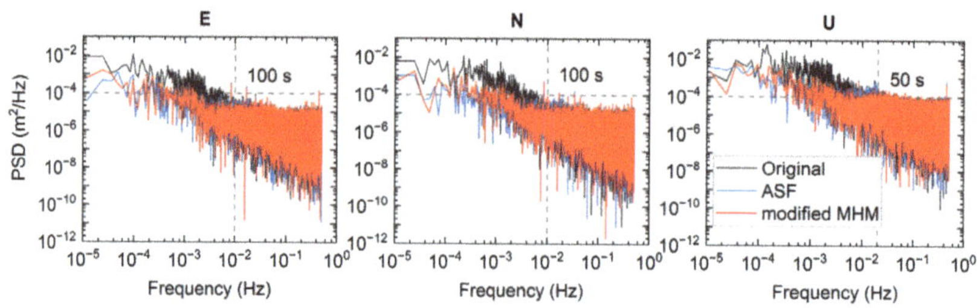

Figure 15. PSD of the original series and the series mitigated by both methods.

Figure 15 illustrates the PSD of the original series and the series mitigated by both methods, and both methods can eliminate most of the middle- to high-frequency multipath. In the frequency below 0.01 Hz, the PSD of the two are more similar on E/N directions, but in the range of 0.01–0.02 Hz, the PSD of the modified MHM is lower than that of the ASF in the U direction. Compared with the ASF, the modified MHM has a positive effect on mitigating the middle-to-high-frequency multipath in the period of 50–100 s.

In order to compare and analyze the differences between our method and the MHM in terms of accuracy and computational efficiency, we added experiments on the MHM.

The multipath depends on the position of the satellite and not on the specific satellite, based on the spatio-temporal correlation of the multipath. The MHM method for multipath mitigation is as follows. The single-difference residuals of GPS, Galileo, and BDS for a total of 10 days are denoised to obtain multipaths, and the multipaths are placed in a grid. The grid is defined by elevation and azimuth angle in a carrier coordinate system [15]. The grids were divided according to certain elevation and azimuth angle intervals. The MHM model is established by taking the mean value of the single-difference residuals in the grid as the multipath of the grid. In order to achieve more accurate multipath error reduction, considering the large amount of data in the multi-GNSS environment, we adopted an MHM grid resolution of $0.1^\circ \times 0.1^\circ$. Such a high resolution helps to improve the refinement of the model, thereby more effectively capturing and reducing the multipath effect. In the subsequent multipath mitigation process, the grid corresponding to the elevation and azimuth angles of the satellite is found in the MHM model, and the multipath of the grid is introduced into the observation equation for multipath mitigation. Comparing the computational efficiency of MHM and the modified MHM, we found that, excluding the time for reading and saving data, the time for establishing the MHM model was 6:30:05 in the range of elevation angles of 30°–90° and azimuth angles of 120°–180°. The grid search times of $20^\circ \times 20^\circ$, $10^\circ \times 10^\circ$, and $1^\circ \times 1^\circ$ were 3:45:23, 0:15:35, and 1:23:15, respectively. Therefore, the k-d number enhanced the search method with a $10^\circ \times 10^\circ$ grid resolution is the most efficient.

The coordinate precision improvement statistics results after multipath mitigation based on the three models are in Figure 16. The coordinate components for 15 consecutive days were statistically analyzed to further distinguish the differences between the ASF and modified MHM methods for multipath mitigation in continuous data. Figure 16 shows the precision improvement of the original components and components after multipath mitigation by both methods from DOY92-DOY106. In the E direction, the precision improvements of the ASF and modified MHM are more similar in DOY92-97, but after DOY92, the precision improvements of the modified MHM are greater than those of the ASF. In the N direction, after DOY95, the precision improvement of the modified MHM was higher than that of the ASF. In the U direction, after DOY96, the precision improvements of the modified MHM are greater than those of the ASF. The mean precision improvement of the modified MHM is 27.94%, 28.06%, and 21.81% in the E, N, and U directions, respectively. The mean precision improvements of the modified MHM over the ASF were 1.40%, 2.00%, and 1.11% in the E, N, and U directions, respectively. The precision improvement of MHM is more variable in the E and N directions, and the precision improvement in the E direction of DOY99-101 and the N direction of DOY100-102 is significantly lower than that of the ASF and the improved MHM. In the U direction, the whole precision performance of MHM is better. The mean precision improvement of MHM in the E, N, and U directions are 24.97%, 23.85%, and 24.768%. The experimental results show that the mean improvement of the modified MHM is 2.97% and 4.21% higher than those of the MHM. However, in the U direction, the average improvement of the MHM is improved by 2.95% higher than that of the modified MHM. Nevertheless, from the overall performance point of view, the modified MHM still performs better than the MHM in multipath mitigation.

SF mainly relies on the satellite repeat period to mitigate the multipath. ASF, as an extension of the SF method, calculates the repeat period of each satellite separately to enhance the multipath mitigation performance. For multi-GNSS multipath mitigation, the calculation of the repeat period of different systems becomes cumbersome. At the same time, the calculation method of the repeat period is inevitably affected by satellite maneuvers. The satellite repeat period changes slowly over time. All of the above reasons will cause the performance of ASF in correcting the multipath to continue to decrease. The modified MHM is not affected by the satellite repeat period, so the modified MHM method performs better than the ASF method in mitigating the multipath.

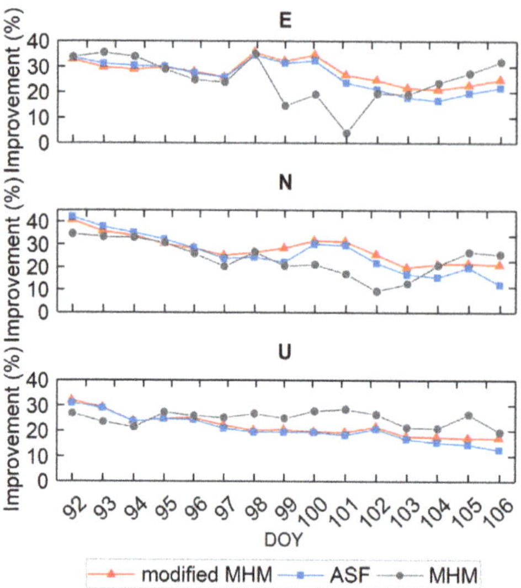

Figure 16. Precision improvement of original components and the components after multipath mitigation by three methods from DOY92-DOY106.

5. Conclusions

We proposed a modified MHM method to mitigate the multipath for single-frequency multi-GNSS tightly combined positioning. In the case of single overlapping frequency, the difference of the multipath between different systems is small; the performance of the multipath model can be improved by interoperation between multi-GNSS systems. Due to the sparsity of satellite orbits and the slight orbit offset, the traditional MHM mitigating multipaths may suffer from omissions. To avoid the shortcomings of multipath mitigation, we use KNN to search for multipaths. We adopt the multi-dimensional data structure of k-d tree to speed up the KNN search operation in the space, and also build a grid with $10^{\circ} \times 10^{\circ}$ resolution to improve the search efficiency while ensuring the precision. The experimental results are as follows:

We used a modified MHM for multipath mitigation of the GPS, BDS, Galileo, and GPS/BDS/Galileo combinations. It can be observed that the precision improvement of the GPS/BDS/Galileo combination was higher than that of a single system. However, the RMS of BDS was higher than that of GPS/BDS/Galileo in the U direction. This indicates that the increased satellite signals may also lead to an increased multipath effect, which in turn decreases the positioning precision.

We mitigated the multipath of the GPS/BDS/Galileo combination using ASF and modified MHM and compared it with the uncorrected values. The mean precision of the modified MHM over the ASF was improved by 10.20%, 10.77%, and 9.29% for the GPS, BDS, and Galileo satellites' single-difference residuals, respectively. The mean precision of the MHM in the E, N, and U directions increased by 1.40%, 2.00%, and 1.11%, respectively, compared with the ASF. The MHM outperforms the ASF for multipath mitigation in the case of the GPS\Galileo\BDS combination. The conclusion is that the modified MHM has better performance and more consistent behavior, and can improve positioning precision in quasi-static environments.

Author Contributions: Methodology, Y.T., C.L. and R.T.; software, Y.T. and C.L.; validation, Y.T., R.T. and J.W.; formal analysis, C.L., R.T., X.Z. and Y.F.; data curation, Y.T., R.T., X.Z. and Y.F.; writing—original draft preparation, Y.T.; writing—review and editing, C.L., X.Z. and J.W. All authors have read and agreed to the published version of the manuscript.

Funding: This research was funded by the Key Project of Natural Science Research in Universities of Anhui Province (grant numbers 2022AH050849, KJ2020A0312, and KJ2021A0443), the Youth Project of Anhui Natural Science (grant number 2008085QD179), the Natural Science Foundation of Beijing (grant number 8222011), and the Science and Technology Research Project of Colleges and Universities in Hebei Province (grant number ZD2021023).

Data Availability Statement: Upon reasonable request, the data that support the findings of this study are available from the corresponding author.

Conflicts of Interest: The authors declare no conflicts of interest.

References

1. Xi, R.; Chen, Q.; Meng, X.; Jiang, W.; An, X.; He, Q. A multi-GNSS differential phase kinematic post-processing method. *Remote Sens.* **2020**, *12*, 2727. [CrossRef]
2. Li, G.; Geng, J.; Guo, J.; Zhou, S.; Lin, S. GPS+ Galileo tightly combined RTK positioning for medium-to-long baselines based on partial ambiguity resolution. *J. Glob. Position. Syst.* **2018**, *16*, 1. [CrossRef]
3. Tong, R.; Liu, C.; Tao, Y.; Fan, Y.; Chen, J. Deformation Information Extraction from Multi-GNSS Coordinate Series Based on EWT-ICA-R. *Sustainability* **2023**, *15*, 4578. [CrossRef]
4. Tian, Y.; Liu, Z.; Lin, M.; Li, K. Modelling and mitigation of GNSS multipath effects by least-squares collocation considering spatial autocorrelation. *J. Geod.* **2023**, *97*, 37. [CrossRef]
5. Radočaj, D.; Plaščak, I.; Jurišić, M. Global navigation satellite systems as state-of-the-art solutions in precision agriculture: A review of studies indexed in the web of science. *Agriculture* **2023**, *13*, 1417. [CrossRef]
6. Han, J.-h.; Park, C.-h.; Park, Y.-J.; Kwon, J.H. Preliminary Results of the Development of a Single-Frequency GNSS RTK-Based Autonomous Driving System for a Speed Sprayer. *J. Sens.* **2019**, *2019*, 4687819. [CrossRef]
7. Tong, R.; Liu, C.; Tao, Y.; Wang, X.; Sun, J. ConvGRU-MHM: A CNN GRU-enhanced MHM for mitigating GNSS multipath. *Meas. Sci. Technol.* **2024**, *35*, 045007. [CrossRef]
8. Tao, Y.; Liu, C.; Chen, T.; Zhao, X.; Liu, C.; Hu, H.; Zhou, T.; Xin, H. Real-Time Multipath Mitigation in Multi-GNSS Short Baseline Positioning via CNN-LSTM Method. *Math. Probl. Eng.* **2021**, *2021*, 6573230. [CrossRef]
9. Zhong, P.; Ding, X.; Yuan, L.; Xu, Y.; Kwok, K.; Chen, Y. Sidereal filtering based on single differences for mitigating GPS multipath effects on short baselines. *J. Geod.* **2010**, *84*, 145–158. [CrossRef]
10. Liu, C.; Lin, P.; Zhao, X.; Gao, J. Reducing GPS carrier phase errors in the measurement and position domains for short-distance static relative positioning. *Acta Geod. Geophys.* **2016**, *51*, 81–93. [CrossRef]
11. Sun, A.; Zhang, Q.; Gao, X.; Meng, X.; Zhang, Y.; Hancock, C. BDS dual-frequency carrier phase multipath hemispherical map model and its application in real-time deformation monitoring. *Sensors* **2023**, *23*, 6357. [CrossRef]
12. Xie, L.; Cui, X.; Zhao, S.; Lu, M. Mitigating multipath bias using a dual-polarization antenna: Theoretical performance, algorithm design, and simulation. *Sensors* **2017**, *17*, 359. [CrossRef]
13. Genrich, J.F.; Bock, Y. Instantaneous geodetic positioning with 10–50 Hz GPS measurements: Noise characteristics and implications for monitoring networks. *J. Geophys. Res.: Solid Earth* **2006**, *111*, B03403. [CrossRef]
14. Zheng, K.; Zhang, X.; Li, P.; Li, X.; Ge, M.; Guo, F.; Sang, J.; Schuh, H. Multipath extraction and mitigation for high-rate multi-GNSS precise point positioning. *J. Geod.* **2019**, *93*, 2037–2051. [CrossRef]
15. Dong, D.; Wang, M.; Chen, W.; Zeng, Z.; Song, L.; Zhang, Q.; Cai, M.; Cheng, Y.; Lv, J. Mitigation of multipath effect in GNSS short baseline positioning by the multipath hemispherical map. *J. Geod.* **2016**, *90*, 255–262. [CrossRef]
16. Wang, M.; Wang, J.; Dong, D.; Chen, W.; Li, H.; Wang, Z. Advanced sidereal filtering for mitigating multipath effects in GNSS short baseline positioning. *ISPRS Int. J. Geo-Inf.* **2018**, *7*, 228. [CrossRef]
17. Su, M.; Zheng, J.; Yang, Y.; Wu, Q. A new multipath mitigation method based on adaptive thresholding wavelet denoising and double reference shift strategy. *GPS Solut.* **2018**, *22*, 1–12. [CrossRef]
18. Ye, S.; Chen, D.; Liu, Y.; Jiang, P.; Tang, W.; Xia, P. Carrier phase multipath mitigation for BeiDou navigation satellite system. *GPS Solut.* **2015**, *19*, 545–557. [CrossRef]
19. Tao, Y.; Liu, C.; Liu, C.; Zhao, X.; Hu, H.; Xin, H. Joint time–frequency mask and convolutional neural network for real-time separation of multipath in GNSS deformation monitoring. *GPS Solut.* **2021**, *25*, 1–13. [CrossRef]
20. Fuhrmann, T.; Luo, X.; Knöpfler, A.; Mayer, M. Generating statistically robust multipath stacking maps using congruent cells. *GPS Solut.* **2015**, *19*, 83–92. [CrossRef]
21. Lu, R.; Chen, W.; Dong, D.; Wang, Z.; Zhang, C.; Peng, Y.; Yu, C. Multipath mitigation in GNSS precise point positioning based on trend-surface analysis and multipath hemispherical map. *GPS Solut.* **2021**, *25*, 119. [CrossRef]
22. Wang, Z.; Chen, W.; Dong, D.; Wang, M.; Cai, M.; Yu, C.; Zheng, Z.; Liu, M. Multipath mitigation based on trend surface analysis applied to dual-antenna receiver with common clock. *GPS Solut.* **2019**, *23*, 1–15. [CrossRef]
23. Lu, R.; Chen, W.; Zhang, C.; Li, L.; Peng, Y.; Zheng, Z. Characteristics of the BDS-3 multipath effect and mitigation methods using precise point positioning. *GPS Solut.* **2022**, *26*, 41. [CrossRef]
24. Ren, H.; Li, G.; Geng, J.; Wang, F.; Li, P. Multipath hemispherical map model with geographic cut-off elevation constraints for real-time GNSS monitoring in complex environments. *GPS Solut.* **2023**, *27*, 188. [CrossRef]

25. Pan, Y.; Möller, G.; Soja, B. Machine learning-based multipath modeling in spatial domain applied to GNSS short baseline processing. *GPS Solut.* **2024**, *28*, 9. [CrossRef]
26. Wu, M.; Liu, W.; Wang, W.; Zhang, X. Differential inter-system biases estimation and initial assessment of instantaneous tightly combined RTK with BDS-3, GPS, and Galileo. *Remote Sens.* **2019**, *11*, 1430. [CrossRef]
27. Paziewski, J.; Wielgosz, P. Accounting for Galileo–GPS inter-system biases in precise satellite positioning. *J. Geod.* **2015**, *89*, 81–93. [CrossRef]
28. Odijk, D.; Teunissen, P.J. Characterization of between-receiver GPS-Galileo inter-system biases and their effect on mixed ambiguity resolution. *GPS Solut.* **2013**, *17*, 521–533. [CrossRef]
29. Li, W.; Zhu, S.; Ming, Z. Estimation of Inter-System Biases between BDS-3/GPS/Galileo and Its Application in RTK Positioning. *Remote Sens.* **2021**, *13*, 3507. [CrossRef]
30. Geng, J.; Li, P.; Li, G. Combining the GPS/Galileo/BDS-3 signals on overlap frequencies for interoperable multipath hemispherical maps. *J. Geod.* **2024**, *98*, 1–16.
31. Lu, R.; Chen, W.; Li, Z.; Dong, D.; Jiang, W.; Wang, Z.; Huang, L.; Duan, X. An improved joint modeling method for multipath mitigation of GPS, BDS-3, and Galileo overlapping frequency signals in typical environments. *J. Geod.* **2023**, *97*, 95. [CrossRef]
32. Bentley, J.L. Multidimensional binary search trees used for associative searching. *Commun. ACM* **1975**, *18*, 509–517. [CrossRef]
33. Liu, C.; Tao, Y.; Xin, H.; Zhao, X.; Liu, C.; Hu, H.; Zhou, T. A single-difference multipath hemispherical map for multipath mitigation in BDS-2/BDS-3 short baseline positioning. *Remote Sens.* **2021**, *13*, 304. [CrossRef]
34. Teunissen, P.; De Jonge, P.; Tiberius, C. The least-squares ambiguity decorrelation adjustment: Its performance on short GPS baselines and short observation spans. *J. Geod.* **1997**, *71*, 589–602. [CrossRef]

remote sensing

MDPI

Article

Assessment of PPP Using BDS PPP-B2b Products with Short-Time-Span Observations and Backward Smoothing Method

Lewen Zhao [1,2,3,*] and Wei Zhai [1]

1 School of Remote Sensing and Geomatics Engineering, Nanjing University of Information Science and Technology, Nanjing 210044, China; zhaiwei@nuist.edu.cn
2 Technology Innovation Center for Integration Applications in Remote Sensing and Navigation, Ministry of Natural Resources, Nanjing 210044, China
3 Key Laboratory of Meteorology Disaster, Ministry of Education (KLME), Joint International Research Laboratory of Climate and Environment Change (ILCEC), Collaborative Innovation Center on Forecast and Evaluation of Meteorological Disasters (CIC-FEMD), Nanjing University of Information Science and Technology, Nanjing 210044, China
* Correspondence: lwzhao@nuist.edu.cn

Abstract: The BeiDou Navigation Satellite System (BDS) offers orbit and clock corrections through the B2b signal, enabling Precise Point Positioning (PPP) without relying on ground communication networks. This capability supports applications such as aerial and maritime mapping. However, achieving high precision during the convergence period remains challenging, particularly for missions with short observation durations. To address this, we analyze the performance of PPP over short periods using PPP-B2b products and propose using the backward smoothing method to enhance the accuracy during the convergence period. Evaluation of the accuracy of PPP-B2b products shows that the orbit and clock accuracy of the BDS surpass those of GPS. Specifically, the BDS achieves orbit accuracies of 0.059 m, 0.178 m, and 0.186 m in the radial, along-track, and cross-track components, respectively, with a clock accuracy within 0.13 ns. The hourly static PPP achieves 0.5 m and 0.1 m accuracies with convergence times of 4.5 and 25 min at a 50% proportion, respectively. Nonetheless, 7.07% to 23.79% of sessions fail to converge to 0.1 m due to the limited availability of GPS and BDS corrections at certain stations. Simulated kinematic PPP requires an additional 1–4 min to reach the same accuracy as the static PPP. Using the backward smoothing method significantly enhances accuracy, achieving 0.024 m, 0.046 m, and 0.053 m in the north, east, and up directions, respectively. For vehicle-based positioning, forward PPP can achieve a horizontal accuracy better than 0.5 m within 4 min; however, during the convergence period, positioning errors may exceed 1.5 m and 3.0 m in the east and up direction. By applying the smoothing method, horizontal accuracy can reach better than 0.2 m, while the vertical accuracy can improve to better than 0.3 m.

Keywords: precise point positioning; B2b; real time; kinematic PPP; vehicle

1. Introduction

With the rapid development of Global Navigation Satellite System (GNSS) technology, the demand for high-precision real-time positioning has increased significantly. Precise Point Positioning (PPP) [1] technology enables high-precision positioning using a single receiver, overcoming the limitations of traditional Real-Time Kinematic (RTK) positioning, which relies on ground communication networks and reference stations. This technology holds broad application prospects in fields such as precise positioning in remote areas [2,3], space weather monitoring [4,5], and high-precision time and frequency transfer [6,7]. PPP relies on augmentation corrections, such as precise orbits and clocks, which are essential for achieving high precision. Additionally, tropospheric and ionospheric corrections can shorten the convergence time of the positioning [8,9]. The application of PPP based on post-processed products has been extensively validated. To meet the requirements of rapid and real-time PPP applications, researchers have conducted in-depth studies using different orbit and clock products. Byram et al. (2012) utilized the rapid and ultra-rapid products provided by the United States Naval Observatory (USNO) for PPP. The results demonstrated that the combination of GPS and GLONASS observation data significantly enhances the position accuracy [10]. Elsobeiey and Al-Harbi (2016) analyzed the real-time PPP accuracy using International GNSS Service (IGS) ultra-rapid products and real-time services. Their results showed that, compared to ultra-rapid products, real-time products achieved a 50% improvement in positioning accuracy [11]. Chun et al. (2018) analyzed the real-time orbit and clock products provided by the Centre National d'Études Spatiales (CNES), which revealed that the user accuracy of real-time products ranges between approximately 0.03 and 0.35 m [12]. Subsequent research has further evaluated the performance of real-time corrections and multi-GNSS PPP on GPS, BDS, and other systems [13–15]. Furthermore, Yu et al. (2023) conducted a comparative analysis of PPP based on 12 different centers and assessed its accuracy. The results indicated that products from Wuhan University showed the best performance, with static PPP over 24 h achieving an accuracy of 1.0 cm in the horizontal and vertical directions and a median convergence time of 12.0 min. The hourly kinematic PPP achieved an accuracy of 10.8 cm and 9.5 cm in the horizontal and vertical directions, respectively. The multi-GNSS combination improved the accuracy to 5.2 cm [16].

However, the realization of the aforementioned real-time service still relies on receiving real-time correction data via ground communication networks, as it is only accessible through the network. PPP corrections are transmitted via the B2b signal (PPP-B2b) from three BDS geosynchronous Earth orbit (GEO) satellites. These corrections include orbit and clock data for both GPS and BDS-3 satellites, along with differential code bias (DCB) corrections. When combined with the GPS LNAV and BDS CNAV1 navigation messages, the precise satellite orbit and clock can be recovered, enabling decimeter to centimeter-level positioning across China and the surrounding areas (75°E–135°E, 10°N–55°N) without relying on internet connectivity [17]. Scholars have conducted thorough evaluations on the latency and accuracy of orbit and clock products, and the positioning performance of PPP-B2b products. Nie et al. (2021) evaluated the accuracy of PPP-B2b orbit and clock corrections against the final precise products obtained from GFZ. The results show that the radial, along-track, and cross-track accuracies for BeiDou-3 satellites are 0.138 m, 0.131 m, and 0.145 m, respectively, with clock offset accuracies reaching centimeter-level precision [18]. Tao et al. (2021) further compared the PPP-B2b with CNES (Centre national d'études spatiales) products. The results demonstrated that BDS-3-only PPP-B2b kinematic PPP can achieve a centimeter-level accuracy comparable to the GPS-only results using CNES [19]. Song et al. (2023) validated the performance of PPP-B2b using different signal combinations and demonstrated that PPP-B2b can meet the positioning performance required in China [20]. Furthermore, the application of PPP-B2b in ocean environments [21], single-frequency scenarios [22], time transfer [23], and its integration with inertial navigation systems [24] has been validated, demonstrating the promising enhancements PPP-B2b offers for real-time users.

The assessments so far have primarily focused on daily solutions and simulated kinematic experiments. However, applications such as aerial mapping [25] and agriculture [26] usually require rapid static positioning during the work preparation phase or high positional accuracy during the operational phase. The convergence period, particularly at the start or after an interruption, tends to exhibit relatively poor performance due to the incomplete elimination of atmospheric errors and the reinitialization of ambiguities. Although some studies investigate the performance of PPP with short observation sessions [27,28], research on real-time PPP-B2b remains relatively limited. To address this issue, this paper evaluates the performance of PPP-B2b products and investigates the positioning accuracy achieved through both static and kinematic PPP with varying session lengths. Additionally, a backward smoothing method is applied to improve the accuracy of PPP during the convergence period, enabling high-precision positioning throughout the process.

2. Methods and Theory

The GNSS raw pseudorange and carrier phase observations across different frequencies are detailed as follows:

$$\begin{aligned} p^s_{r,j} &= \rho^s_r + c\cdot(dt_r + d_{r,j}) - c\cdot\left(dt^s + d^s_j\right) + I^s_{r,j} + T^s_r + \varepsilon^s_{r,p,j} \\ \varphi^s_{r,j} &= \rho^s_r + c\cdot(dt_r + \delta_{r,j}) - c\cdot\left(dt^s + \delta^s_j\right) - I^s_{r,j} + T^s_r + \lambda_j\cdot N^s_{r,j} + \varepsilon^s_{r,\varphi,j} \end{aligned} \tag{1}$$

where superscripts s and r denote a specific satellite and receiver, respectively; p and φ represent the pseudorange and carrier phase observations; and j denotes the frequency and the corresponding wavelength λ_j. Furthermore, ρ is the geometric distance from the satellite to receiver, while $\mathrm{d}t_r$ and $\mathrm{d}t^s$ are clock offsets for the receiver and satellite, respectively. c represents the speed of light in a vacuum; I^s_r is the ionospheric delay; T is the tropospheric delay; N stands for ambiguity; ε represents the observation noise and multipath effects; $d_{r,j}$ and d^s_j are the receiver-dependent and satellite-dependent instrumental delay biases in the pseudorange, respectively; and $\delta_{r,j}$ and δ^s_j are the receiver-dependent and satellite-dependent instrumental delay biases in the carrier phase observations, respectively.

2.1. Recovery of Precise Orbits and Clocks Products Using PPP-B2b Correction

According to the ICD (Interface Control Document) [29] of the BDS PPP-B2b signal, orbit corrections are provided in the radial, along-track, and cross-track directions within the satellite orbit coordinate system and denoted as δO_{RAC}. The corrections should be translated to the Earth Center Earth Fixed (ECEF) system using the following equation:

$$\delta\mathbf{X}_{ECEF} = -\left[\frac{\mathbf{X}}{|\mathbf{X}|} \quad \frac{\mathbf{X}}{|\mathbf{X}|}\times\frac{\mathbf{X}\times\mathbf{V}}{|\mathbf{X}\times\mathbf{V}|} \quad \frac{\mathbf{X}\times\mathbf{V}}{|\mathbf{X}\times\mathbf{V}|}\right]\cdot\delta O_{RAC} \tag{2}$$

where $\mathbf{X}$ and $\mathbf{V}$ are the satellite position and velocity vectors calculated by the broadcast ephemeris, respectively; $\delta\mathbf{X}_{ECEF}$ is the transmitted satellite orbit correction. After calculating the initial position $\boldsymbol{dX}_{brdc}$ with the broadcast ephemerides, the precise satellite orbit $\boldsymbol{dX}_{prec}$ can be calculated with the following:

$$\mathbf{X}_{prec} = \boldsymbol{dX}_{brdc} + \delta\mathbf{X}_{ECEF} \tag{3}$$

In addition, the precise clock $\mathrm{d}t_{\mathrm{prec,B2}b}$ can be calculated using the following:

$$\mathrm{d}t_{\mathrm{prec,B2}b} = \mathrm{d}t_{brdc} - C_0/C_{light} \tag{4}$$

where $\mathrm{d}t_{brdc}$ is the clock correction calculated from the broadcast ephemeris in meters; C_0 is the clock corrections derived from the PPP-B2b corrections in meters and $\mathrm{C_{light}}$ is the velocity of light.

The PPP-B2b service provides clock and orbit corrections with update intervals of 6 s and 48 s, respectively. To ensure the proper synchronization of these corrections, the "IOD Corr" parameter acts as a version identifier, aligning orbit and clock data. Additionally, the "IOD SSR" (Issue of Data, State Space Representation) parameter is used to match the PPP-B2b corrections with the broadcast ephemerides. Although the synchronization of these corrections is typically reliable, discrepancies can arise due to missing correction products for certain satellites or changes in the ephemeris IOD. Such mismatches may result in the incorrect recovery of orbit and clock information or a reduction in the number of usable satellites, thereby degrading PPP performance.

To address these challenges, a predictive correction strategy is employed, leveraging the nearest available correction data to compensate for missing corrections. This method effectively handles gaps in orbit corrections within a 300 s window and clock corrections within a 60 s window. In this way, the integrity of PPP positioning is maintained even in the presence of correction gaps from certain satellites.

Apart from the satellite orbit and clock errors, satellite and receiver biases are another main factor affecting the accuracy of PPP. The receiver code hardware delays can be absorbed by the clock parameter, and the satellite and receiver phase biases can be absorbed by the float ambiguity parameters. However, the satellite code biases persist and impact the convergence of PPP. Code biases corrections are also provided in PPP-B2b corrections and can be corrected using the following equation:

$$\widetilde{P_f} = P_f - DCB_f \tag{5}$$

where P_f is the original pseudorange observations and $\widetilde{P_f}$ is the corrected observations; DCB_f is the PPP-B2b DCB corrections on the B1Cp and B2Ap signals for corresponding signals of BDS-3 in meters.

The B1I and B3I ionosphere-free (IF) combination are generally used for the satellite orbit and clock estimation, in addition to the IF PPP at the user end. However, the real-time precise product clocks calculated with PPP-B2b corrections and the matching broadcast ephemeris are referenced to the B3 frequency. For the PPP using the IF model, the satellite hardware delay correction should be applied to convert the clocks from B3I to the B1I/B3I [30], which can be described as follows:

$$dt_{prec} = \mathrm{d}t_{\mathrm{prec,B2}b} - \frac{f_{B1I}^2}{f_{B1I}^2 - f_{B3I}^2}\mathrm{DCB}_{B1I}^s \tag{6}$$

where dt_{prec} is the precise satellite clock offset referred to the B1I/B3I IF combination, which can be used in the same manner as the IGS standard clock products; f_{B1I}^2 and f_{B3I}^2 represent the frequencies of the B1I and B3I signals, respectively; and DCB_{B1I}^s is the hardware delays for the B1I signal.

2.2. SISRE Assessment

To assess the accuracy of the satellite orbits and clocks, the Signal-in-Space Ranging Error (SISRE) is calculated [31], which can be expressed as follows:

$$SISRE = \sqrt{(\alpha \cdot dR - c \cdot dt)^2 + \frac{(dA^2 + dC^2)}{\beta}} \tag{7}$$

where dR, dA, dC are the radial, along- track, and cross-track orbit errors in the satellite orbit coordinate system, respectively; dt is the clock bias; c is the speed of light in a vacuum; and α and β are coefficients determined by the satellite orbit altitudes.

The satellite orbits calculated from PPP-B2b refer to the satellite's antenna phase center, while the precise orbit products provided by the MGEX analysis centers refer to the satellite's center of mass. Therefore, prior to comparing broadcast ephemeris and precise orbits, Phase Center Offset (PCO) correction must be applied according to the following formula:

$$\delta \boldsymbol{X}_{B2b} = \boldsymbol{X}_{pre,final} - \left[\boldsymbol{X}_{pre,B2b} + \boldsymbol{A} \cdot \delta \boldsymbol{X}_{pco}\right] \tag{8}$$

where $\delta \boldsymbol{X}_{B2b}$ is the orbit error vector; $\boldsymbol{X}_{pre,final}$ is the reference orbit vector obtained from final products; $\boldsymbol{X}_{pre,B2b}$ is the real-time PPP-B2b precise orbit vector; A is the satellite attitude matrix; and $\delta \boldsymbol{X}_{pco}$ is the satellite PCO correction vector obtained from the latest "igs14.atx" file released by IGS.

3. Data and Experiments

The objectives of this section are to comprehensively analyze the quality of BDS PPP-B2b corrections and their accuracy for the static and kinematic PPP.

3.1. Accuracy of PPP-B2b Orbits and Clocks

To verify the accuracy of real-time PPP-B2b products, the final products provided by Wuhan University (WUM) were selected as the reference. The reported accuracy of these final products for BDS-3 satellites is 3 to 4 cm [32,33]. Figure 1 shows the error time series of the PPP-B2b precise orbit products for days 075 to 092 of 2024. Notably, in the GPS orbit error series, the radial orbit errors for satellites G22 and G28 exhibit systematic biases of approximately 0.4 m and 1 m, respectively. These biases persist until day 087, after which the errors return to levels consistent with those of other satellites. Apart from this, no significant anomalies are detected in the GPS and BDS products across different days, indicating the stable quality of PPP-B2b products. Overall, the radial component, which significantly impacts positioning errors, shows the smallest errors among the three components, with most satellites exhibiting errors within 0.3 m. In the along-track direction, the GPS products exhibit relatively larger errors, whereas in the cross-track direction, GPS and BDS exhibit comparable accuracy.

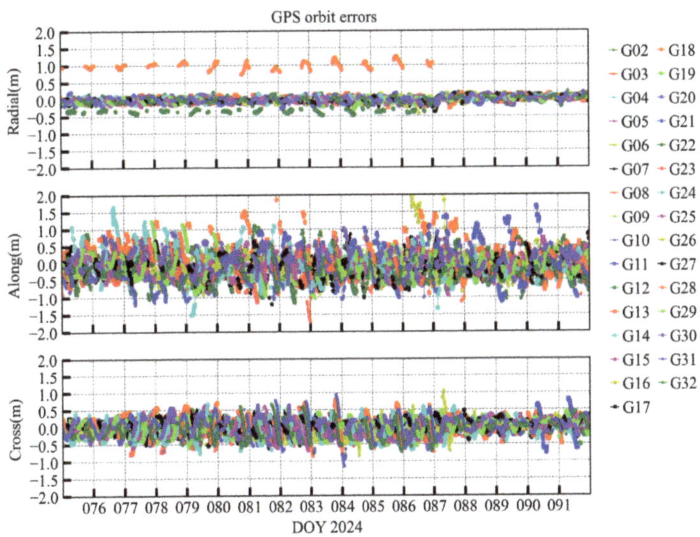

Figure 1. *Cont.*

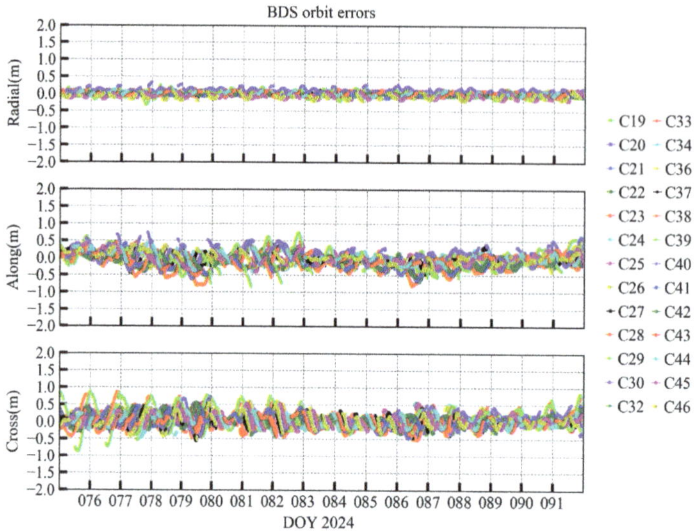

Figure 1. Time series of GPS and BDS orbit errors in the radial, along-track, and cross-track directions for PPP, referenced using WUM products.

Figure 2 presents the Root Mean Square (RMS) of PPP-B2b real-time orbits errors. The average values are 0.098 m, 0.368 m, and 0.214 m for the radial, along-track, and cross-track components for GPS, respectively. In comparison, the average RMS values for the BDS-3 MEO satellites are 0.059 m, 0.178 m, and 0.186 m, respectively. These results confirm that BDS real-time orbits are more accurate than GPS. This improved accuracy is due to the additional observations from the inter-satellite link (ISL) terminals on the BDS-3 satellites, which enhance orbit determination, particularly in the along-track direction [34].

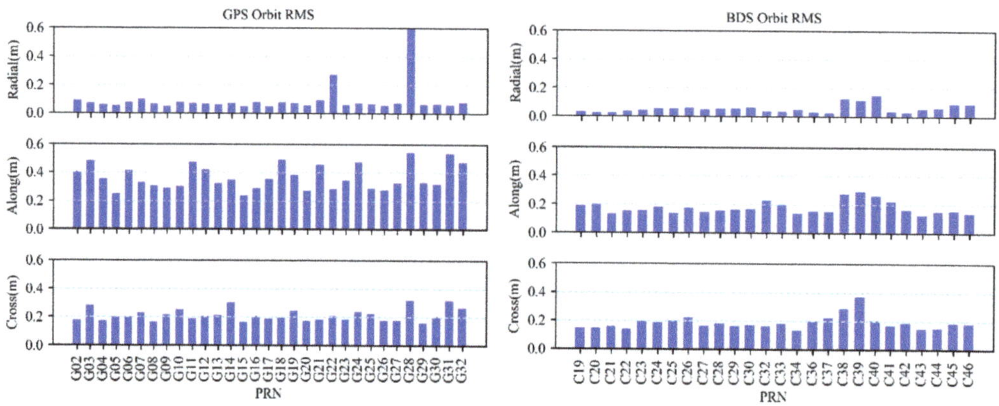

Figure 2. The statistics of GPS and BDS orbit errors in the radial, along-track, and cross-track directions, using WUM products as the reference.

The accuracy of satellite clocks is evaluated using the double-difference method [35]. This method involves first calculating the single difference between the real-time clock and final clock. The mean of these differences at each epoch is used as the reference clock bias. Double differences are then computed by subtracting this mean from the clock error of each satellite. Since the PPP-B2b products are estimated based on regional stations, interruptions can cause discontinuities in the clock datum. To address this issue, the quality of satellite clocks is evaluated using the average standard deviation (STD) and RMS on an hourly basis, as presented in Figures 3 and 4 for GPS and BDS, respectively. Overall, the averaged STD and RMS for the GPS are 0.33 ns and 2.76 ns, respectively, which are significantly higher than the corresponding values of 0.13 ns and 1.82 ns for the BDS. Several GPS satellites exhibit clock RMS values exceeding 6 ns, which degrades the precision of the pseudorange observations and increases the convergence time of PPP. However, this does not impact the accuracy after convergence, as the systematic bias is absorbed into the observation residuals and ambiguities during PPP [36].

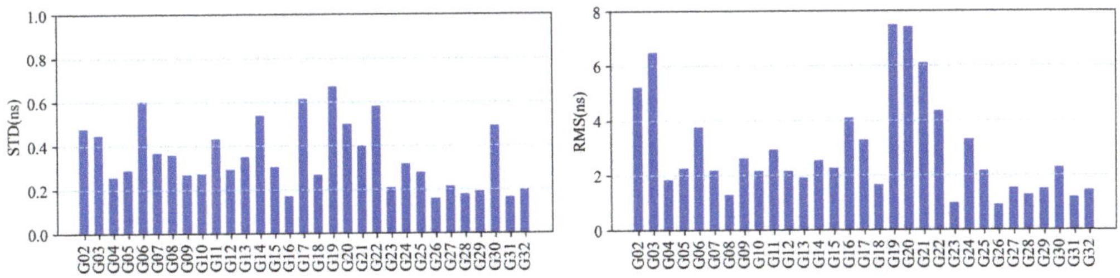

Figure 3. The clock STD and RMS for the GPS products of PPP-B2b, using the "WUM" as reference.

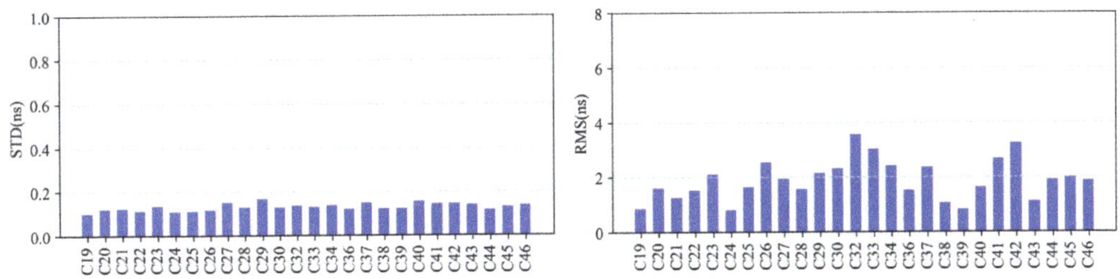

Figure 4. The clock STD and RMS for the BDS products of PPP-B2b, using the "WUM" as reference.

Figure 5 illustrates the time series of SISRE for the GPS and BDS-3 satellites. The SISRE of BDS-3 outperforms that of the GPS, which is consistent with the orbit and clock accuracy assessment presented in the previous analysis. The GPS SISRE exhibits significant errors and a substantial discrepancy on DOY 087, which can be easily identified through the positioning outlier screening process in PPP due to its relatively large magnitude. Moreover, no notable discrepancies are observed among different satellites, differing from the characteristics seen in the GPS orbit evaluation results. This can be attributed to the use of consistent orbit and clock products, where radial errors are effectively compensated [37].

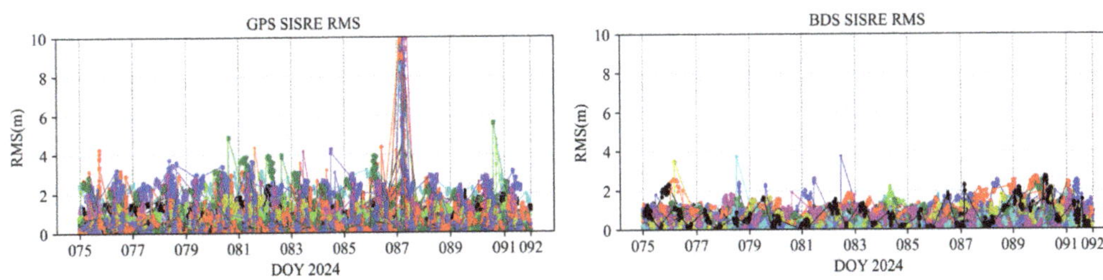

Figure 5. The SISRE time series for the GPS and BDS satellite from PPP-B2b products, with different colors representing the different satellites.

3.2. Experiments and Validation

Figure 6 illustrates the distribution of stations used for data processing in this study. Data from eight MGEX stations for days of year (DOY) 075 to 092 in 2024 were selected, with a data sampling rate of 5 s. The observation data were processed using the in-house developed software, NavEngine V1.1. Both GPS and BDS measurements were processed using PPP-B2b real-time orbits, clocks, and code biases, as well as the rapid products from Wuhan University (WHR). All data were processed using a dual-frequency combination and an ionosphere-free model. The receiver coordinates in the forward Kalman filter [38] were initialized using the position obtained from single-point positioning, with a variance of 30 m. Afterwards, the coordinates were estimated as white noise at 30 m for kinematic mode and as constant in the static mode. The backward smoothing method [39] utilizes the estimated parameters and variance from the forward filter as constraints, thereby enhancing the accuracy of the estimated parameters during the convergence process. Details of the processing strategies are outlined in Table 1.

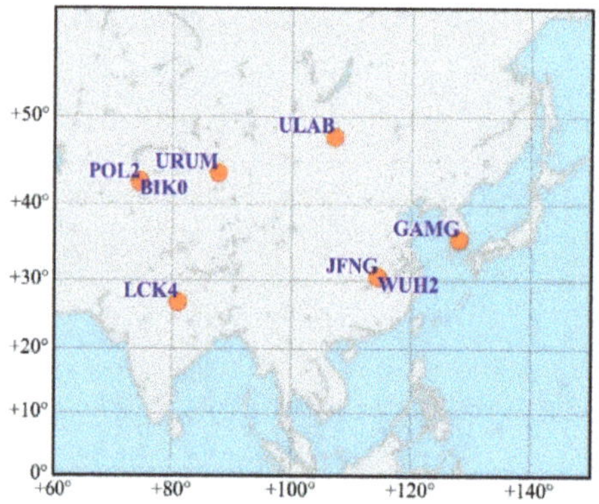

Figure 6. Distribution of the MGEX stations.

Table 1. Processing strategies of PPP using the PPP-B2b and WHR products.

Items	Strategies
Observation	GPS: L1/L2; BDS: B1I/B3I
Elevation mask	7°
Weight for observations	Elevation-dependent weighting
Noise for observations	0.003 m for phase and 0.6 m for code
Satellite orbit/clock	PPP-B2b and WHR
Satellite antenna phase center	Using igs20.atx for WHR PPP
Filter method	Forward Kalman and backward smoothing
Tides correction	IERS 2010 [40]
Troposphere	Zenith wet delay is estimated as a random walk
Ionosphere	Ionosphere-free combination
Ambiguity	Estimated as constant with float solution
Receiver coordinate	Constant in static processing and white noise in the kinematic processing
Receiver clocks	Estimated as white noise

3.2.1. Static PPP Processing

Static PPP solutions were calculated in 1 h intervals. A total of 8 stations were used over 15 days, resulting in 24 periods per day, which corresponds to 2880 h of data. Figure 7 illustrates the convergence of positioning errors over time in the horizontal direction. The green dots represent the original positioning error sequence, while the red, blue, and purple curves represent the 50%, 68%, and 95% percentile positioning error at corresponding times. The three gray lines indicate horizontal positioning accuracies of 0.5 m, 0.2 m, and 0.1 m, respectively. A comparison of the positioning errors during initialization shows that the errors for "WHR" products are around 2 m, while the initial positioning error for PPP-B2b is greater than 4 m. After initialization, the overall positioning errors for "WHR" products are significantly smaller than those for PPP-B2b. At some stations, the errors exceed 1 m even after 1 h of convergence, primarily due to differences in the availability of GPS and BDS corrections across the PPP-B2b service region.

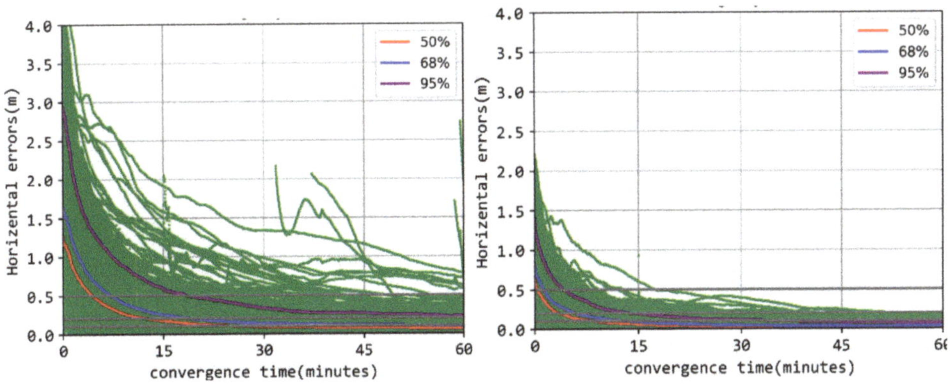

Figure 7. Convergence time for static PPP using with PPP-B2b products (**left**) and WHR products (**right**) with green dots representing the original positioning errors.

For further quantitative analysis, Table 2 compares the time required for different proportions of positioning errors to reach a desired accuracy threshold of 50 cm, 20 cm, and 10 cm. At each sampling interval, the positioning errors from all processed stations are sorted. The times at which the proportions of errors within the 50%, 68%, and 95% confidence intervals first meet these thresholds are recorded as the convergence times. The analysis reveals that for static positioning with PPP-B2b, 4.5 min, 12.7 min, and 26.4 min are required for 50% of positioning errors to converge to 0.5 m, 0.2 m, and 0.1 m, respectively. The time required for high-accuracy convergence to 0.1 m increases significantly compared to 0.2 m, primarily because achieving high-accuracy positioning requires more precise atmospheric delay information, which takes time to converge. To meet 68% of positioning errors within 0.5 m, 0.2 m, and 0.1 m, it takes 7.3 min, 18.9 min, and 55.5 min, respectively, with significant increases in time. Using "WHR" rapid orbit products, 50% of the positioning errors converge to an accuracy better than 0.1 m within 8.6 min, and it takes 13.0 min for 68% of positioning errors to reach this accuracy.

Due to the fact that PPP-B2b corrections primarily cover the Asia–Pacific region, different stations experience variations in the number of visible satellites and DOP values, which in turn affect the positioning accuracy. To further analyze this impact, we examined the station-specific convergence time needed for PPP to reach 0.5 m and 0.1 m accuracy, as presented in Figure 8. The average convergence time needed to achieve a 0.5 m positioning accuracy is 6 min for PPP-B2b corrections and 2 min for WHR products, with the average difference between stations being within 2 min. To achieve a 0.1 m positioning accuracy, the average convergence time required is 24.8 min for PPP-B2b corrections and 16.6 min for WHR products, with the average difference between stations being within 5 min and 8 min, respectively. However, it is important to note that, in some cases, static PPP fails to converge to 0.1 m even after the mean convergence time or beyond 50 to 60 min, as indicated by the station-specific statistics in Table 3. This is primarily attributed to the observation geometry and the limited number of available GPS and BDS corrections from PPP-B2b.

Table 2. The average convergence time required for horizontal positioning errors to reach specified thresholds at different proportions.

Product Type	Error Threshold (m)	Time for 50% Percentile (min)	Time for 68% Percentile (min)	Time for 95% Percentile (min)
PPP-B2B	0.5	4.5	7.3	19.0
	0.2	12.7	18.9	60.0
	0.1	26.4	55.5	60.0
WHR	0.5	0.4	1.3	4.2
	0.2	3.6	5.9	14.3
	0.1	8.6	13.0	34.5

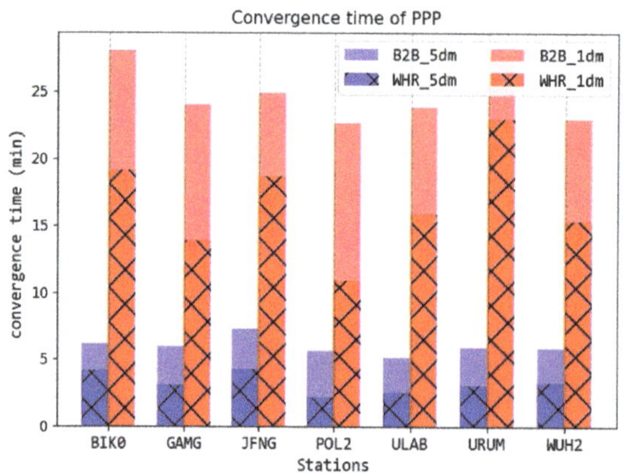

Figure 8. Station-specific convergence time using different products to achieve 0.5 m and 0.1 m.

Table 3. Proportion of convergence time exceeding the average, 50 min and 60 min for each station.

Station	BIK0	GAMG	JFNG	POL2	ULAB	URUM	WUH2
>average	39.23%	36.66%	39.02%	38.46%	37.18%	44.48%	42.32%
>50 min	27.33%	9.65%	11.74%	19.55%	12.50%	27.05%	18.26%
>60 min	23.79%	7.07%	7.57%	16.98%	10.26%	19.57%	12.45%

Figure 9 further depicts the station-specific hourly three-dimensional positioning RMS and its corresponding average number of satellites. The average RMS is approximately 0.1 m with hourly data at different stations, with the number of satellites ranging from 12.5 to 14.5. Stations with a lower average number of satellites typically exhibit longer convergence times. The statistical analysis reveals that 95% of the positions fall within 0.065 m, 0.143 m, and 0.148 m for the hourly positioning RMS in the N/E/U directions, respectively.

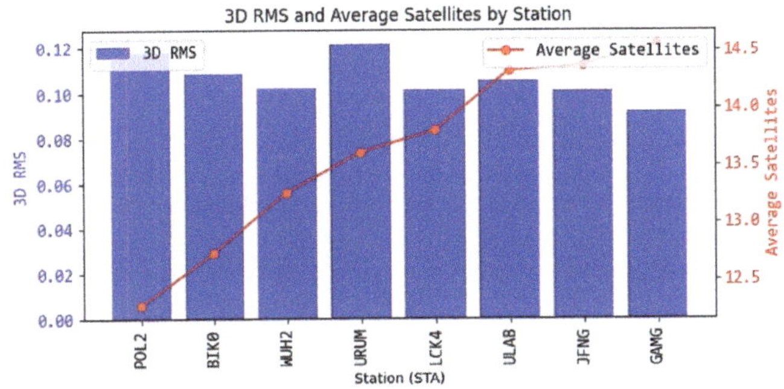

Figure 9. Station-specific three-dimensional positioning RMS and its corresponding average number of satellites.

3.2.2. Simulated Kinematic PPP Processing

Based on selected data, static stations are estimated using the simulated kinematic mode, where receiver coordinates are re-initialized each epoch, and ambiguities are continuously estimated unless a cycle slip is detected. Figure 10 shows the convergence times needed for the hourly dynamic PPP to achieve accuracies of 0.5 m and 0.1 m using PPP-B2b and WHR products. The average times needed to reach a 0.5 m accuracy are 9.9 min with PPP-B2b and 2.1 min with WHR, while for a 0.1 m accuracy, the times are 25.9 min and 13.2 min, respectively. Compared to static PPP, dynamic PPP using PPP-B2b products requires an additional 1–4 min to achieve the same level of accuracy.

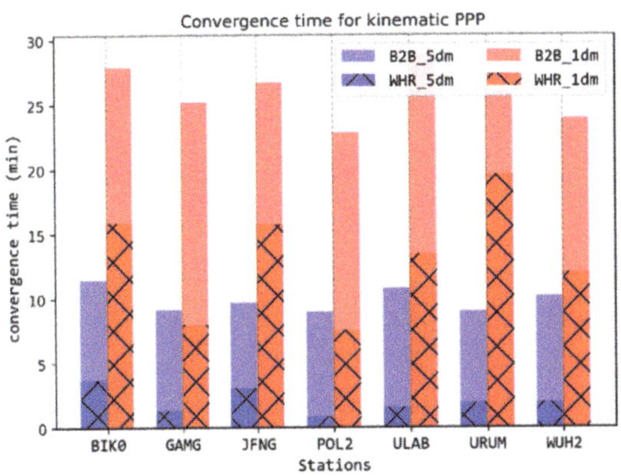

Figure 10. Convergence time of kinematic PPP to reach 0.1 m and 0.5 m accuracy using PPP-B2b and WHR products.

To further characterize the station-specific positioning accuracy, Figure 11 shows the relationship between the RMS of 3D positioning errors after convergence to 0.1 m accuracy and the number of visible satellites. The blue data points represent the RMS for each hourly session, plotted against an average number of visible satellites, while the red line shows the RMS average grouped by the integer number of visible satellites. The results indicate that the RMS of 3D positioning errors varies significantly after convergence to 0.1 m accuracy, ranging from 0.05 m to 0.3 m. This can be attributed to factors such as the number of visible satellites, the orbit accuracy of different satellites, and the data quality of different stations. Some stations experience periods with fewer than 10 usable satellites in the combined GPS + BDS system. As the average number of visible satellites exceeds 10, the average 3D positioning error gradually decreases.

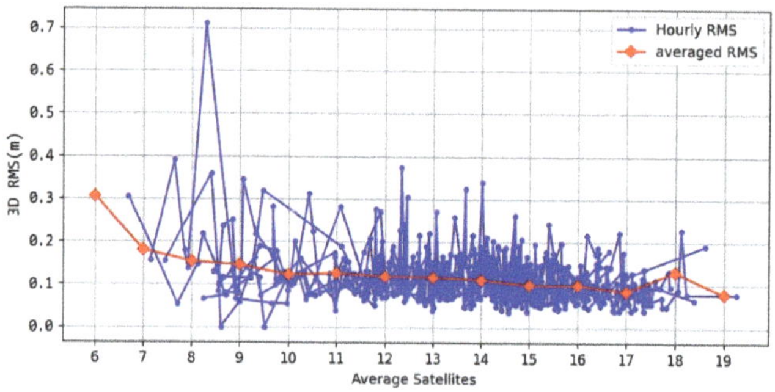

Figure 11. Variation in 3D positioning errors with respect to the number of visible satellites for hourly PPP.

To eliminate the inaccuracies during the convergence period and fully utilize observations from dynamic scenarios over a short time span, backward filtering was conducted based on forward Kalman filtering. The average positioning RMS values in the N/E/U directions using PPP-B2b corrections are 0.024 m, 0.046 m, and 0.053 m, respectively, while using WHR products, they are reduced to 0.009 m, 0.021 m, and 0.027 m. Figure 12 shows the distribution of positioning error in the N/E/U directions using PPP-B2b products (left) and WHR products (right) for each station. Specifically, the proportions of hourly PPP-B2b solutions with positioning errors less than 0.1 m in the N/E/U directions are 94.03%, 73.00%, and 60.79%, respectively, while those with errors less than 0.2 m are 98.53%, 91.00%, and 88.89%, respectively. When using WHR products, the proportions of positioning errors less than 0.1 m in the N/E/U directions are 100.00%, 97.61%, and 96.88%, respectively; the proportions with positioning errors less than 0.2 m are 100.00%, 100.00%, and 99.82%, respectively. The comparison reveals that the larger positioning errors obtained with PPP-B2b products are mainly due to the accuracy of the PPP-B2b products rather than the quality of the station observation.

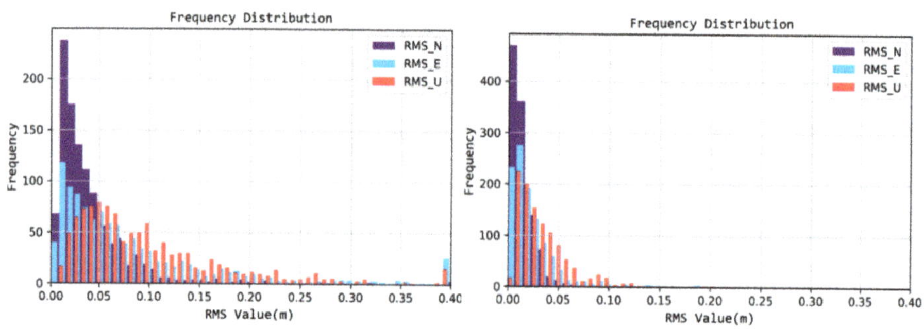

Figure 12. Distribution of hourly PPP positioning errors in the N/E/U directions calculated using PPP-B2b (**left**) and WHR (**right**) products.

Overall, the results indicate that the positioning accuracy in the north direction is better than that in the east direction, with a higher proportion of smaller errors. However, some stations still exhibit hourly positioning errors exceeding 0.4 m in the E and U directions. Figure 13 further illustrates the positioning time series of station BIK0 on DOY 075 in the north, east, and up directions, along with the variations in the number of satellites (Nsat) and Geometric Dilution of Precision (GDOP) values. It confirms that the backward smoothing PPP achieves better consistency in the north direction, while the east and up directions exhibit greater variations across different sessions. Notably, the number of satellites decreases to fewer than 10 from 12:00 to 13:00, during which the positioning errors of the forward PPP increase to approximately 0.6 m in all three components. Although the backward smoothing PPP improves the positioning accuracy during this session, it still performs worse than in other sessions. As only the dual-system GPS and BDS-3 can be used with PPP-B2b, despite the availability of multi-GNSS observations from the receiver, a high positioning accuracy cannot be maintained when the number of satellites is insufficient. A reasonable solution is to partially combine the PPP-B2b with other correction sources, such as those from the Galileo HAS service [41], using a proper weighting model or an equivalent transformation model.

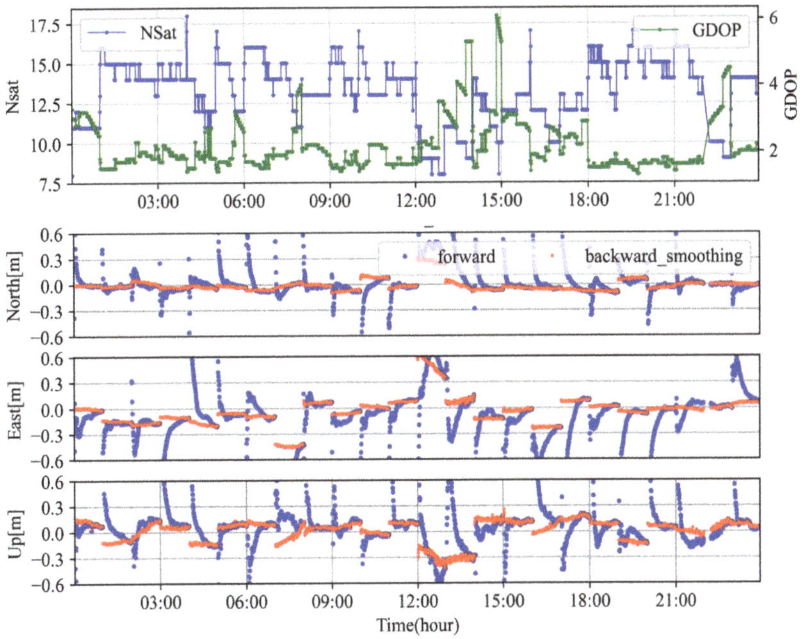

Figure 13. Positioning time series of station BIK0 on DOY 075 in the north, east, and up directions, along with the variations in the number of satellites (Nsat) and GDOP values.

3.2.3. Vehicle-Based Dynamic PPP

To further validate the accuracy of dynamic PPP, we conducted a vehicle dynamic experiment. The vehicle's trajectory, as shown in the left panel of Figure 14, begins at the point marked "1" and ends at the point marked "2". Data were collected using a multi-constellation, multi-frequency GNSS positioning module (UM980) manufactured by Unicore, along with a corresponding multi-frequency GNSS antenna mounted on the roof of the vehicle. The experiment took place on the school campus from 08:07:44 to 08:31:00 (GPS time) on 15 July 2024, with a sampling interval of 0.5 s. The corresponding time series of positioning errors in the N/E/U directions, which are converted into a local coordinate system based on the coordinates at the final epoch, are shown in the right panel of Figure 14. The vehicle remained stationary for approximately five minutes before initiating dynamic motion, followed by another stationary period from 08:26:21 to 08:28:58, with brief movement thereafter. Due to the 40 km distance between the base station and the rover, the RTK ambiguity resolution rate was only 93.2%, as indicated by the float ambiguities represented by the yellow points. The positioning errors observed during stationary periods show that the RTK positioning accuracy achieved was better than 0.05 m, which is sufficient for assessing the accuracy of the PPP results.

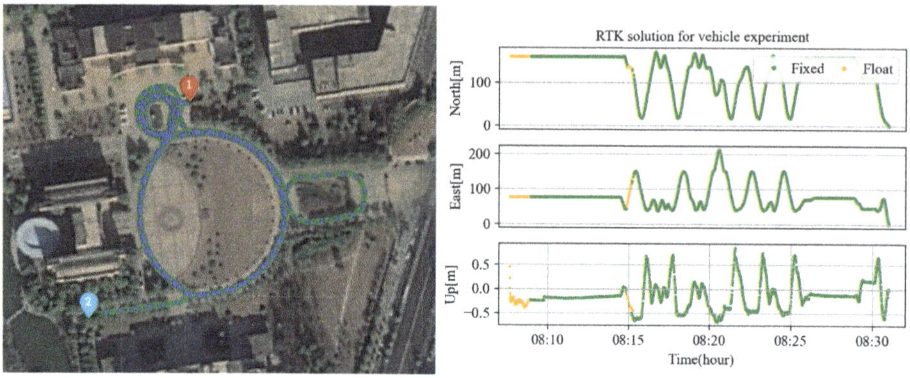

Figure 14. Vehicle trajectory for the dynamic experiment starting at point '1' and ending at point '2' (left), and time series of RTK positioning, along with its ambiguity fixing status, where the green color represents the fixed solution.

First, simulated kinematic PPP using observations from the reference station was used to evaluate the accuracy of PPP using two separate software tools, namely PRIDE PPP-AR [42] and NavEngine. PRIDE utilizes a total least squares algorithm, while NavEngine uses forward Kalman filtering combined with backward smoothing. Both tools used rapid products provided by Wuhan University. Figure 15 compares the time series of positioning errors, revealing that both software tools produced stable error sequences. However, systematic biases of varying magnitudes were observed across the three components. Table 4 summarizes the statistical results of positioning errors. The analysis shows that the accuracy, measured by the STD, is similar across both software, achieving millimeter-level precision, while the vertical precision is slightly lower, with errors around 1 cm. The RMS analysis reveals systematic errors of 0.1 m in the east direction and 0.06 m in the north direction. This discrepancy is primarily due to the short observation duration of the reference station's data, which lasted only two hours, leading to the incomplete convergence of the dynamic solution's positioning accuracy.

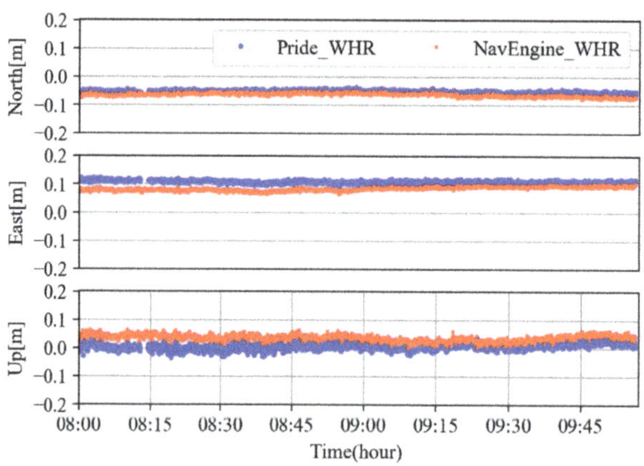

Figure 15. Time series of positioning errors at the base station using PrideLab and NavEngine software with WHR products.

Table 4. Averaged RMS and STD using different processing modes.

	RMS (m)			STD (m)		
	North	East	Up	North	East	Up
Pride_WHR	0.051	0.111	0.014	0.004	0.004	0.011
NavEngine_WHR	0.061	0.086	0.039	0.005	0.008	0.010
NavEngine_B2b	0.055	0.076	0.049	0.006	0.024	0.045

Furthermore, a comparison of the positioning errors observed using NavEngine with WHR and PPP-B2b products, as shown in Figure 16, reveals that the systematic biases exhibit good consistency, with RMS differences of less than 0.01 m, which is reasonable within the kinematic positioning noise of PPP. However, the STD accuracy of the PPP-B2b solution is larger than that of WHR, particularly in the east and up directions. Overall, the positioning errors obtained using PPP-B2b products at the reference station can achieve centimeter-level accuracy when utilizing the backward smoothing method.

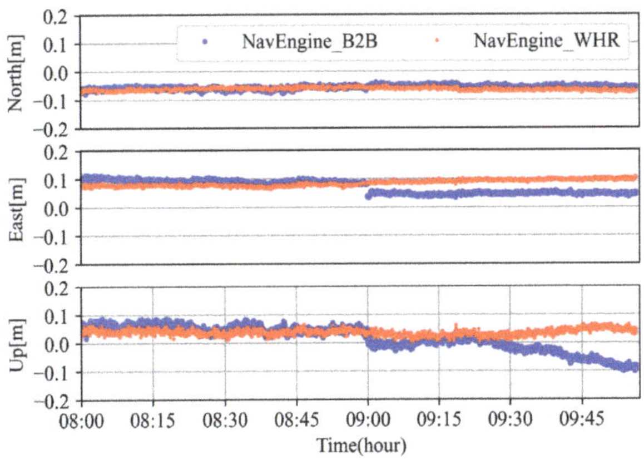

Figure 16. Time series of positioning errors at the base station using WHR and PPP-B2b products processed with NavEngine.

Following the simulated kinematic validation, the positioning accuracy of the vehicle dynamic station was analyzed by comparing it against the RTK solution. Figure 17 presents the positioning errors of the forward PPP using WHR and PPP-B2b products in NavEngine software. The results indicate that PPP achieves a horizontal accuracy of 0.5 m within 4 min of convergence. Afterward, the horizontal positioning errors continue to decrease, with WHR and PPP-B2b products converging to horizontal accuracies of 0.05 m and 0.23 m, respectively. The vertical positioning error requires about 12 min to converge to a 0.5 m accuracy, and a constant bias persists even after the convergence.

In comparison, Figure 18 shows the positioning error time series using the WHR and PPP-B2b products with backward smoothing. Although the trends are similar due to variations in the reference RTK solution, significant biases exist between the products. The WHR products exhibit better accuracy in the horizontal direction, with average RMS values of 0.062 m and 0.152 m for the north and east directions, respectively, compared to 0.139 m and 0.163 m for the PPP-B2b products. Conversely, PPP-B2b products demonstrate superior accuracy in the vertical direction, achieving an RMS value of 0.137 m compared to 0.354 m for the WHR products, respectively. This discrepancy arises because the solution using PPP-B2b products continues to converge vertically during the forward filter, whereas the WHR products reach a stable state earlier. Although backward filtering can enhance the

PPP accuracy during the convergence period, the overall accuracy largely depends on the parameter convergence precision achieved through the forward filtering. The convergence period is partially influenced by the quality control strategy implemented in the software, as well as the accuracy of the products.

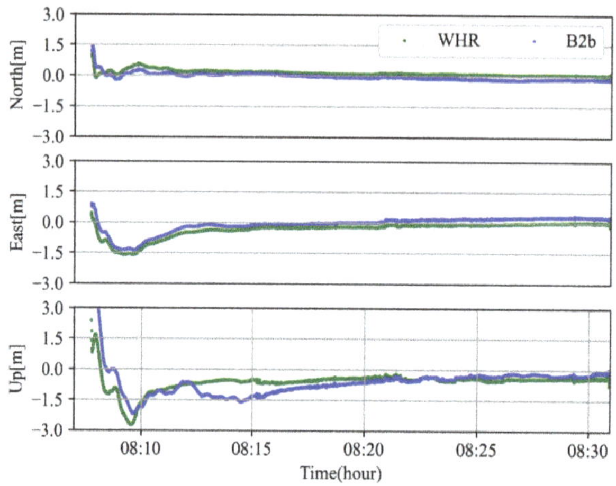

Figure 17. Time series of positioning errors for forward PPP using WHR and PPP-B2b products.

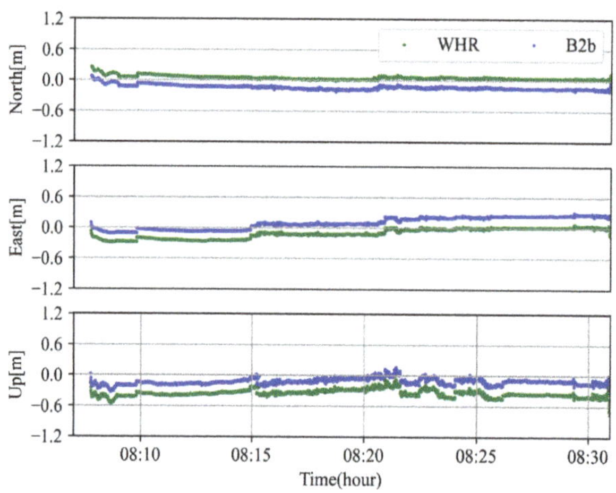

Figure 18. Positioning error time series of backward PPP using the WHR and PPP-B2b products.

4. Conclusions

BDS provides precise orbit, clock, and bias corrections through PPP-B2b signals, enabling PPP in the Asia–Pacific region without the need for internet connectivity. To achieve a high positioning accuracy with short observation periods, the accuracy of static PPP using PPP-B2b was evaluated, and a backward smoothing method was proposed to improve the accuracy during the initialization period. The performance was validated using the static and vehicle observations. The following conclusions can be drawn:

(1) The accuracy of the orbits and clocks recovered from the PPP-B2b is assessed, indicating that BDS products outperform GPS. The improvement is largely due to the BDS inter-satellite link ranging, which enhances orbit performance.

(2) The static hourly PPP results show that the average convergence times required to achieve horizontal accuracies better than 0.5 m and 0.1 m are approximately 4.5 min and 25 min, respectively. However, a proportion of the sessions, ranging from 7.07% to 23.79%, fail to converge to 0.1 m due to the limited availability of GPS and BDS satellites. For sessions that do converge, the average 3D RMS is 0.1 m.
(3) The simulated kinematic PPP results indicate that the average times required to achieve horizontal accuracies of 0.5 m and 0.1 m are approximately 9.9 min and 25.9 min, respectively. The average positioning RMS values in the N/E/U directions using backward smoothing PPP are 0.024 m, 0.046 m, and 0.053 m. Similar to the case of static PPP, convergence to 0.1 m accuracy cannot always be achieved due to variations in the observation quality at different stations. Using the backward smoothing method, the proportions of positioning errors less than 0.1 m in the north, east, and up directions are 94.03%, 73.00%, and 60.79%, respectively.
(4) The vehicle experiments show that forward PPP can achieve a horizontal accuracy better than 0.5 m within 4 min, with steady improvement over longer observation periods. However, large positioning errors of 1.5 m in the east direction and 3.0 m in the vertical direction are observed during the convergence period. Using the backward smoothing method, an RMS of 0.139 m, 0.163 m, and 0.137 m is achieved in the north, east, and up directions, respectively.

Overall, the results demonstrate that using PPP-B2b can achieve decimeter-level positioning with different observation durations, without relying on an external base station. The backward smoothing method further enhances accuracy during the convergence period. However, the accuracy observed in the vehicle experiments is lower than that of the simulated real-time results due to the effects of the observation environment. For applications such as Unmanned Aerial Vehicles (UAVs) and marine navigation, where the observation environment typically offers open-sky conditions, the results are expected to be comparable. Furthermore, at the boundary of the PPP-B2b service region, the number of available satellites decreases, which may degrade the positioning accuracy. The integration of PPP-B2b with the High-Accuracy Service (HAS) could be further explored to improve global positioning accuracy.

Author Contributions: Conceptualization and software development, L.Z.; validation and formal analysis, W.Z.; writing—original draft preparation, W.Z.; writing—review and editing, L.Z. All authors have read and agreed to the published version of the manuscript.

Funding: This work is supported by the Open Fund of Technology Innovation Center for Integrated Applications in Remote Sensing and Navigation, Ministry of Natural Resources, (No. TICIARSN-2023-01), National Natural Science Foundations of China (No. 42104018), China Postdoctoral Science Foundation (2022M711669).

Data Availability Statement: The GNSS observations used in this study can be accessed at https://cddis.nasa.gov/archive (accessed on 23 December 2024). Wuhan University provides GPS/Galileo/BDS phase clock/bias products at ftp://igs.gnsswhu.cn/pub/whu/phasebias/ (accessed on 23 December 2024). The vehicle data and PPP-B2b data will be made available on request.

Acknowledgments: The authors gratefully acknowledge IGS MGEX for providing the GNSS data. We also acknowledge the Wuhan University for providing real-time and rapid precise products.

Conflicts of Interest: The authors declare no conflicts of interest.

References

1. Zumberge, J.F.; Heflin, M.B.; Jefferson, D.C.; Watkins, M.M.; Webb, F.H. Precise point positioning for the efficient and robust analysis of GPS data from large networks. *J. Geophys. Res. Solid Earth* **1997**, *102*, 5005–5017. [CrossRef]
2. Geng, J.; Teferle, F.N.; Meng, X.; Dodson, A.H. Kinematic precise point positioning at remote marine platforms. *GPS Solut.* **2010**, *14*, 343–350. [CrossRef]
3. Su, K.; Jin, S.; Jiao, G. Assessment of multi-frequency global navigation satellite system precise point positioning models using GPS, BeiDou, GLONASS, Galileo and QZSS. *Meas. Sci. Technol.* **2020**, *31*, 064008. [CrossRef]
4. Shi, J.; Xu, C.; Guo, J.; Gao, Y. Real-time GPS precise point positioning-based precipitable water vapor estimation for rainfall monitoring and forecasting. *IEEE Trans. Geosci. Remote Sens.* **2014**, *53*, 3452–3459.
5. Tunalı, E.; Özlüdemir, M.T. GNSS PPP with different troposphere models during severe weather conditions. *GPS Solut.* **2019**, *23*, 82. [CrossRef]
6. Zhang, R.; Li, L.; Li, X.; Ma, H.; Xiao, G.; Zhang, J. Research on Quad-Frequency PPP-B2b Time Transfer. *IEEE Instrum. Meas. Mag.* **2024**, *27*, 57–62. [CrossRef]
7. He, Z.; Li, L.; Zhang, R.; Hou, J.; Xiao, G.; Guang, W.; Zhang, J.; He, X. Research on Single-Frequency PPP-B2b Time Transfer. *IEEE Instrum. Meas. Mag.* **2023**, *26*, 42–47. [CrossRef]
8. Hou, P.; Zha, J.; Liu, T.; Zhang, B. Recent advances and perspectives in GNSS PPP-RTK. *Meas. Sci. Technol.* **2023**, *34*, 051002. [CrossRef]
9. Cai, C.; Liu, G.; Yi, Z.; Cui, X.; Kuang, C. Effect analysis of higher-order ionospheric corrections on quad-constellation GNSS PPP. *Meas. Sci. Technol.* **2019**, *30*, 024001. [CrossRef]
10. Byram, S.; Hackman, C. High-precision GNSS orbit, clock and EOP estimation at the United States Naval Observatory. In Proceedings of the 2012 IEEE/ION Position, Location and Navigation Symposium, Myrtle Beach, SC, USA, 23–26 April 2012.
11. Elsobeiey, M.; Al-Harbi, S. Performance of real-time Precise Point Positioning using IGS real-time service. *GPS Solut.* **2016**, *20*, 565–571. [CrossRef]
12. Cheng, C.; Zhao, Y.; Li, L.; Cheng, J.; Sun, X. Preliminary analysis of URA characterization for GPS real-time precise orbit and clock products. In Proceedings of the 2018 IEEE/ION Position, Location and Navigation Symposium (PLANS), Monterey, CA, USA, 23–26 April 2018.
13. Wang, L.; Li, Z.; Ge, M.; Neitzel, F.; Wang, Z.; Yuan, H. Validation and Assessment of Multi-GNSS Real-Time Precise Point Positioning in Simulated Kinematic Mode Using IGS Real-Time Service. *Remote Sens.* **2018**, *10*, 337. [CrossRef]
14. Abdi, N.; Ardalan, A.A.; Karimi, R.; Rezvani, M.-H. Performance Assessment of Multi-GNSS Real-Time PPP over Iran. *Adv. Space Res.* **2017**, *59*, 2870–2879. [CrossRef]
15. Wang, A.; Zhang, Y.; Chen, J.; Wang, H. Improving the (re-)convergence of multi-GNSS real-time precise point positioning through regional between-satellite single-differenced ionospheric augmentation. *GPS Solut.* **2022**, *26*, 39. [CrossRef]
16. Yu, C.; Zhang, Y.; Chen, J.; Chen, Q.; Xu, K.; Wang, B. Performance assessment of multi-GNSS real-time products from various analysis centers. *Remote Sens.* **2022**, *15*, 140. [CrossRef]
17. Yang, Y.; Ding, Q.; Gao, W.; Li, J.; Xu, Y.; Sun, B. Principle and performance of BDSBAS and PPP-B2b of BDS-3. *Satell. Navig.* **2022**, *3*, 5. [CrossRef]
18. Nie, Z.; Xu, X.; Wang, Z.; Du, J. Initial assessment of BDS PPP-B2b service: Precision of orbit and clock corrections, and PPP performance. *Remote Sens.* **2021**, *13*, 2050. [CrossRef]
19. Tao, J.; Liu, J.; Hu, Z.; Zhao, Q.; Chen, G.; Ju, B. Initial assessment of the BDS-3 PPP-B2b RTS compared with the CNES RTS. *GPS Solut.* **2021**, *25*, 131. [CrossRef]
20. Song, W.; Zhao, X.; Lou, Y.; Sun, W.; Zhao, Z. Performance Evaluation of BDS-3 PPP-B2b Service. *Geomat. Inf. Sci. Wuhan Univ.* **2023**, *48*, 408–415.
21. Geng, T.; Li, Z.; Xie, X.; Liu, W.; Li, Y.; Zhao, Q. Real-time ocean precise point positioning with BDS-3 service signal PPP-B2b. *Measurement* **2022**, *203*, 111911. [CrossRef]
22. Zhou, H.; Wang, L.; Fu, W.; Han, Y.; Li, T.; Li, W.; Chen, R. Real-time single-frequency precise point positioning using BDS-3 PPP-B2b corrections. *Measurement* **2022**, *205*, 112178. [CrossRef]
23. He, Q.; Chen, L.; Liu, L.; Zhao, D.; Gong, X.; Lou, Y.; Guan, Q. Long-term performance evaluation of BeiDou PPP-B2b products and its application in time service. *Remote Sens.* **2023**, *15*, 1358. [CrossRef]
24. Li, M.; Chai, H. Real-time marine PPP-B2b/SINS integrated navigation based on BDS-3. *Meas. Sci. Tech.* **2023**, *34*, 105113. [CrossRef]
25. Erol, B.; Turan, E.; Erol, S.; Kuçak, R.A. Comparative performance analysis of precise point positioning technique in the UAV based mapping. *Measurement* **2024**, *233*, 114768. [CrossRef]
26. Kowalczyk, W.Z.; Hadas, T. A comparative analysis of the performance of various GNSS positioning concepts dedicated to precision agriculture. *Rep. Geod. Geoinform.* **2024**, *117*, 11–20. [CrossRef]

27. Gandolfi, S.; Tavasci, L.; Poluzzi, L. Study on GPS–PPP precision for short observation sessions. *GPS Solut.* **2016**, *21*, 887–896. [CrossRef]
28. Kurtz, B.M.; Gómez, D.D.; Bevis, M.G. Characterization of the precision of PPP solutions as a function of latitude and session length. *J. Geod. Sci.* **2024**, *14*, 20220176. [CrossRef]
29. China Satellite Navigation Office. *BeiDou Navigation Satellite System Signal in Space Interface Control Document Precise Point Positioning Service Signal PPP-B2b Version 1.0*; China Satellite Navigation Office: Beijing, China, 2020.
30. Guo, F.; Zhang, X.; Wang, J. Timing group delay and differential code bias corrections for BeiDou positioning. *J. Geod.* **2015**, *89*, 427–445. [CrossRef]
31. Montenbruck, O.; Steigenberger, P.; Hauschild, A. Multi-GNSS signal-in-space range error assessment–Methodology and results. *Adv. Space Res.* **2018**, *61*, 3020–3038. [CrossRef]
32. Zhao, Q.; Guo, J.; Wang, C.; Lyu, Y.; Xu, X.; Yang, C.; Li, J. Precise orbit determination for BDS satellites. *Satell. Navig.* **2022**, *3*, 2. [CrossRef]
33. Guo, J.; Wang, C.; Chen, G.; Xu, X.; Zhao, Q. BDS-3 precise orbit and clock solution at Wuhan University: Status and improvement. *J. Geod.* **2023**, *97*, 15. [CrossRef]
34. Xie, X.; Geng, T.; Zhao, Q.; Cai, H.; Zhang, F.; Wang, X.; Meng, Y. Precise orbit determination for BDS-3 satellites using satellite-ground and inter-satellite link observations. *GPS Solut.* **2019**, *23*, 40. [CrossRef]
35. Yao, Y.; He, Y.; Yi, W.; Song, W.; Cao, C.; Chen, M. Method for evaluating real-time GNSS satellite clock offset products. *GPS Solut.* **2017**, *21*, 1417–1425. [CrossRef]
36. Zhao, L.; Dousa, J.; Ye, S.; Vaclavovic, P. A flexible strategy for handling the datum and initial bias in real-time GNSS satellite clock estimation. *J. Geod.* **2020**, *94*, 3. [CrossRef]
37. Lou, Y.; Zhang, W.; Wang, C.; Yao, X.; Shi, C.; Liu, J. The impact of orbital errors on the estimation of satellite clock errors and PPP. *Adv. Space Res.* **2014**, *54*, 1571–1580. [CrossRef]
38. Simon, D. Kalman filtering. *Embed. Syst. Program.* **2001**, *14*, 72–79.
39. Vaclavovic, P.; Dousa, J. Backward smoothing for precise GNSS applications. *Adv. Space Res.* **2015**, *56*, 1627–1634. [CrossRef]
40. Petit, G.; Luzum, B. *IERS Conventions (2010)*; IERS Tech. Note 36; IERS: Frankfurt am Main, Germany, 2010; Volume 1.
41. Galileo GNSS Agency. *Galileo High Accuracy Service (HAS)*; Publications Office of the EU, Galileo GNSS Agency: Paris, France, 2020.
42. Geng, J.; Chen, X.; Pan, Y.; Mao, S.; Li, C.; Zhou, J.; Zhang, K. PRIDE PPP-AR: An open-source software for GPS PPP ambiguity resolution. *GPS Solut.* **2019**, *23*, 91. [CrossRef]

remote sensing

MDPI

Article

An Improved Fading Factor-Based Adaptive Robust Filtering Algorithm for SINS/GNSS Integration with Dynamic Disturbance Suppression

Zhaohao Chen [1], Yixu Liu [2], Shangguo Liu [1,*], Shengli Wang [3] and Lei Yang [2]

1 College of Geodesy and Geomatics, Shandong University of Science and Technology, Qingdao 266590, China; 202383020037@sdust.edu.cn

2 National Deep-Sea Base Management Center, Qingdao 266237, China; lyx@ndsc.org.cn (Y.L.); yangsir@ndsc.org.cn (L.Y.)

3 College of Ocean Science and Engineering, Shandong University of Science and Technology, Qingdao 266590, China; shlwang@sdust.edu.cn

* Correspondence: sdkdlsg@sdust.edu.cn; Tel.: +86-0532-80681158

Abstract: Aiming at the problem of nonlinear observation model mismatch and insufficient anti-interference ability of SINS/GNSS integrated navigation system in complex dynamic environment, this paper proposes an adaptive robust filtering algorithm with improved fading factor. Aiming at the problem that the traditional Kalman filter is easy to diverge in severe heave motion and abnormal observation, a multi-source information fusion framework integrating satellite positioning geometric accuracy factor (PDOP), solution quality factor (Q value), effective satellite observation number (Satnum), and residual vector is constructed. The dynamic weight adjustment mechanism is designed to realize the real-time optimization of the fading factor. Through the collaborative optimization of robust estimation theory and adaptive filtering, a dual robust mechanism is constructed by combining the sequential update strategy. In the measurement update stage, the observation weight is dynamically adjusted according to the innovation covariance, and the fading memory factor is introduced in the time update stage to suppress the error accumulation of the model. The experimental results show that compared with EKF, Sage-Husa adaptive filtering and robust filtering algorithms, the three-dimensional positioning accuracy is improved by 47.12%, 35.26%, and 9.58%, respectively, in the vehicle strong maneuvering scene. In the scene of ship-borne heave motion, the corresponding increase is 19.44%, 10.47%, and 8.28%. The research results provide an effective anti-interference solution for navigation systems in high dynamic and complex environments.

Keywords: SINS/GNSS integrated navigation; robust adaptive filtering kalman filtering; optimization algorithm; dynamic disturbances; positioning accuracy

1. Introduction

Positioning, Navigation, and Timing (PNT) is regarded as the cornerstone and important infrastructure of national economy and national defense security [1]. In a complex environment, a single PNT service system has the risk of discontinuity, unavailability, or unreliability, and even completely loses its service capability [2]. Therefore, how to make the carrier system generate efficient, stable, and reliable PNT information under complex interference environment is an important research direction.

The noise in the marine environment usually presents a non-Gaussian distribution and has time-varying characteristics. It is difficult for traditional filtering methods to accurately

https://doi.org/10.3390/rs17081449

describe its statistical characteristics, resulting in distortion of state estimation results [3]. At the same time, the marine navigation system is highly nonlinear. Especially in the case of severe heave motion, the linearization of the traditional filtering method may introduce significant errors, which may lead to the performance degradation or even divergence of the filter. GNSS is widely used in the fields of smart agriculture, public transportation, aviation, and navigation. Because GNSS signal is a form of radio propagation, it is prone to inaccuracy or signal loss in complex environments such as occlusion and interference [4]. Relying solely on the global satellite navigation and positioning system to achieve navigation and positioning cannot meet the requirements of high precision, high stability, and high reliability. It is a reasonable choice to make full use of multi-sensors to obtain multi-source PNT information. In order to obtain more accurate and stable results, the information obtained by SINS and GNSS is usually fused. As an independent navigation method, the strapdown inertial navigation system (SINS) is not easily affected by the external environment, but the position error will accumulate over time. In addition, due to the influence of vibration and other factors, the inertial measurement unit (IMU) data may have abnormal values. Especially in the ocean, a low speed and high vibration heave environment, the sensor output is unstable [5]. When the GNSS signal is disturbed by a multipath effect, occlusion and other non-Gaussian noise, or the navigation information output by the inertial sensor in a complex environment is unreliable, it is easy to cause the Kalman filter to diverge, resulting in inaccurate positioning information. In view of the above situation, researchers have proposed a variety of improved Kalman Filter algorithms based on robust and adaptive Kalman Filter to improve the adaptability and reliability of integrated navigation.

In view of GNSS signal anomaly and noise distribution uncertainty, robust and adaptive filters have become research hotspots. Yang, Y, improved the abnormal observation and state disturbance control in the filtering process through adaptive estimation. A method of using adaptive factor to control the abnormal influence of prior prediction state is proposed. The adaptive factor is constructed by a variety of error statistics, such as prediction residual statistics and state inconsistency statistics. In addition, they use a two-or three-segment function model to implement the adaptive factor, thereby enhancing the adaptability and robustness of the filter to dynamic system noise and observation anomalies [6]. In the unknown time-varying noise environment, Song uses the improved square root cubic Kalman filter (ISRCKF) to estimate the motion state of the target vehicle. It is proposed to combine the Sage-Husa noise statistical estimator with the decay memory index weighting method to construct a time-varying noise statistical estimator suitable for nonlinear systems, so as to improve the filtering accuracy under the condition of uncertain noise characteristics, and solve the problem of low accuracy or even divergence of traditional filtering methods in high-dimensional nonlinear target tracking [7]. Chun Ma proposed the Variational Bayesian-based Robust Adaptive Kalman Filtering (VBRAKF) method for GNSS/INS tightly coupled positioning in urban environments. It combines robust estimation and variational Bayesian adaptive filtering to handle measurement outliers and inaccurate noise statistics, improving positioning accuracy. Tested in urban vehicle experiments, VBRAKF outperformed traditional methods, showing significant improvements in accuracy and ambiguity resolution, especially under GNSS signal outages [8]. Jun Xiong proposed an adaptive hybrid robust filter (AHRF) for multi-sensor relative navigation systems. This method integrates KF-RAIM (FDE-based method) and Huber's M-estimation-based Kalman filter (Huber-KF) using an adaptive interactive multiple model (AIMM) framework. The AHRF improves positioning by efficiently switching between the two filters to handle various error conditions, such as slowly growing errors, step errors, and random biases. The proposed method outperforms conventional systems, particularly in scenarios with sharp system model transitions, achieving better robustness and positioning accuracy [9].

The Sage-Husa adaptive filter is mainly used to estimate the statistical characteristics of constant coefficient noise in linear systems, but it does not perform well for nonlinear or time-varying noise systems [10,11]. In addition, it cannot compare the abrupt changes that rely on historical observation data and cannot respond quickly to noise. Robust filtering suppresses the influence of abnormal data, but it may also introduce bias to a certain extent. In some extreme cases, the filter may ignore the drastic change in the real value, resulting in the result deviating from the actual state [12].

In order to solve this problem, a SINS/GNSS adaptive robust filtering algorithm based on an improved fading factor is proposed in this paper. An adaptive filtering algorithm for dynamically adjusting the fading factor is designed. Firstly, the measurement variance is artificially limited to prevent extreme values from affecting the positioning accuracy. The construction of the fading adaptive factor comprehensively considers the satellite positioning geometric precision factor (PDOP value), the solution quality factor (Q value), the number of effective satellite observations and the observation vector residual, dynamically adjusts the measurement noise covariance matrix, and enhances the adaptability to the nonlinear observation model. The measurement is updated by sequential filtering to prevent the negative determination of the measurement noise matrix due to the excessive input state noise. The robust factor is based on the standardized residual vector, and the weight function is used to assign weights to suppress the interference of abnormal observations on the filtering results. In order to further optimize the navigation state estimation, this paper combines the improved adaptive filtering algorithm with the robust filtering algorithm to obtain the optimal estimation value of the navigation, which suppresses the accumulation of model error and reduces the influence of nonlinear error on the state estimation. By designing vehicle-borne and ship-borne experiments, the proposed algorithm is compared with the extended Kalman filter (EKF), robust filter (RKF), and Sage-Husa adaptive filter (AKF) in terms of position, speed, and attitude. The algorithm shows better accuracy and smoother performance. The experimental results show that the algorithm can significantly improve the stability and accuracy of navigation and positioning in a complex dynamic environment.

The structure of this paper is as follows: Section 2 summarizes the proposed method. This section introduces the overall framework and process of the adaptive robust filtering algorithm with improved fading factor and focuses on the implementation steps of adaptive filtering and the dynamic weight adjustment mechanism of real-time optimization of fading factor. Through the collaborative optimization of robust estimation theory and adaptive filtering, combined with sequential update strategy, a dual-factor robust method is constructed. In Section 3, the performance of the proposed method is verified by vehicle experiments under high maneuvering environment and ship-borne experiments under heave and sloshing, and the experimental results are discussed. Section 4 of the article discusses the advancements over other algorithms and the potential benefits and limitations of this paper's algorithm. It also makes an exploration of future research based on the limitations Finally, Section 5 summarizes this article.

2. Methods

2.1. Research Design Framework

This study aims to solve the problem of nonlinear observation model mismatch and insufficient anti-interference ability of SINS/GNSS integrated navigation system in complex dynamic environment, and proposes an improved adaptive robust filtering algorithm (ARKF). The research framework follows the closed-loop research system of 'problem-driven → theoretical modeling → algorithm design → experimental verification'. The specific framework design is as follows:

- Problem-driven: In complex dynamic environments (such as urban canyons, marine high-vibration scenarios), the SINS/GNSS integrated navigation system faces the following core challenges: (1) Model structure mismatch: The traditional EKF uses first-order Taylor expansion linearization. When encountering high-frequency vibrations, the second-order truncation error accumulates, resulting in a non-positive state estimation covariance matrix, causing the risk of filtering divergence. (2) Dissimilation of noise characteristics: GNSS observation noise exhibits pulse characteristics due to multipath effects and occlusion; the colored noise generated by IMU in the vibration environment breaks through the classical Gaussian white noise hypothesis, which leads to the inaccurate estimation of the noise model by the system and affects the filtering effect.
- Theoretical modeling: Based on the SINS/GNSS loosely coupled model, the 15-dimensional error state vector and the 6-dimensional observation vector are defined and the state equation and measurement equation are constructed.
- Algorithm design: The dynamic fading factor is constructed by multi-source information fusion, and the sequential filtering is used to update the measurement to avoid the negative determination of the measurement noise covariance matrix. At the same time, the standardized residual and IGG-III weight function are combined to suppress the influence of abnormal observation on state update, and the two-factor robust mechanism is realized.
- Experimental verification: In order to verify the performance of the algorithm, a set of vehicle-mounted experiments was designed. The experimental time was about 20 min. Due to the high-rise occlusion, the data had a GNSS signal of about 200 s. A set of ship-borne experiments, three-level sea conditions, were designed, and the experiment lasted about 1 h. After the 2000s, the heave of the ship was significantly increased due to the wind and waves. The emergence of these abnormal conditions increases the uncertainty of navigation and can better verify the stability of the algorithm in a disturbed environment.

2.2. Combined SINS/GNSS Navigation Models

In this paper, based on the observation of position and velocity by GNSS, the GNSS/INS loose combination is studied to solve the nonlinear problem of the system [13]. The state space model and measurement model of SINS and GNSS integrated navigation system are as follows [14]:

$$\begin{cases} \boldsymbol{X}_k = \boldsymbol{\Phi}_{k/k-1}\boldsymbol{X}_{k-1} + \boldsymbol{\Gamma}_{k-1}\boldsymbol{W}_{k-1} \\ \boldsymbol{Z}_k = \boldsymbol{H}_k\boldsymbol{X}_k + \boldsymbol{v}_k \end{cases}$$
$$\begin{cases} E[\boldsymbol{W}_k] = 0,\ E[\boldsymbol{W}_k\boldsymbol{W}_j^{\mathrm{T}}] = \boldsymbol{Q}_k\delta_{kj}, \\ E[\boldsymbol{V}_k] = 0,\ E[\boldsymbol{V}_k\boldsymbol{V}_j^{\mathrm{T}}] = \boldsymbol{R}_k\delta_{kj}, \\ E[\boldsymbol{W}_k\boldsymbol{V}_j^{\mathrm{T}}] = 0 \end{cases} \quad (1)$$

where $\boldsymbol{\Phi}_{k/k-1}$ is the state transfer matrix of the system; $\boldsymbol{\Gamma}_{k-1}$ is the noise driver matrix; $\boldsymbol{W}_{k-1}$ is the process noise vector; $\boldsymbol{Q}_k$ is the state noise covariance matrix; $\boldsymbol{R}_k$ is the observation noise covariance matrix; meanwhile, $\boldsymbol{Q}_k$, $\boldsymbol{R}_k$ should be positive definite matrices; $\boldsymbol{X}_k$ is the state vector; $\boldsymbol{Z}_k$ denotes the measurement vector of the system output; $\boldsymbol{H}_k$ is the measurement matrix; v_k denotes the measurement noise, where the 21-dimensional state vector contains the navigation state error and the sensor error, defined as [15].

$$X_k = \begin{bmatrix} \delta r^n & \delta v^n & \phi & b_g & b_a & s_g & s_a \end{bmatrix}^T \tag{2}$$

where δr^n is the inertial position error vector; δv^n is the inertial velocity error vector; ϕ is the attitude error vector; b_g is the three-axis gyroscope bias vector; b_a is the three-axis accelerometer bias vector; s_g is the gyroscope scale factor error; s_a is the accelerometer scale factor error vector.

2.3. SINS/GNSS Adaptive Robust Filtering Algorithm Based on Improved Fading Factor (ARKF)

The SINS/GNSS adaptive robust filtering algorithm based on improved fading factor solves the problem that the traditional filtering algorithm is susceptible to non-Gaussian noise and abnormal observation in complex dynamic environment. The core of the algorithm is to dynamically adjust the filter by constructing a fading adaptive factor and a robust factor to improve the filtering performance and navigation accuracy.

The key to improving adaptive filtering is to construct a fading adaptive factor and adjust the measurement noise covariance matrix in real time, so as to enhance the adaptability of the filter to dynamic environmental changes. This paper comprehensively considers the four parameters of satellite geometric precision factor (PDOP value), solution quality factor (Q value), effective satellite observation number and observation vector residual, and dynamically adjusts the size of the adaptive factor. And the gross error detection is performed before the measurement update, and the upper and lower limits of the measurement variance are artificially set to avoid the influence of extreme values. When the PDOP value and Q value are low, and the number of effective satellite observations is sufficient, the measurement environment is better, the adaptive factor is close to 1, and the filter trusts the current observation data more. When the PDOP value and Q value are large, the number of effective satellite observations decreases or the observation residual increases, the adaptive fading factor decreases, and the filter reduces the weight of abnormal observations, thereby reducing the impact of abnormal data on state estimation. According to the evaluation results of the normalized vector residual mean, if a certain error standard C is met, the weight of the filter will be adjusted accordingly. Secondly, the robust factor is constructed, and the right function is used to determine the weight, so as to ensure the resistance of the robust filter to abnormal data. Finally, the fusion of these two filtering methods can further optimize the dynamic estimation accuracy of the system, especially in the case of complex environment or strong interference, to ensure the stable operation of the navigation system.

The flow chart of SINS/GNSS adaptive robust filtering algorithm based on improved fading factor is shown in Figure 1:

2.4. Improved Fading Factor Adaptive Filtering Algorithm (IAKF)

Since some information about the system model is implied in the measurement output, if the system model parameters are not accurate enough, some of the parameters can be modeled by adaptive estimation based on the measurement output [16,17].

When $E[W_k]$ and $E[V_k]$ do not satisfy equal to 0, using the traditional Kalman filter is prone to filter divergence and does not allow the system to reach the optimal estimation [18]. When the observation is in an abnormal state, the measurement noise will have a great impact on the filtering results. Therefore, the adaptive filtering is introduced to estimate the measurement noise while estimating the parameters. The system noise variance is difficult to adapt, and the measurement noise variance matrix is relatively easy to adapt. Therefore, the designed adaptive filter uses the output edge as the state estimation and the measurement noise identification. The adaptive filtering algorithm of the improved fading factor is as follows:

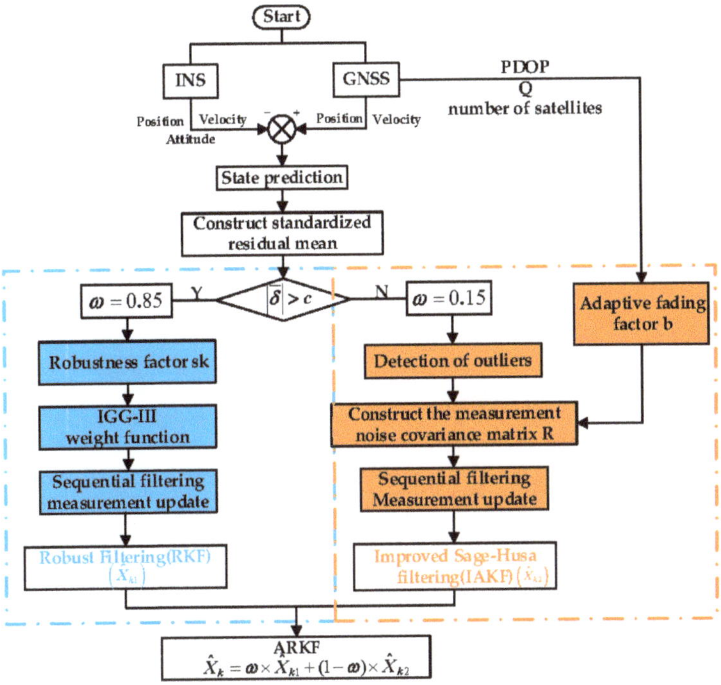

Figure 1. SINS/GNSS adaptive robust filtering algorithm based on an improved fading factor.

The measurement noise covariance matrix is adaptive by innovation. The measurement innovation is as follows [19]:

$$\hat{V}_k = Z_k - \hat{Z}_{k/k-1} = H_k X_k + V_k - H_k \hat{X}_{k/k-1} = H_k \widetilde{X}_{k/k-1} + V_k \tag{3}$$

where $\hat{V}_k$ is the measurement innovation; Z_k denotes the observation vector; $\hat{X}_{k/k-1}$ denotes the predicted state vector; H_k is a measurement matrix [20,21]:

$$\hat{V}_k \sim \mathrm{N}(0, H_k P_{k/k-1} H_k^{\mathrm{T}} + R_k) \tag{4}$$

R_k is the measurement noise covariance matrix. where the measurement innovation is considered to be white noise with 0 as the mean and $H_k P_{k/k-1} H_k^{\mathrm{T}} + R_k$ represents the variance when the system is working normally [22]; and $P_{k/k-1}$ is the covariance matrix of the prediction error. The measurement noise variance matrix is estimated as follows [23]:

$$\hat{R}_k = \frac{1}{k}\sum_{i=1}^{k}(\widetilde{V}_{i/i-1}\widetilde{V}_{i/i-1}^{\mathrm{T}} - H_i P_{i/i-1} H_i^{\mathrm{T}}) \tag{5}$$

$$\hat{R}_k = \frac{1}{k}[\sum_{i=1}^{k-1}(\widetilde{V}_{i/i-1}\widetilde{V}_{i/i-1}^{\mathrm{T}} - H_i P_{i/i-1} H_i^{\mathrm{T}}) + (\widetilde{V}_{k/k-1}\widetilde{V}_{k/k-1}^{\mathrm{T}} - H_k P_{k/k-1} H_k^{\mathrm{T}})] \tag{6}$$

$$\hat{R} = (1-\beta_k)\hat{R}_{k-1} + \beta_k(\widetilde{V}_{k/k-1}\widetilde{V}_{k/k-1}^{\mathrm{T}} - H_k P_{k/k-1} H_k^{\mathrm{T}}) \tag{7}$$

In Equation (7):

$$\beta_k = \frac{\beta_{k-1}}{\beta_{k-1}+b} \tag{8}$$

where $\beta_0 = 1, \beta_\infty = 1 - b$, b is called the adaptive fading factor $(0 < b < 1)$.

In Equation (5), when the actual noise is relatively small, it is easy to have negative determination of the measurement variance, which is handled here by using sequential filtering.

$$\rho_k^{(i)} = (\widetilde{V}_{k/k-1}^{(i)})^2 - H_k^{(i)} P_{k/k-1}^{(i)} (H_k^{(i)})^{\mathrm{T}} \tag{9}$$

In order to prevent the filter from becoming unstable due to inaccurate measurement noise estimation. By limiting the update range of the estimated measurement noise variance, the stability of the filter is improved. The upper and lower limits of $R_k^{(i)}$ are set artificially. In the experiment, the lower limit R_{min} is 0.3 times the nominal variance of GNSS position noise and velocity noise, and the upper limit R_{max} is 3 times the nominal variance of GNSS position noise and velocity noise. This setting can prevent extreme values from causing filter divergence, while retaining the space for dynamic adjustment of noise [24]. There are as follows:

$$\hat{R}_k^{(i)} = \begin{cases} R_{\min}^{(i)} & \rho_k^{(i)} \le R_{\min}^{(i)} \\ (1-\beta_k)\hat{R}_{k-1}^{(i)} + \beta_k \rho_k^{(i)} & R_{\min}^{(i)} < \rho_k^{(i)} < R_{\max}^{(i)} \\ R_{\max}^{(i)} & \rho_k^{(i)} \ge R_{\max}^{(i)} \end{cases} \tag{10}$$

In this paper, we also add fault judgment on the basis of adaptive filtering, according to Equation (10) when the system is fault-free $\widetilde{\boldsymbol{V}}_{k/k-1} \sim N(0, \boldsymbol{H}_k \boldsymbol{P}_{k/k-1} \boldsymbol{H}_k^{\mathrm{T}} + \boldsymbol{R}_k)$, based on the residual detection technique, the filtering dispersion judgment is based on the following [25]:

$$\widetilde{\boldsymbol{V}}_{k/k-1}^{\mathrm{T}} \widetilde{\boldsymbol{V}}_{k/k-1} > 3tr(\boldsymbol{H}_k \boldsymbol{P}_{k/k-1} \boldsymbol{H}_k^{\mathrm{T}} + \boldsymbol{R}_k) \tag{11}$$

If the above equation is valid, it means that the actual error value is more than 3 times that of the theoretical value, according to the principle that it is anomalous value, and the use of these data for filtering can easily lead to unstable filtering dispersion, at this time, let $\hat{R}_k^{(i)} = R_{\max}^i$, through the increase in the large number of measurement of the noise variance array to make the anomalous value of the filtering system weighting is reduced to reduce the impact on the results.

In order to obtain the adaptive fading factor b, the adaptive fading factor is constructed by the PDOP value, Q value, effective satellite observation number and measurement vector residual obtained by satellite observation. Firstly, the measurement scale factors α, β and λ are obtained to adjust the sensitivity of the system to the measurement error. The system can evaluate the quality of the current environment and observation data in real time and dynamically adjust the weight of the filter. The following is the formula:

$$\begin{cases} \alpha = \frac{PDOP - PDOP_{\min}}{PDOP_{\max} - PDOP_{\min}} \\ \beta = \frac{Q - Q_{\min}}{Q_{\max} - Q_{\min}} \\ \lambda = 1 - \frac{satnum - satnum_{\min}}{satnum_{\max} - satnum_{\min}} \end{cases} \tag{12}$$

In this formula, the factor α is used to adjust the influence of PDOP. PDOP represents the geometric distribution of the satellite. The lower PDOP indicates a better satellite geometric distribution and usually has better positioning accuracy, while the higher PDOP indicates a worse geometric distribution. When PDOP is poor, this factor is at a lower value, which makes the filtering system weaken the influence of measurement innovation on positioning, thus improving the reliability of the filtering. The Q value measures the quality of the satellite positioning solution and represents the accuracy and reliability of the GNSS positioning data. The β factor ensures that when Q is high, it usually indicates that the quality of the positioning solution is poor. At this time, the filter is dynamically adjusted to reduce the dependence on the data. The λ factor is adjusted according to the number of effective satellites used for positioning. When the number of satellites is large, it means that the number of satellites involved in the solution is large, and the positioning is usually more accurate. The λ factor helps the system to adapt to changes in the number of visible satellites. When the number of satellites is small, it can still maintain stability.

The construction formula for the fading adaptive factor is as follows:

$$\begin{cases} \mathrm{u} = \frac{\widetilde{V}_k^T \widetilde{V}_k}{3 \times \left| \mathrm{tr}(\boldsymbol{HPH}^T + \boldsymbol{R}) \right|} \\ \omega_1 + \omega_2 + \omega_3 = 1 \\ \mathrm{b} = (1 - \mathrm{u}) \times [0.999 - (\omega_1 \times \alpha + \omega_2 \times \beta + \omega_3 \times \lambda) * \mathrm{A}] \end{cases} \tag{13}$$

where b is the adaptive fading factor, the measurement noise weight is adjusted by the adaptive fading factor. The parameter u becomes larger, so that the adaptive fading factor b decreases, and by reducing the value of b, the system's response to the measurement noise can be accelerated, so as to improve the tracking accuracy, especially in the case of large noise fluctuations, so that it can be better adjusted and adapted to the current environmental changes.

A represents the amplitude of the amplitude adjustment coefficient of the adaptive fading factor, and the value is 0.07. Its core significance is to balance the dynamic response speed of the noise estimation, so that the value of the fading factor is kept in the range of [0.9–0.999] as much as possible [26]. If the fading factor is too small, the noise estimation result will jump too violently.

For the selection of the adaptive fading factor amplitude adjustment coefficient A, we re-compare the performance of the adaptive fading factor when A = 0.05, A = 0.07, and A = 0.09. It can be seen from the Figure 2 that when A = 0.05, the adaptive fading factor b is insensitive to changes in the disturbance environment, which weakens the system's adaptive ability and reduces the response speed. When A = 0.09, the known adaptive fading factor b should be between [0.9–0.999]. When b is too small, it is easy to cause instability of the system. Therefore, when A = 0.07, the response speed and stability of the filter achieve the best balance.

It is the weight coefficient of the satellite geometric accuracy factor PDOP, the weight coefficient of the solution factor Q, and the weight coefficient of the number of effective satellite observations. The selection and distribution of weight coefficients adhere to the principle of significant influence of geometric accuracy factor, satellite solution factor and effective satellite observation number on satellite positioning error.

- Satellite geometric precision factor (PDOP): it directly reflects the influence of satellite geometric distribution on positioning accuracy [27], and its weight ω_1 is set to 0.4.
- Satellite solution factor (Q): used to quantify the quality of GNSS solution. Table 1 shows the relationship between Q value and 3D positioning accuracy. The Q value directly reflects the error range of GNSS positioning. Its weight ω_2 is also set to 0.4.
- The number of effective satellite observations (Satnum): The number of effective satellite observations does not directly reflect the advantages and disadvantages of GNSS positioning solutions, but it can reflect the redundancy of GNSS positioning. Therefore, this paper sets its weight ω_3 to 0.2.

In the filtering process, when the GNSS observation results are abnormal, the PDOP and Q values will increase significantly, which represents the deterioration of the geometric distribution of the satellite and the decrease in the solution quality [28]. This change means that the system faces greater uncertainty, which may lead to a decrease in the confidence of the filter in the observed data. At the same time, the observation residuals of GNSS will also increase, and the system will increase the adaptability to the measurement noise according to the change in these abnormal residuals, and it will gradually improve the response ability to the noise estimation. In addition, the parameter u will also increase, thus prompting the adaptive fading factor b to decrease within a reasonable range. The dynamic adjustment of the fading factor b enables the filtering system to flexibly adjust the measurement noise covariance matrix according to the actual observation conditions, thereby improving the robustness of the system in the face of complex disturbances. It effectively suppresses the influence of abnormal data on the filtering results and prevents abnormal observations from adversely affecting the navigation system.

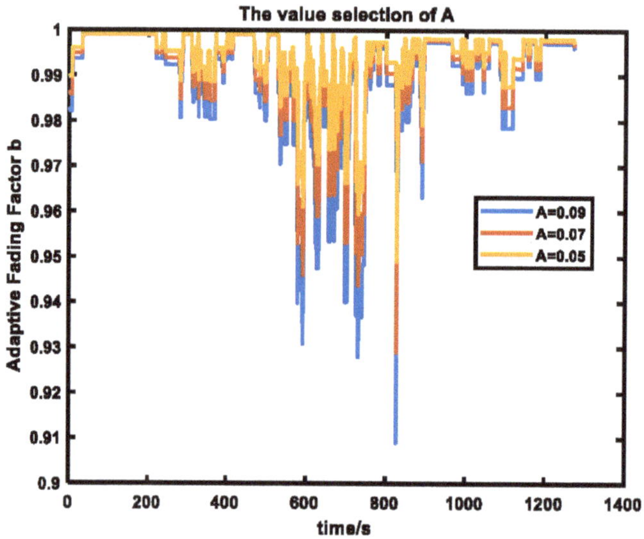

Figure 2. Selection of adaptive fading factor amplitude adjustment coefficient A.

Table 1. Solution quality factor.

Q	Description	3D Accuracy (m)
1	Fixed integer	0.00–015
2	Converged float or noisy fixed integer	0.05–0.40
3	Converging float	0.20–1.00
4	Converging float	0.50–2.00
5	DGPS	1.00–5.00

The value of fading factor b should not be too small. If b is set too small, the response of the system to noise estimation will change dramatically. This change may lead to unstable noise estimation results and jump phenomenon, thus affecting the stability and positioning accuracy of the filter. Therefore, the fading factor should be kept within a reasonable range, usually set between $[0.9, 0.999]$ to ensure that the filtering system can respond smoothly during dynamic adjustment [29,30].

- When the GNSS observations are reliable, when the satellite geometry is well distributed and the number of effective satellite observations is sufficient, the PDOP value and Q value are usually low, and the system can obtain high-quality positioning results [31]. In this case, the fading factor b will increase, the confidence of the filter to the observation data will increase, and the system believes that the measurement data are more reliable. The update of the measurement noise covariance matrix will become more stable, and the filter will give greater weight when processing the observation data, so that the estimation results are more dependent on the observation data.

- When the GNSS observations are unreliable, when the satellite geometric distribution is poor and the number of effective satellite observations is insufficient, the PDOP value and Q value will increase, and the system will face higher uncertainty and positioning error. In this case, the quality of the observation data is poor, and there may be a large measurement error or noise, which will lead to a decrease in the trust of the filter to the observation data [31]. As the fading factor decreases, the update of the measurement noise covariance matrix becomes more conservative, the trust of the filter to the measurement data decreases, and the system depends more on the inertial navigation system (SINS). This dynamic adjustment can effectively avoid the estimation error caused by measurement noise or abnormal data.

Therefore, the dynamic adjustment mechanism of the adaptive fading factor designed in this paper can make the system flexibly adapt to the changes under different environmental conditions. Under harsh environmental conditions, a large number of measurement noise matrices are obtained by reducing the b value, and the sensitivity to measurement noise is enhanced. When the observation conditions are improved, the fading factor approaches 1, which helps the system to make more accurate use of historical data, so as to maintain efficient noise suppression and positioning accuracy during long-term operation. In the filtering system, with the dynamic adjustment of the fading factor, the system can quickly reduce the dependence on abnormal observations in harsh environments, and restore a high degree of trust in valid data when the observation conditions are good. Through this adaptive mechanism, the system can automatically adjust the trust degree of different measurements according to the real-time observation quality and achieve higher positioning accuracy and robustness. Especially in the environment of GNSS signal occlusion or large external vibration, it can effectively avoid the attenuation of filter performance and the expansion of positioning error.

In summary, the overall flow chart of the improved adaptive filtering algorithm is shown in Figure 3:

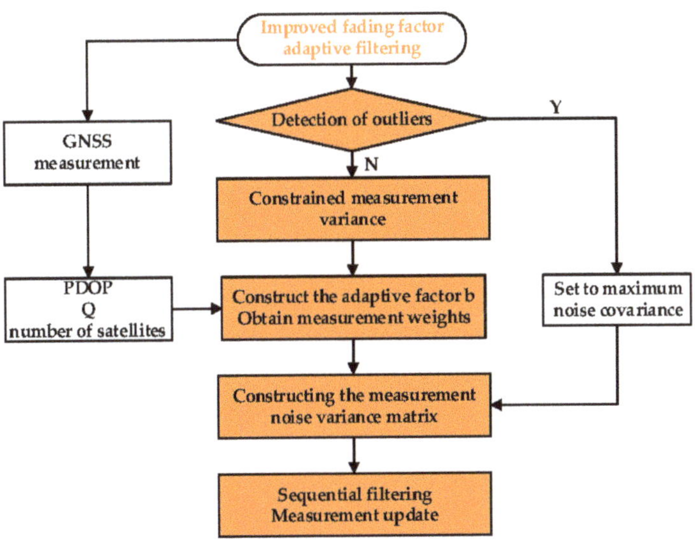

Figure 3. Improved fading factor adaptive fading algorithm (IAKF).

2.5. Robust Filtering (RKF)

Robust filtering can solve the problem of outliers and errors in navigation systems. By reducing the influence of outliers on the filtering results, robust filtering can avoid filtering divergence and ensure the stable operation of the integrated navigation system in various complex environments [32]. The robust filtering process is as follows:

The prediction residual vector can better respond to the anomalies of state perturbations [33]. Then, the weight function is constructed based on the standardized residual vector. The construction process of standardized vector residual construction and anti-differential factor sk is as follows:

$$\delta_k = \frac{\left|\widetilde{\boldsymbol{V}}_{\boldsymbol{k}}(i)\right|}{\sqrt{\boldsymbol{R}_{\boldsymbol{k}}[i,i]}} \tag{14}$$

Sorting δ_k, we have $\delta_k^{\text{sorted}}[j] \geq \delta_k^{\text{sorted}}[j+1]$, $\forall j \in [0, n-2]$. Since the observation vector is 6-dimensional, $n = 6$, is taken:

$$\delta_k^{mid} = \frac{\delta_k^{\text{sorted}}[3] + \delta_k^{\text{sorted}}[4]}{2} \tag{15}$$

The vector of resistance factor is as follows:

$$sk = \frac{\left|\widetilde{\boldsymbol{V}}_{\boldsymbol{k}}(i)\right|}{\delta_k^{mid}\sqrt{\boldsymbol{R}_{\boldsymbol{k}}[i,i]}} \tag{16}$$

Here, we use a three-stage weight reduction factor, IGGIII weight function [27,33]:

$$\overline{p_i} = \begin{cases} p_i & sk \leq k_0 \\ p_i\omega(\widetilde{V}_i) & k_0 < sk \leq k_1 \\ 0 & sk > k_1 \end{cases} = \begin{cases} p_i & sk \leq k_0 \\ p_i\frac{k_0}{\left|\widetilde{V}_i\right|}\left(\frac{k_1-\left|\widetilde{V}_i\right|}{k_1-k_0}\right)^2 & k_0 < sk \leq k_1 \\ 0 & sk > k_1 \end{cases} \tag{17}$$

where $\left|\widetilde{V}_i\right|$ denotes the standardized residual; k_0 and k_1 are constants, usually chosen as $k_0 = 1.0 \sim 1.5$, $k_1 = 2.5 \sim 8.0$.

Sequential filtering is used to update measurements. In the robust filtering framework, only the data at the current time and the estimation results at the previous time are needed, and the weight can be dynamically adjusted according to the weighting factor and residual of the current observation without storing all the historical data [34].

$$\begin{cases} \boldsymbol{K}_k^{(N)} = \overline{p_i}\boldsymbol{p}_k^{(N-1)}(\boldsymbol{H}_k^{(N)})^T[\boldsymbol{H}_k^{(N)}\boldsymbol{p}_k^{(N-1)}(\boldsymbol{H}_k^{(N)})^T + R_k^{(N)}]^{-1} \\ \boldsymbol{X}_k^{(N)} = \boldsymbol{X}_k^{(N-1)} + \boldsymbol{K}_k^{(N)}(\boldsymbol{Z}_k^{(N)} - \boldsymbol{H}_k^{(N)}\boldsymbol{X}_k^{N-1}) \\ \boldsymbol{p}_k^{(N)} = (\boldsymbol{I} - \boldsymbol{K}_k^{(N)}\boldsymbol{H}_k^{(N)})\boldsymbol{p}_k^{(N-1)}(\boldsymbol{I} - \boldsymbol{K}_k^{(N)}\boldsymbol{H}_k^{(N)})^T + \boldsymbol{K}_k^{(N)}\boldsymbol{R}_k^{(N)}\boldsymbol{K}_k^{(N)T} \end{cases} \tag{18}$$

In summary, the overall flow chart of the robust filtering algorithm is as follows: In summary, the overall flow chart of the robust filtering algorithm is shown in Figure 4:

2.6. Data Collection and Processing Analysis Strategy

The inertial navigation of SINS/GNSS integrated navigation vehicle data adopts Honeywell HG4930, and its sampling frequency is 100 Hz. GNSS uses a single antenna with a sampling frequency of 5 Hz. The vehicle line is a circle around Shandong University of Science and Technology. The experiment lasted about 20 min. The ship-borne data also uses HG4930 with a frequency of 100 Hz, and the GNSS sampling frequency is 10 Hz. The ship-borne route revolves around the offshore wind farm in Changyi, and the experiment time is about 1 h. The time synchronization of GNSS and SINS is shown in Figure 5. According to the satellite data, the UTC and PPS signals are obtained. The initial data of IMU is timed by PPS, and the subsequent data are recursively timed according to the initial time of IMU, and the subsequent data are measured by timer. The final IMU data with time information is obtained by weighted fusion of recursive time and measurement time [35].

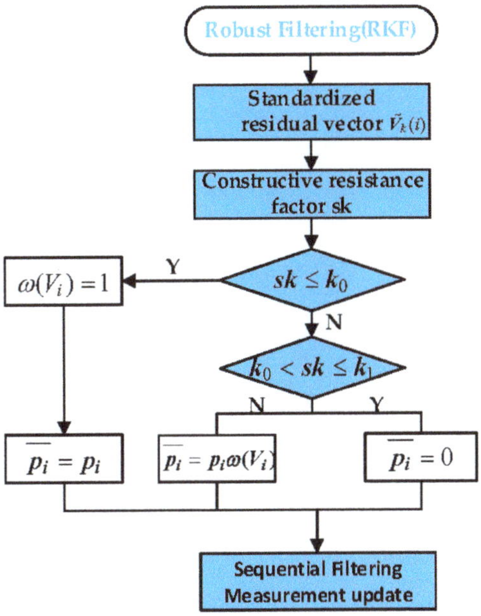

Figure 4. Robust filtering algorithm.

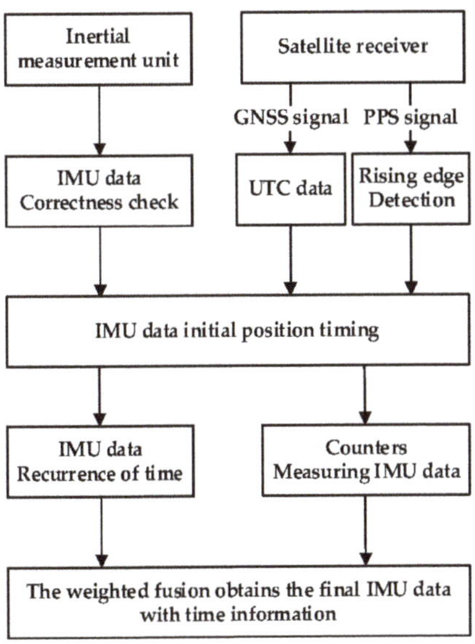

Figure 5. Hardware time synchronization strategy.

3. Experimental Demonstration and Analysis of Results

3.1. Experimental Environment

In order to evaluate the performance of the SINS/GNSS adaptive robust filtering algorithm based on the improved fading factor, a set of vehicle-mounted measured data collected in the complex environment of Qingdao City in 2024 was processed and analyzed (Figure 6a); a set of ship-borne measured data collected in the sea area near Changyi, Weifang, in 2024 under heave, low frequency, and high vibration environment are analyzed (Figure 6b).

(**a**) (**b**)

Figure 6. Experimental environment: (**a**) in-vehicle experiments; (**b**) shipboard experiments.

The INS data of the following two sets of experiments were collected by HG4930, and the performance index parameters are shown in Table 2.

Table 2. HG4930 performance parameters.

	Performance Parameters	HG4930
gyros	Range/(deg/s)	±400
	Bias instability/(deg/h)	0.35
	ARW/(deg/$\sqrt{h}$)	0.05
accelerometer	Range/g	±20
	Bias instability/mg	0.05
	VRW/(m/s/$\sqrt{h}$)	0.05

3.2. In-Vehicle Experiments

GNSS uses the differential positioning solution, and the GNSS base station is set up in the comprehensive building of Shandong University of Science and Technology Science Park, with a baseline length of about 1 km. The experiment lasted for about 1200 s. The experimental scene was selected as a complex urban environment, and the sports car experiment was carried out around Shandong University of Science and Technology.

The GNSS differential sports car trajectory is as follows Figure 7 shows that the differential signal is lost during the experiment. The satellite signal will be interfered or reflected due to the occlusion of the building; similarly, satellite signals may also be reflected by surrounding buildings or objects, resulting in multipath effects. The signal incorrectly receives the reflected wave, affecting the accuracy of positioning [36].

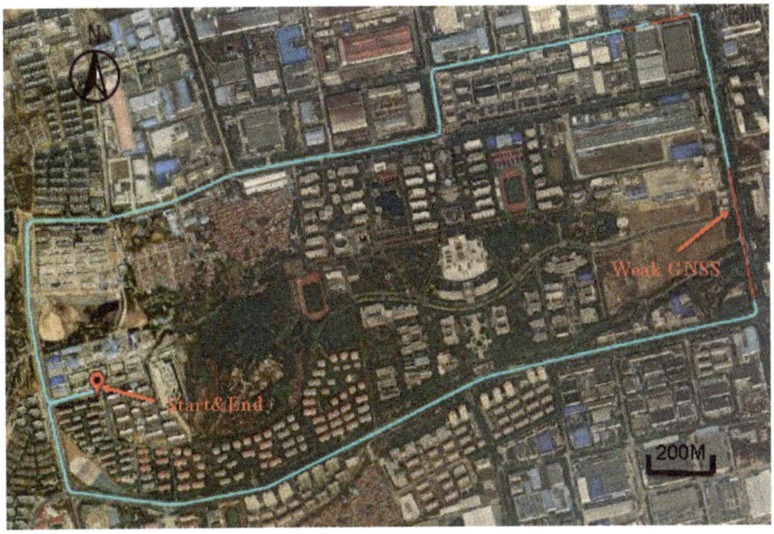

Figure 7. Vehicle experiment running trajectory.

Four different schemes are designed for this data set: method one uses extended Kalman filter (EKF); the second method is the Sage-Husa adaptive Kalman filter (AKF); the third method is robust filtering (RKF); method 4 uses the filtering method (ARKF) mentioned in this paper. The high-precision navigation results obtained by NovAtel's post-processing software Inertial Explorer 8.70 are used as the true value of the reference ground [37].

Figure 8A shows the geometric accuracy factor of the vehicle-borne experimental satellite, which directly reflects the influence of the distribution of the satellite geometry on the positioning error. When the vehicle passes under the tall buildings between 600 s and 800 s, most of the satellite signals are blocked, resulting in poor satellite geometry, which affects the observation effect of differential positioning. Figure 8B represents the satellite solution factor (Q), which is used to quantify the GNSS solution quality. The Q value can directly reflect the problem of satellite positioning difference between 600 and 800 s. Figure 8C shows the number of satellites involved in the solution in the vehicle experiment. There are only four satellites involved in the solution between 600 and 800 s, and the solution quality is poor. Figure 8D shows the change in adaptive fading factor b in the vehicle experiment. In the case of poor satellite solution quality, the filtering system can quickly adapt to environmental changes by adjusting the adaptive fading factor to obtain the optimal measurement noise matrix in the current situation. When the observation conditions are good and the measurement data are reliable, the value of the adaptive fading factor approaches 1, so that the measurement noise is closer to the historical value. This adjustment helps to improve the smoothness and stability of measurement noise, effectively reduce the impact of noise fluctuations on system performance, and ensure the stability of positioning accuracy.

The position, velocity, and attitude errors obtained by using the above four methods and professional solution software are shown in Figures 9 and 10.

Figure 9A,C,E represent the comparison of the position errors of the four algorithms in the northeast direction. Figure 9B,D,F represent the comparison of the velocity errors of the four algorithms in the northeast direction, respectively. In the 600–800 s period with poor satellite observation conditions, the algorithm proposed in this paper shows more stable error characteristics than the other three algorithms. Specifically, the error value approaches

0, indicating that the algorithm can effectively deal with the unfavorable observation environment and reduce the influence of external factors on the positioning accuracy.

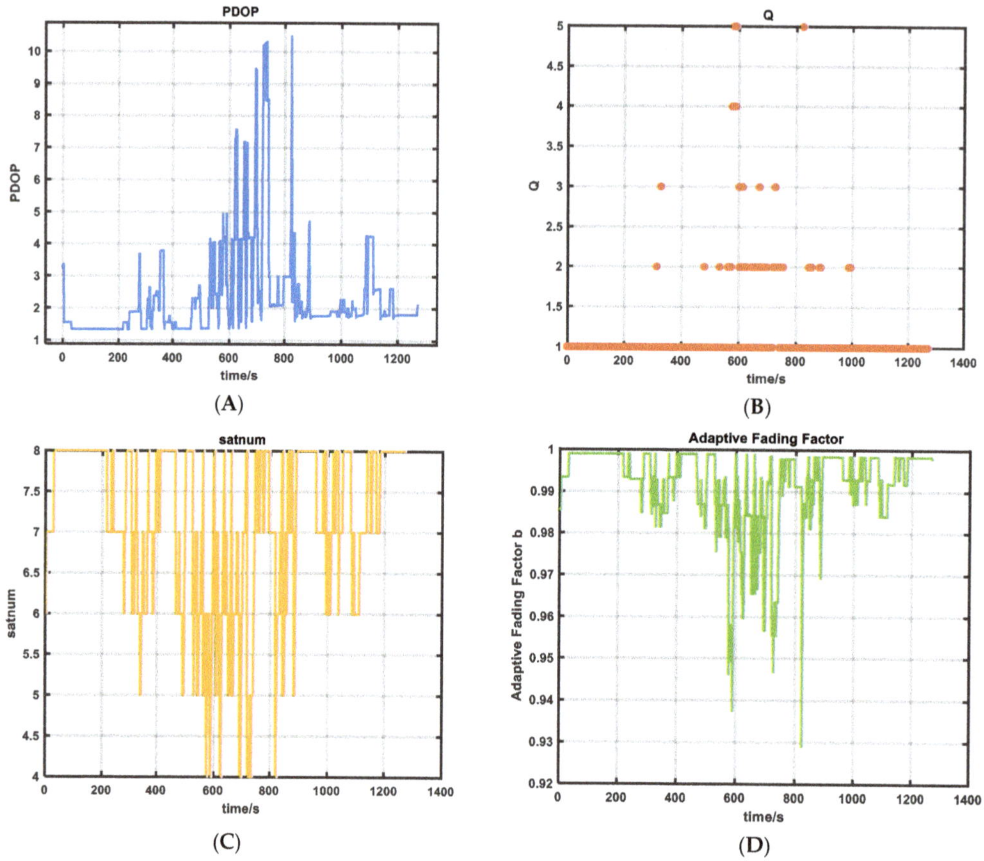

Figure 8. Vehicle experiment (**A**) satellite geometric accuracy factor PDOP, (**B**) satellite solution quality Q, (**C**) the number of effective satellite observations, (**D**) adaptive fading factor.

Figure 10A–C represent the error changes in the four algorithms in the roll, pitch, and heading directions with time. According to the figure, because the algorithm in this paper has adaptive characteristics, it can be adjusted according to different environmental conditions. The proposed algorithm (ARKF) shows good stability and smoothness in the estimation of roll, pitch, and heading angles.

In summary, EKF shows large transient fluctuations in multiple images, especially in areas with high GNSS occlusion. This shows that EKF has poor response to nonlinear and dynamic complex systems. The algorithm in this paper can quickly track the system state in areas with large dynamic changes, and there is no significant deviation or oscillation. The algorithm in this paper has high tracking accuracy for changes and can quickly recover to a stable state. In summary, the algorithm in this paper shows obvious smoothness and has the smallest fluctuation compared with other algorithms. In order to further compare the accuracy of the position, velocity, and attitude of the four schemes, the position and velocity error (RMS) are plotted as a histogram (Figure 11), and the corresponding results are also summarized in Table 3.

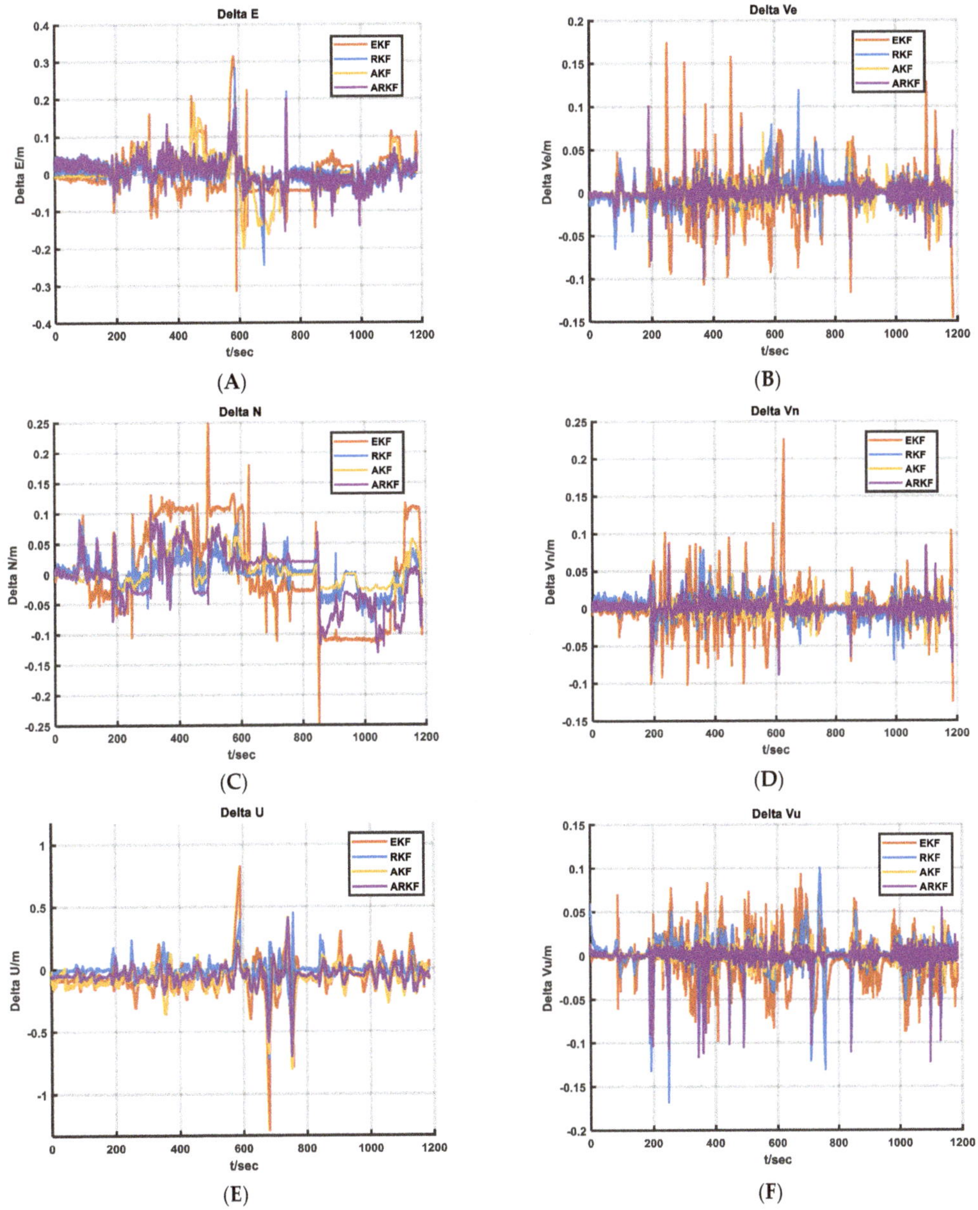

Figure 9. In the vehicle experiment, (**A**,**C**,**E**) represent the comparative analysis of the position errors of the four algorithms in the three directions of the northeast day. In the vehicle experiment, (**B**,**D**,**F**) represent the speed error comparison and analysis of the four algorithms in the three directions of the northeast day.

For vehicle experiments, it can be seen from Table 3 that the accuracy of the algorithm in the three-dimensional position is 47.12% higher than that of EKF, 35.26% higher than that of Sage-Husa adaptive filtering, and 9.58% higher than that of robust filtering.

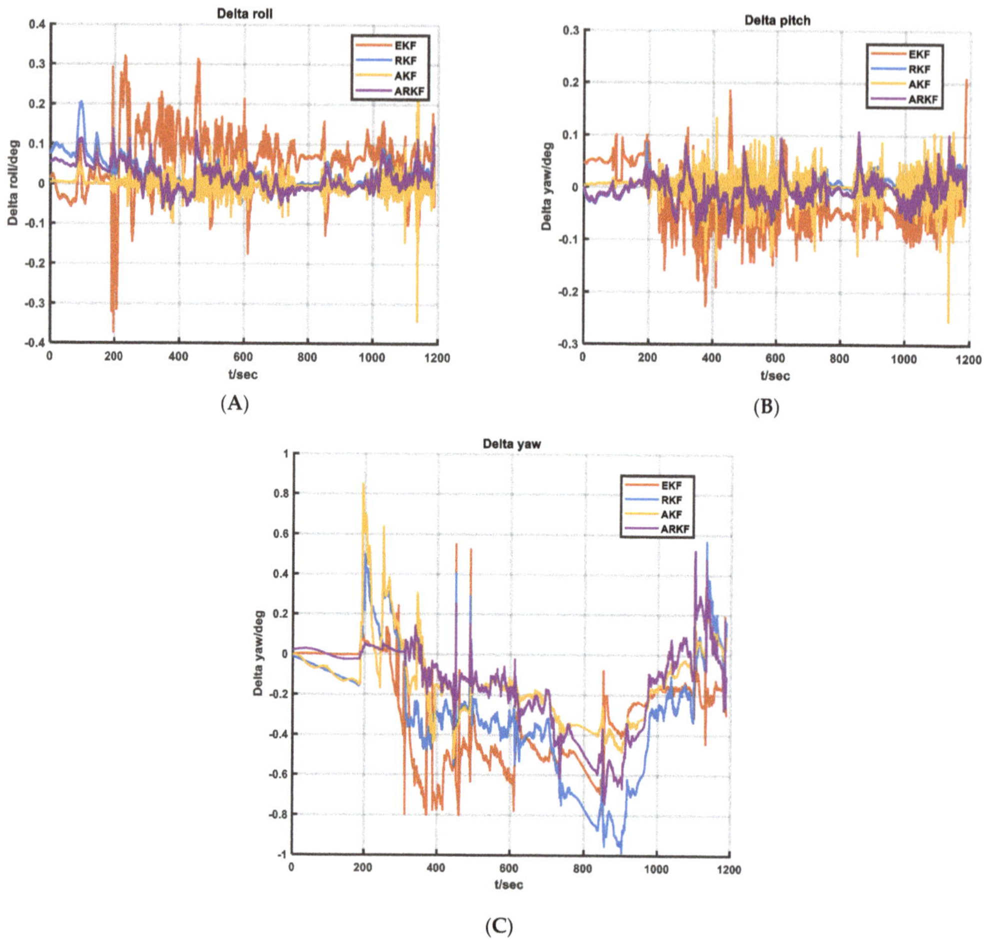

Figure 10. In the vehicle experiment, (**A–C**) are the error comparison of the four algorithms in roll, pitch, and yaw, respectively.

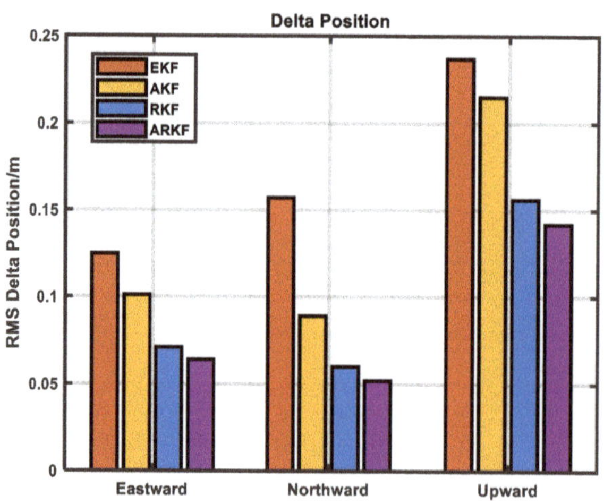

Figure 11. Vehicle Experiment Rms: position error.

Table 3. Vehicle-mounted experiments. Comparative analysis table of position, velocity, and attitude errors.

ERROR		EKF	AKF	RKF	ARKF
Velocity (m/s)	Eastward	0.191	0.068	0.024	0.011
	Northward	0.215	0.075	0.015	0.012
	Upward	0.078	0.040	0.054	0.055
Attitude (deg)	Roll	0.098	0.125	0.049	0.038
	Pitch	0.054	0.131	0.022	0.026
	Heading	0.715	0.308	0.575	0.313
Position (m)	Eastward	0.125	0.101	0.071	0.064
	Northward	0.157	0.089	0.060	0.052
	Upward	0.237	0.215	0.156	0.142
	Improved accuracy (%)	47.12%	35.26%	9.58%	

3.3. Shipboard Experiments

The experiment was carried out in the sea area around Changyi, Weifang, China. The experimental scene was a low-speed and high-vibration marine environment. The experiment lasted about 1 h. The ship is equipped with four GNSS antennas, and a main antenna is used as the GNSS solution result. The base station is set up on the coast, and the baseline length is about 20 km, so as to obtain the differential GNSS positioning solution. The specific ship-borne experimental route is shown in Figure 12.

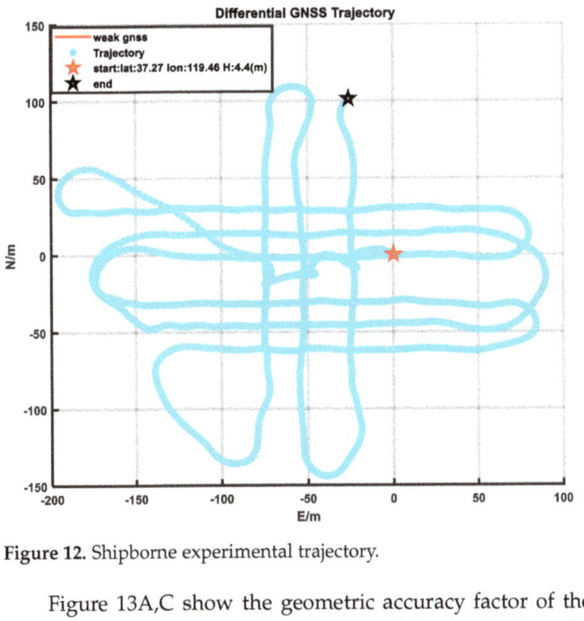

Figure 12. Shipborne experimental trajectory.

Figure 13A,C show the geometric accuracy factor of the shipborne experimental satellite and the number of satellites participating in the solution. Because there is no satellite occlusion problem under the sea observation condition, the observation condition is better. Figure 13B represents the satellite solution factor (Q) of the sea experiment. Figure 13D shows the change in the adaptive fading factor b in the ship-borne experiment.

Figure 14 shows the change in the ship's heave with time. For the marine sloshing environment, the influence of the ship's heave on the positioning accuracy cannot be ignored, and the maximum heave amplitude reaches 1 m. After 2500, the heave amplitude of the ship increases significantly. The SINS is prone to abnormal measurement in the low-speed and high-vibration heave environment, which has a great impact on the estimation of the filter.

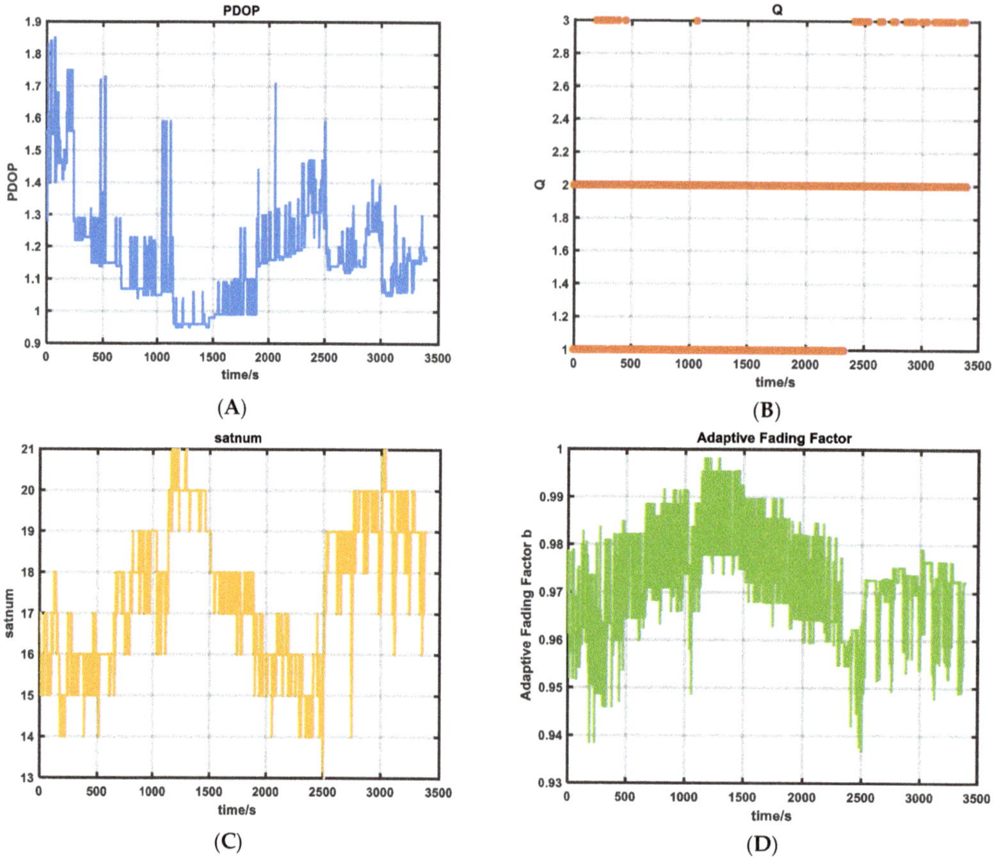

Figure 13. Shipborne experiment. (**A**) Satellite geometric precision factor PDOP, (**B**) satellite solution quality Q, (**C**) the number of effective satellite observations, (**D**) adaptive fading factor.

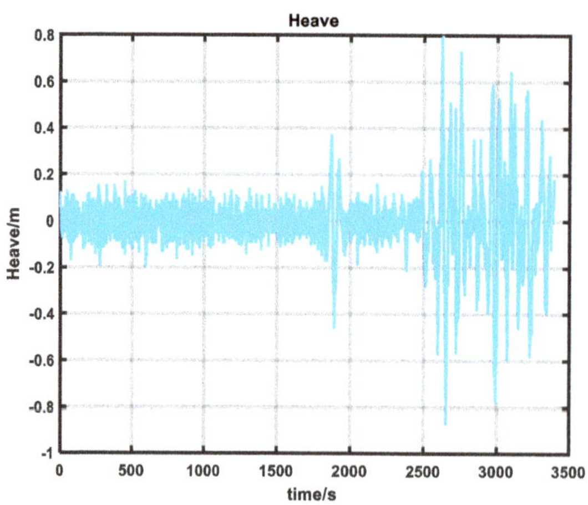

Figure 14. Shipboard heave.

In order to verify the accuracy of the algorithm in this paper, four algorithms are selected for analysis. The position, velocity and attitude errors obtained by using the above four methods and professional solution software are shown in Figures 15 and 16.

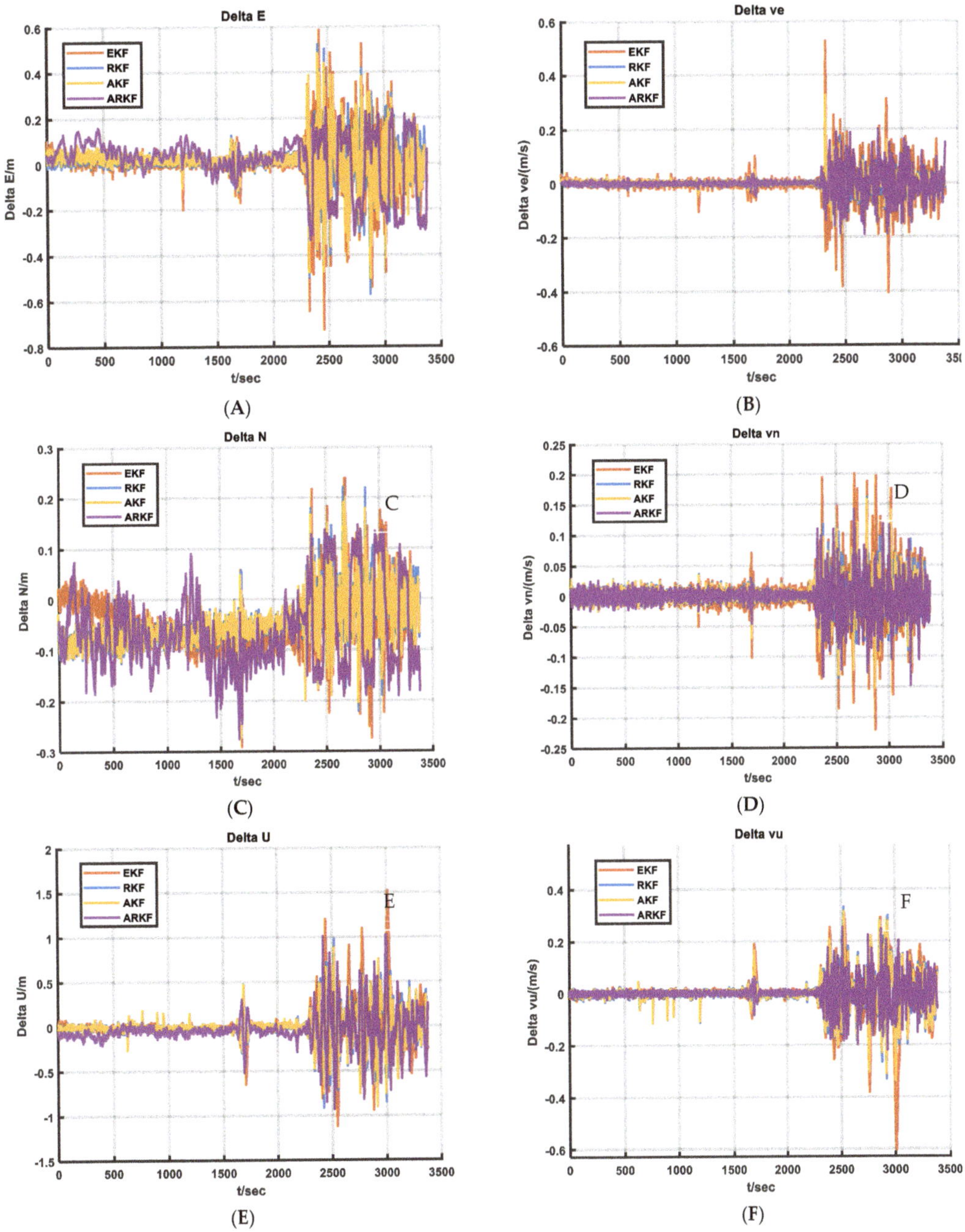

Figure 15. In the ship-borne experiment, (**A**,**C**,**E**) represent the comparative analysis of the position errors of the four algorithms in the three directions of the northeast sky. In the ship-borne experiment, (**B**,**D**,**F**) represent the comparative analysis of the velocity errors of the four algorithms in the three directions of the northeast sky.

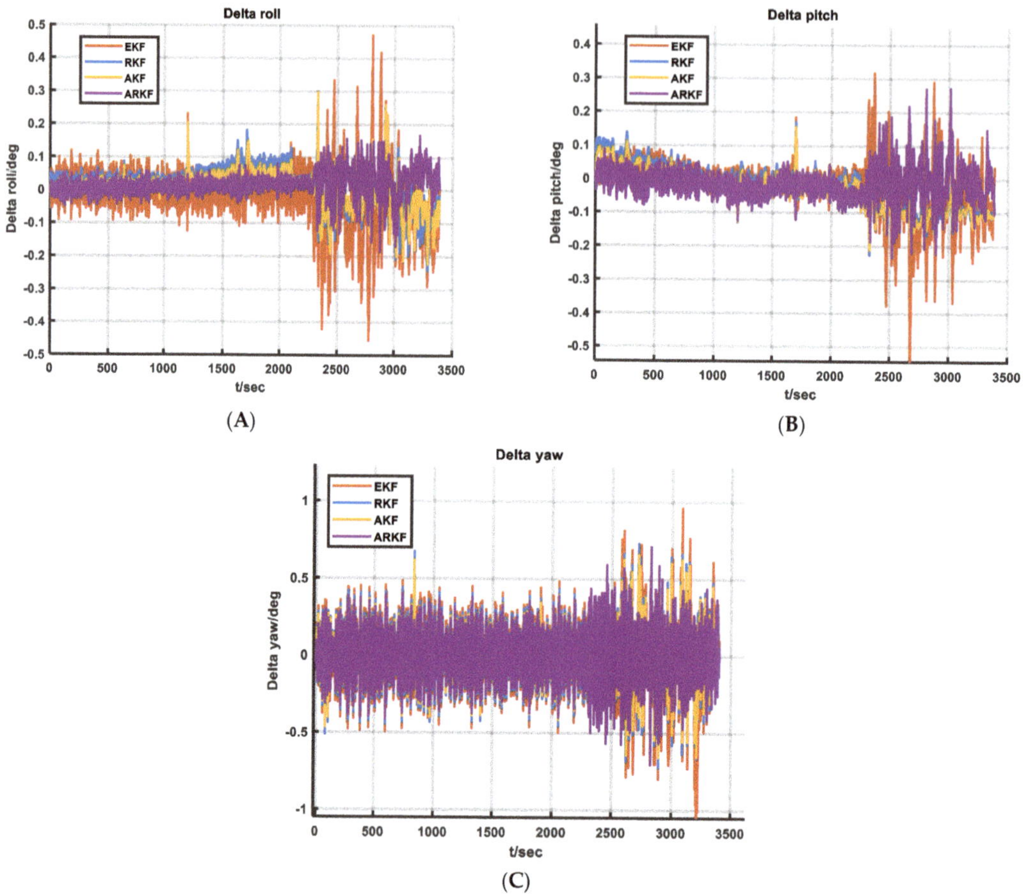

Figure 16. In the ship-borne experiment, (**A**–**C**) are the error comparison of the four algorithms in roll, pitch and, yaw, respectively.

At around 2000 s, the interference of the ship's heave motion to the positioning system increases significantly. When the heave motion is enhanced, EKF has insufficient ability to deal with the interference of nonlinear and non-Gaussian distribution, and the error fluctuates greatly. Robust filtering needs to be improved and adapted to the offshore heave vibration environment. The error of the algorithm in this paper can still maintain low amplitude and good stability. Compared with all the algorithms in this paper, this algorithm has the smallest error in the sky direction and the smallest fluctuation of the velocity error curve, which shows that it has better stability in the strong interference environment.

In summary, other algorithms have poor adaptability to dynamic environments. Especially when the heave motion is intensified, the attitude estimation error increases significantly, and even diverges. Whether it is roll angle, pitch angle, or heading angle, the error curve of the algorithm in this paper maintains a small fluctuation under the influence of heave motion (around 2000 s).

In order to further compare the accuracy of the position, velocity, and attitude of the four schemes, the position and velocity error (RMS) is plotted as a histogram (Figure 17), and the corresponding results are also summarized in Table 4.

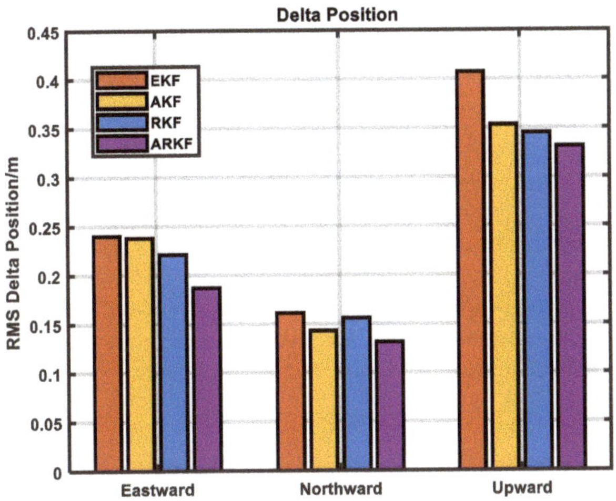

Figure 17. Shipboard experiment Rms: position error.

Table 4. Shipboard experiment position, velocity, and attitude error comparison and analysis table.

ERROR		EKF	AKF	RKF	ARKF
Velocity (m/s)	Eastward	0.101	0.091	0.084	0.080
	Northward	0.112	0.094	0.086	0.071
	Upward	0.254	0.216	0.209	0.173
Attitude (deg)	Roll	0.181	0.140	0.151	0.107
	Pitch	0.204	0.163	0.141	0.128
	Heading	0.355	0.279	0.321	0.223
Position (m)	Eastward	0.240	0.238	0.221	0.187
	Northward	0.161	0.143	0.156	0.134
	Upward	0.407	0.353	0.345	0.331
	Improved accuracy (%)	19.44%	10.47%	8.28%	

For ship-borne experiments, it can be seen from Table 4 that the accuracy of the algorithm in three-dimensional position is improved by 19.44% compared with EKF, 10.47% compared with Sage-Husa adaptive filtering, and 8.28% compared with robust filtering. These results show that although the system faces great uncertainty and noise interference in the marine environment with low speed, high vibration and severe heave, the algorithm proposed in this paper can still effectively maintain a high accuracy level.

4. Discussion

The ARKF algorithm can maintain stable and accurate positioning performance in the vehicle environment with poor satellite positioning and the complex marine environment with severe heave. Compared with other algorithms, ARKF algorithm has stronger robustness, better anti-interference ability, and higher application value.

The quantitative correlation between environmental perception (PDOP value, Q value) and state prediction credibility breaks through the coupling problem of traditional filtering noise statistics and disturbance suppression, and it provides a new paradigm for nonlinear filtering theory. Two-factor robust theoretical framework: Through the collaborative optimization of fading factor and residual weight, the stability criterion in the dynamic noise scene is established, and the related results can be extended to the field of multi-sensor fusion. In terms of practical value, high-precision navigation of unmanned systems can be achieved. Vehicle experiments show that the three-dimensional positioning error of ARKF in complex urban areas is less than 0.1 m. Meet the sub-meter precision requirements of unmanned distribution vehicles. In the ship-borne experiment, the heading angle error is less than 0.223, which can support the accurate berthing of autonomous ships.

Limitations of the algorithm in this paper. In the existing research, the collaborative optimization of robust filtering and adaptive filtering can partially suppress the influence of non-Gaussian noise. However, if the statistical characteristics of noise are highly deviated from the preset model, the adaptive mechanism of the algorithm may face the challenge of insufficient convergence speed or lagging weight adjustment. For example, in the vehicle experiment of this paper, when the satellite signal is completely occluded and the inertial sensor noise is suddenly abnormal, the positioning error of ARKF is better than other algorithms mentioned in this paper, but there are still short-term fluctuations. This indicates that extreme non-Gaussian noise may impose higher requirements on the dynamic adjustment range of the fading factor.

Aiming at the limitations, the deep learning method is introduced, and the filtering parameters are adjusted by the online noise characteristic identification network to enhance the real-time modeling ability of non-stationary noise. The weight design is improved, and the robustness factor is optimized by combining more flexible distribution assumptions to improve the robustness to impulse noise.

5. Conclusions

In this study, a dynamic robust filtering framework based on multi-source information fusion is constructed to solve the problem of navigation accuracy attenuation in low-speed and high-vibration environment. The real-time optimization of the fading factor is realized by PDOP-Q-effective satellite number-residual joint criterion, and the double weighting strategy constructed by standardized residuals breaks through the coupling problem of traditional algorithms in measurement noise estimation and dynamic disturbance suppression. The experimental results show that in the urban vehicle scene, the proposed algorithm reduces the RMS value of three-dimensional positioning error by 47.12%, 35.26%, and 9.58%, respectively, compared with EKF, RKF, and AKF, and improves the heading angle accuracy by 42.3%. The mean square error of the velocity vector is reduced by 19.44%, the roll angle tracking delay is shortened by 40 ms, and the continuous failure time of the system is extended from 8.2 s to 23.7 s under strong disturbance conditions.

By establishing a closed-loop correction mechanism for the dynamic noise covariance matrix, the collaborative optimization of robust estimation and adaptive filtering is successfully realized, and the nonlinear mapping law between the observation quality quantification index and the state prediction credibility is revealed. It provides a new solution for the fault-tolerant control of multi-source navigation systems in complex dynamic environments, especially for the high-precision pose estimation requirements of unmanned vehicles in low-frequency and high-vibration coupling scenarios.

Author Contributions: Conceptualization, Z.C. and Y.L.; data curation, Z.C., Y.L. and S.L.; formal analysis, Z.C.; funding acquisition, Y.L.; investigation, Z.C. and L.Y.; methodology, Z.C. and Y.L.; project administration, Y.L., S.L. and L.Y.; software, Z.C.; validation, Z.C.; writing—original draft, Z.C.; writing—review and editing, S.L. and S.W. All authors have read and agreed to the published version of the manuscript.

Funding: This research was funded by National Key Research and Development Program of China (NO. 2024YFC2814400); Supported by the China Postdoctoral Science Foundation under Grant Number 2024M750683; Funded by Shandong Postdoctoral Science Foundation SDCX-ZG-202400179; Funded by the key research and development plan of Shandong Province: the development and application demonstration of key core integrated navigation components of marine stereo observation equipment, No. 2019JZZY010809; Supported by Shandong Province Technology Innovation Guidance Plan: Key Technology and Transfer Transformation Demonstration Project of Polar Intelligent Unmanned Ship, No. YpZX202A093.

Data Availability Statement: Data available on request due to restrictions privacy or ethical. The data presented in this study are available on request from the corresponding author.

Conflicts of Interest: The authors declare no conflicts of interest.

References

1. Zhang, Z.; Xiao, G.; Nie, Z.; Ferreira, V.; Casula, G. High-Precision and High-Reliability Positioning, Navigation, and Timing: Opportunities and Challenges. *Remote Sens.* **2024**, *16*, 4403. [CrossRef]
2. Yang, Y.; Yao, Z.; Mao, Y.; Xu, T.; Wang, D. Resilient satellite-based PNT system design and key technologies. *Sci. China Earth Sci.* **2025**, *68*, 669–682. [CrossRef]
3. Xu, B.; Wang, X.; Guo, Y.; Zhang, J.; Razzaqi, A.A. A novel adaptive filter for cooperative localization under time-varying delay and non-Gaussian noise. *IEEE Trans. Instrum. Meas.* **2021**, *70*, 1–15. [CrossRef]
4. Delamer, J.-A.; Watanabe, Y.; Chanel, C.P. Safe path planning for UAV urban operation under GNSS signal occlusion risk. *Robot. Auton. Syst.* **2021**, *142*, 103800. [CrossRef]
5. Yang, Y.; Dai, Z.; Chen, Y.; Yuan, Y.; Yalikun, Y.; Shang, C. Emerging MEMS sensors for ocean physics: Principles, materials, and applications. *Appl. Phys. Rev.* **2024**, *11*, 021320. [CrossRef]
6. Yang, Y.; Cui, X. Adaptively robust filter with multi adaptive factors. *Surv. Rev.* **2008**, *40*, 260–270. [CrossRef]
7. Song, S.; Wu, J. Motion State Estimation of Target Vehicle under Unknown Time-Varying Noises Based on Improved Square-Root Cubature Kalman Filter. *Sensors* **2020**, *20*, 2620. [CrossRef]
8. Ma, C.; Pan, S.; Gao, W.; Wang, H.; Liu, L. Variational Bayesian-based robust adaptive filtering for GNSS/INS tightly coupled positioning in urban environments. *Measurement* **2023**, *223*, 113668. [CrossRef]
9. Xiong, J.; Cheong, J.W.; Xiong, Z.; Dempster, A.G.; Tian, S.; Wang, R. Adaptive hybrid robust filter for multi-sensor relative navigation system. *IEEE Trans. Intell. Transp. Syst.* **2021**, *23*, 11026–11040. [CrossRef]
10. Narasimhappa, M.; Mahindrakar, A.D.; Guizilini, V.C.; Terra, M.H.; Sabat, S.L. MEMS-based IMU drift minimization: Sage Husa adaptive robust Kalman filtering. *IEEE Sens. J.* **2019**, *20*, 250–260. [CrossRef]
11. Zou, T.; Zeng, W.; Yang, W.; Ong, M.C.; Wang, Y.; Situ, W. An adaptive robust cubature Kalman filter based on Sage-Husa estimator for improving ship heave measurement accuracy. *IEEE Sens. J.* **2023**, *23*, 10089–10099. [CrossRef]
12. Petrus, P. Robust Huber adaptive filter. *IEEE Trans. Signal Process.* **1999**, *47*, 1129–1133. [CrossRef]
13. Mao, Y.; Sun, R.; Wang, J.; Cheng, Q.; Kiong, L.C.; Ochieng, W.Y. New time-differenced carrier phase approach to GNSS/INS integration. *GPS Solut.* **2022**, *26*, 122. [CrossRef]
14. Li, Q.; Zhang, L.; Wang, X. Loosely coupled GNSS/INS integration based on factor graph and aided by ARIMA model. *IEEE Sens. J.* **2021**, *21*, 24379–24387. [CrossRef]
15. Zhang, Q.; Dai, Y.; Zhang, T.; Guo, C.; Niu, X. Road Semantic-Enhanced Land Vehicle Integrated Navigation in GNSS Denied Environments. *IEEE Trans. Intell. Transp. Syst.* **2024**, *25*, 20889–20899. [CrossRef]
16. Jiang, C.; Zhang, S.; Li, H.; Li, Z. Performance evaluation of the filters with adaptive factor and fading factor for GNSS/INS integrated systems. *GPS Solut.* **2021**, *25*, 130. [CrossRef]
17. Song, J.; Li, J.; Wei, X.; Hu, C.; Zhang, Z.; Zhao, L.; Jiao, Y. Improved multiple-model adaptive estimation method for integrated navigation with time-varying noise. *Sensors* **2022**, *22*, 5976. [CrossRef]
18. Yan, G.; Weng, J. *Integrated Inertial Navigation Algorithms and Principles of Integrated Navigation*; Northwestern Polytechnical University: Xi'an, China, 2019; ISBN 978-7-5612-6547-5.
19. Liu, Y.; Wang, L.; Hu, L.; Cui, H.; Wang, S. Analysis of the Influence of Attitude Error on Underwater Positioning and Its High-Precision Realization Algorithm. *Remote Sens.* **2022**, *14*, 3878. [CrossRef]
20. Guo, F.; Zhang, X. Adaptive robust Kalman filtering for precise point positioning. *Meas. Sci. Technol.* **2024**, *25*, 105011. [CrossRef]
21. Yin, Z.; Yang, J.; Ma, Y.; Wang, S.; Chai, D.; Cui, H. A Robust Adaptive Extended Kalman Filter Based on an Improved Measurement Noise Covariance Matrix for the Monitoring and Isolation of Abnormal Disturbances in GNSS/INS Vehicle Navigation. *Remote Sens.* **2023**, *15*, 4125. [CrossRef]
22. Jiang, Y.; Pan, S.; Meng, Q.; Zhang, M.; Ma, C.; Yu, H.; Gao, W. Multiple faults detection algorithm based on REKF for GNSS/INS integrated navigation. In Proceedings of the International Conference on Guidance, Navigation and Control, Harbin, China, 5–7 August 2022; Springer Nature Singapore: Singapore, 2022; pp. 886–894. [CrossRef]
23. Liu, J.; Pu, J.; Sun, L.; He, Z. An approach to robust INS/UWB integrated positioning for autonomous indoor mobile robots. *Sensors* **2019**, *19*, 950. [CrossRef] [PubMed]

24. Chhabra, A.; Venepally, J.R.; Kim, D. Measurement noise covariance-adapting Kalman filters for varying sensor noise situations. *Sensors* **2021**, *21*, 8304. [CrossRef] [PubMed]
25. Cheng, Y.; Zhang, S.; Wang, X.; Wang, H.; Yang, H. Kalman Filter with Adaptive Covariance Estimation for Carrier Tracking under Weak Signals and Dynamic Conditions. *Electronics* **2024**, *13*, 1288. [CrossRef]
26. Li, H.; Medina, D.; Vilà-Valls, J.; Closas, P. Robust variational-based Kalman filter for outlier rejection with correlated measurements. *IEEE Trans. Signal Process.* **2020**, *69*, 357–369. [CrossRef]
27. Biswas, S.K.; Qiao, L.; Dempster, A.G. Effect of PDOP on performance of Kalman Filters for GNSS-based space vehicle position estimation. *GPS Solut.* **2017**, *21*, 1379–1387. [CrossRef]
28. Wang, Z.; Song, S.; Jiao, W.; Wei, W.; Zhou, W.; Jiao, G.; Liu, J. A unified PDOP evaluation method for multi-GNSS with optimization grid model and temporal–spatial resolution. *GPS Solut.* **2024**, *28*, 36. [CrossRef]
29. Chen, C.; Zhu, J.; Bo, Y.; Chen, Y.; Jiang, C.; Jia, J.; Duan, Z.; Karjalainen, M.; Hyyppä, J. Pedestrian Smartphone Navigation Based on Weighted Graph Factor Optimization Utilizing GPS/BDS Multi-Constellation. *Remote Sens.* **2023**, *15*, 2506. [CrossRef]
30. Jiang, S.; Chen, Y.; Liu, Q. Advancements in Buoy Wave Data Processing through the Application of the Sage–Husa Adaptive Kalman Filtering Algorithm. *Sensors* **2023**, *23*, 7298. [CrossRef]
31. Rahemi, N.; Mosavi, M.R.; Martín, D. A new variance-covariance matrix for improving positioning accuracy in high-speed GPS receivers. *Sensors* **2021**, *21*, 7324. [CrossRef]
32. Wang, D.; Dong, Y.; Li, Z.; Li, Q.; Wu, J. Constrained MEMS-based GNSS/INS tightly coupled system with robust Kalman filter for accurate land vehicular navigation. *IEEE Trans. Instrum. Meas.* **2019**, *69*, 5138–5148. [CrossRef]
33. Knight, N.L.; Wang, J. A comparison of outlier detection procedures and robust estimation methods in GPS positioning. *J. Navig.* **2009**, *62*, 699–709. [CrossRef]
34. Dong, Y.; Wang, D.; Zhang, L.; Li, Q.; Wu, J. Tightly coupled GNSS/INS integration with robust sequential Kalman filter for accurate vehicular navigation. *Sensors* **2020**, *20*, 561. [CrossRef] [PubMed]
35. Li, B.; Rizos, C.; Lee, H.K.; Lee, H.K. A GPS-slaved time synchronization system for hybrid navigation. *GPS Solut.* **2006**, *10*, 207–217. [CrossRef]
36. Chai, D.; Sang, W.; Chen, G.; Ning, Y.; Xing, J.; Yu, M.; Wang, S. A novel method of ambiguity resolution and cycle slip processing for single-frequency GNSS/INS tightly coupled integration system. *Adv. Space Res.* **2022**, *69*, 359–375. [CrossRef]
37. Instructions for the Software (Inertial Explorer 8.70) Used as the Reference Truth Value in the Comparative Experiment. Available online: https://hexagondownloads.blob.core.windows.net/public/Novatel/assets/Documents/Waypoint/Downloads/GrafNav-GrafNet-User-Manual-870/GrafNav-GrafNet-User-Manual-870.pdf (accessed on 1 April 2025).

remote sensing

MDPI

Article

Breaking the Cyclic Prefix Barrier: Zero-Padding Correlation Enables Centimeter-Accurate LEO Navigation via 5G NR Signals

Lingyu Deng [1], Yikang Yang [1,*], Jiangang Ma [2], Tao Wu [3,4], Xingyou Qian [1] and Hengnian Li [5]

1 School of Electronic and Information Engineering, Xi'an Jiaotong University, Xi'an 710049, China; dengly0625@stu.xjtu.edu.cn (L.D.); qxy000320@stu.xjtu.edu.cn (X.Q.)
2 China Academy of Space Technology (Xi'an), Xi'an 710100, China; jiangangma@126.com
3 Space Engineering University, Beijing 101416, China; ttc_wt98@hgd.edu.cn
4 Key Laboratory of Intelligent Space TTC&O (Space Engineering University), Ministry of Education, Beijing 101416, China
5 State Key Laboratory of Astronautic Dynamics, Xi'an Satellite Control Center, Xi'an 710043, China; henry_xscc@mail.xjtu.edu.cn
* Correspondence: yangyk74@xjtu.edu.cn

Abstract: Low Earth orbit (LEO) satellites offer a revolutionary potential for positioning, navigation, and timing (PNT) services due to their stronger signal power and rapid geometric changes compared to traditional global navigation satellite systems (GNSS). However, dedicated LEO navigation systems face high costs, so opportunity navigation based on LEO satellites is a potential solution. This paper presents an orthogonal frequency division multiplexing (OFDM)-based LEO navigation system and analyzes its navigation performance. We use 5G new radio (NR) as the satellite transmitting signal and introduce the NR signal components that can be used for navigation services. The LEO NR system and a novel zero-padding correlation (ZPC) are introduced. This ZPC receiver can eliminate cyclic prefix (CP) and inter-carrier interference, thereby improving tracking accuracy. The power spectral density (PSD) for the NR navigation signal is derived, followed by a comprehensive analysis of tracking accuracy under different NR configurations (bandwidth, spectral allocation, and signal components). An extended Kalman filter (EKF) is proposed to fuse pseudorange and pseudorange rate measurements for real-time positioning. The simulations demonstrate an 80% improvement in ranging precision (3.0–4.5 cm) and 88.3% enhancement in positioning accuracy (5.61 cm) compared to conventional receivers. The proposed ZPC receiver can achieve centimeter-level navigation accuracy. This work comprehensively analyzes the navigation performance of the LEO NR system and provides a reference for LEO PNT design.

Keywords: low Earth orbit (LEO); new radio (NR); OFDM; navigation; navigation receiver; integrated communication and navigation

1. Introduction

During the past decades, global navigation satellite systems (GNSS) have served as the foundation for positioning, navigation, and timing (PNT) services that are critical for agriculture, industry, and daily life [1]. Existing GNSS satellites are typically deployed in the medium Earth orbit (MEO) to provide extensive signal coverage [2]. Consequently, due to their extremely weak power levels, GNSS signals are highly susceptible to interference and spoofing while suffering severe degradation in urban canyons and indoor environments [3,4].

The burgeoning low Earth orbit (LEO) constellations present a new opportunity to overcome the limitations of existing GNSS and enhance the quality of PNT service [5–7].

Numerous studies have demonstrated the advantages of LEO-based PNT services. (1) Compared to the existing MEO GNSS, the received signal power of LEO satellites enhances by 25~30 dB [4,8,9]. (2) Rapid orbital motion of LEO satellites induces faster geometric changes, reducing the convergence time of precise point positioning (PPP) from 30 to just 1 min [10].

Some studies focus on dedicated LEO satellites for PNT services. The first dedicated LEO navigation satellite was the TRANSIT system designed by the US Navy [11]. Luojia-1A, developed by Wuhan University, broadcasts a dual-band ranging signal with its orbital data, significantly improving GNSS positioning performance [12]. The Centispace system, an LEO navigation augmentation system designed for GNSS, operates in the same radio frequency bands as GNSS (L1 1575.42 MHz and L5 1176 MHz) [13]. The Pulsar constellation, designed by Xona Space, not only provides enhanced GNSS services but also offers encrypted signals to support authentication [14]. Furthermore, by retrofitting existing Iridium satellites, the satellite time and location (STL) service aims to provide assured time and location estimates, even in challenging and strongly attenuated environments [15].

The limited coverage of LEO satellites [16] requires massive constellations to ensure continuous navigation services, imposing prohibitively high costs that render dedicated LEO navigation systems economically unfeasible. As most existing LEO constellations are designed for broadband communications [17], integrating navigation functions into communication signals has become a feasible solution. Three multiplexing methodologies are used to integrate communication and navigation: *time division multiplexing (TDM)*, *frequency division multiplexing (FDM)*, and *code division multiplexing (CDM)* [18]. In TDM, navigation signals are inserted into communication signal as a frame [19,20]; FDM employs binary offset carrier (BOC) modulation to achieve spectral separation between communication and navigation signals [21–24]; and CDM directly superimposes spread-spectrum navigation signals onto communication waveforms. But these multiplexing approaches inherently introduce mutual interference between communication and navigation signals. This fundamental conflict motivates the exploration of opportunity navigation, which directly leverages original reference signals in communication waveforms as navigation signal [25–27].

Due to the spectral efficiency and anti-multipath capabilities [28], orthogonal frequency division multiplexing (OFDM) has been adopted as the core physical modulation method in Starlink [29,30] and 5G NTN networks [31,32] The OFDM-based opportunity navigation is an increasingly vital and attractive paradigm [33–35]. Studies in [36,37] investigate the feasibility of OFDM-based positioning; for non-cooperative LEO OFDM signals, cognitive approaches have been adopted for localization [38,39].

However, there are some limitations for LEO OFDM navigation. (1) Cyclic prefixes (CPs) have been universally adopted in OFDM, and they are a part replica of the OFDM signal. These CP signals are used to mitigate inter-symbol interference caused by multipath effects, but no work analyzes their adverse impact on the navigation performance. (2) The OFDM system has reconfigurable bandwidth and spectrum allocation [31]. In addition, the communication signal includes various channels (such as synchronization, broadcast, and reference signals), and numerous signal components can be used for navigation service. These NR and receiver configurations significantly influence their tracking accuracy. But no study systematically evaluated the impact of different configurations and signal components on tracking accuracy. (3) Many works research opportunity navigation based on actual signals from the LEO constellations like Starlink, and the experiment systems are complicated [30,38–44]. These studies also lack quantitative analyses of error sources, such as clock bias and ionospheric and tropospheric delays. Providing simple, quantitative, and low-cost analysis for LEO navigation research is essential.

This paper builds a simulation system of OFDM opportunity navigation to overcome these limitations. The 5G new radio (NR) is an open communication standard based on OFDM, and it is used as an OFDM signal in the simulation. A novel zero-padding correlation (ZPC) is proposed for the navigation receiver to eliminate interference from cyclic prefixes (CP) and inter-carriers (including communication and reference signals). We derive the power spectral density of NR reference signals and comprehensively analyze how different NR configurations and signal components affect tracking accuracy. Moreover, an extended Kalman filter (EKF) is designed for the LEO position. The experimental results demonstrate an 80% improvement in tracking precision and a 88.3% enhancement in positioning accuracy compared to existing NR receivers. This paper has three contributions:

1. A system architecture for LEO NR opportunity navigation is presented in detail. We design an NR navigation receiver, and the zero-padding correlator (ZPC) is proposed to address the influence of OFDM CP and inter-carrier interference on navigation performance. For the proposed ZPC receiver, all the reference signals in NR can be utilized to provide navigation service.
2. A experiment validates that ZPC eliminates the influence caused by CP and inter-carriers and significantly enhances tracking accuracy. Compared with other works, the ZPC receiver achieves higher tracking accuracy. Moreover, the relationship between the tracking accuracy and NR configurations, including bandwidth, spectrum location, and signal components combination, is rigorously investigated.
3. We discuss the positioning performance of LEO NR based on an EKF, and simulation results show it can achieve centimeter-level accuracy. Additionally, the clock bias and ionospheric and tropospheric effects on positioning performance are quantitatively analyzed.

The remainder of this paper is organized as follows. Section 2 introduces related works. Section 3 details the NR signal model, OFDM modulation, and power spectral density (PSD). Section 4 introduces the LEO NR navigation system and proposes the ZPC receiver and EKF algorithm. Section 5 discusses the advantages of ZPC in tracking accuracy. Section 6 analyzes ZPC's tracking performance under varying NR and receiver configurations. Section 7 evaluates positioning accuracy via simulations. Section 8 summarize this paper.

2. Related Works

To further explore LEO OFDM-based opportunity navigation, we survey relevant research, including cooperative and non-cooperative, OFDM-based, and LEO navigation.

2.1. Non-Cooperative and Cooperative Opportunity Navigation

Opportunity navigation utilizes existing radio signals, such as cellular communication signals and television broadcast signals, to provide navigation services [45]. Opportunity navigation can be categorized into two paradigms: cooperative and non-cooperative. Non-cooperative navigation typically operates without base station assistance, and the navigation receiver does not subscribe to base stations. Thereby, non-cooperative navigation offers enhanced privacy preservation. However, non-cooperative navigation suffers from several limitations. (1) The transmitter positions are either unknown or low-accuracy, resulting in degraded navigation accuracy. For terrestrial communication base stations, estimating transmitters' locations requires sophisticated systems and the estimated results often remain unsatisfactory [46,47]. Although third parties may provide LEO satellite ephemerides, their accuracy is low [48,49]. (2) Unauthorized users cannot precisely acquire signal structures [44,50]. (3) Communication signals are not consistently active.

These challenges in non-cooperative navigation can be effectively addressed through the cooperative method [8,25]. For cooperative receivers, precise orbital data and signal structures can be readily obtained from the transmitter, and transmitting signals can be activated on demand according to receiver requirements. These features enhance navigation accuracy, simplify receiver processing, and ensure continuous navigation services [51]. Hence, cooperative opportunity navigation is adopted in this paper.

2.2. OFDM-Based Opportunity Navigation

OFDM has been widely adopted in LTE, 5G NR, and LEO satellites due to its spectral efficiency and multipath resilience. It is a promising modulation scheme for integrated space-air-ground communications [29,30,33,34]. Therefore, OFDM-based opportunity navigation has become a prominent research domain. LTE opportunity navigation focuses on the time-of-arrival (TOA) delay estimation and interference mitigation. Shamaei [52] pioneered theoretical performance bounds for LTE opportunity positioning through joint TOA and direction-of-arrival (DOA) estimation, while Yang [53,54] experimentally characterized the impact of fading environments and antenna ports on TOA estimation. Wang [55] innovatively modeled inter-cell interference as Gaussian perturbations to quantify TOA accuracy degradation. Techniques such as multipath estimating delay lock loop (MEDLL) and synthetic aperture beamforming [56–58] have been developed to address multipath challenges. Additionally, channel fingerprint machine learning [59] and differential positioning frameworks [60] are used to achieve meter-level accuracy.

The evolution of 5G NR has further expanded OFDM-based opportunity navigation capabilities. Abdallah [61] explored carrier-phase-based localization by NR signals, whereas Xhafa [62] and Koivisto [63] achieved joint uplink time-difference-of-arrival (UTDoA) and angle-of-arrival (AoA) estimation through antenna arrays. Lapin [64] enhanced multipath robustness via ESPRIT-based joint delay-phase estimation. Tensor-based channel estimation [65], millimeter-wave beam fingerprinting [66], and machine learning-driven TOA estimation systems [67,68] collectively constitute intelligent positioning architectures that enhance both accuracy and operational reliability.

However, the CP is an essential part of OFDM signals, and its influence on navigation performance is seldom addressed in these existing studies. Furthermore, the relationship between navigation performance and OFDM configurations remains largely unexplored.

2.3. LEO Opportunity Navigation

The burgeoning development of LEO satellites provides new opportunities for opportunity navigation. Khalife [41] pioneered the feasibility verification of Starlink positioning using carrier-phase observations, establishing a framework for LEO-based opportunity navigation. Building upon this, an efficient integer ambiguity resolution algorithm was developed via LEO differential carrier-phase processing [69]. While Doppler frequency can offer alternative navigation solutions, its tracking accuracy remains constrained [40,42–44]. Signal coginitvie methodology is proposed to identify configurations of Starlink's OFDM signal [30,38]. Moreover, Kozhaya [70] achieved positioning through blind estimation of OneWeb constellation beacons, whereas Zhao [71] enhanced navigation robustness by fusing Doppler measurements from Iridium NEXT and Orbcomm constellations. Fundamental research investigates the impact of LEO satellite oscillator stability on navigation performance [72]. The differential Doppler positioning method is proposed to improve navigation accuracy for orbit-unknown LEO satellites [73].

However, the existing works rarely quantitatively evaluate the impact of different error sources, such as clock bias, ionosphere, and tropospheric delay, on navigation performance. This paper assumes that the LEO constellation transmits NR signals to quantitatively analyze the error source's impact on opportunity navigation performance. The experimental results provide critical insights to future LEO PNT services.

3. Signal Model

This section first introduces the downlink physical channels of 5G NR, emphasizing the signal components that can be used in opportunity navigation. Subsequently, the OFDM modulation with CP is elaborated. The power spectral density (PSD) of the NR reference signal is derived, serving as the foundation for tracking performance analysis.

3.1. NR Signal Components

NR specifies different physical channels to utilize radio resources fully [31]. As shown in Figure 1, an NR frame lasts 10 ms and consists of 10 subframes, each lasting 1 ms. Furthermore, each subframe is divided into 2^{μ} slots and numerology $\mu \in \{0, \ldots, 4\}$ is an NR configuration parameter. Every slot contains 14 (normal CP) or 12 (extended CP) symbols and the subcarrier spacing follows $\Delta f = 2^{\mu} \cdot 15$ kHz. The resource grid further illustrates the allocation of NR physical channels. Each resource element (RE) in the resource grid carries a modulation symbol $c_{n,k}$, where n and k denote the symbol and subcarrier indices, respectively. These modulation symbols $c_{n,k}$, obtained from higher-layer signaling, can be categorized into two types: (1) unknown communication data and (2) predefined reference signals or predictable broadcast signals. The latter category can serve as navigation signals.

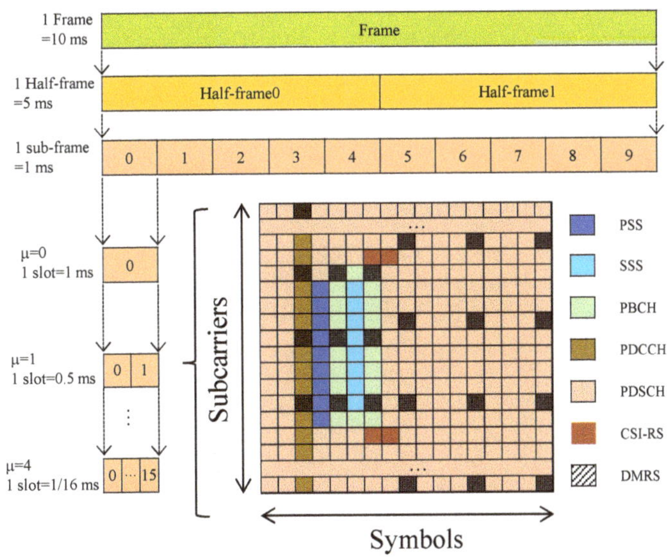

Figure 1. New radio (NR) physical channel structure and resource grid.

NR downlink signals conclude multiple physical channels: a synchronization signal and physical broadcast channel (SSB), physical downlink shared channel (PDSCH), physical downlink control channel (PDCCH), and channel state information reference signal (CSI-RS). The SSB facilitates synchronization of user equipment (UE) and provides system message broadcasting. It comprises three signal components: primary synchronization signal (PSS), secondary synchronization signal (SSS), and physical broadcast channel (PBCH). In the time domain, the SSB occupies four OFDM symbols with specific allocation: one for PSS, one for SSS, and two for PBCH. Its transmission periodicity is configurable with options $\{5, 10, 20, 40, 80, 160\}$ ms. Both PSS and SSS are m-sequences and their combination uniquely determines the physical layer cell identity (PCI):

$$PCI = N_{ID}^{SSS} \times 3 + N_{ID}^{PSS} \tag{1}$$

where $N_{ID}^{PSS} \in \{0, 1, 2\}$ denotes the PSS sequence index, and $N_{ID}^{SSS} \in \{0, 1, \ldots, 335\}$ represents the SSS sequence index. The PSS and SSS are fixed for transmitters and can be used for navigation service.

The PBCH comprises a master information block (MIB) and demodulation reference signals (DMRS). UE uses the DMRS, with predefined positions and content, for channel estimation. The MIB carries system information about the base station: system frame number (SFN), subcarrier spacing configurations, subcarrier offset parameters, and PDCCH configurations. The SFN increases over time, and other MIB parameters typically remain static. To enhance interference resistance, the MIB undergoes polar coding followed by scrambling. As a consequence, the coded MIB signal is time-varying. Nevertheless, cooperative receivers can effectively estimate these dynamic MIB components through predefined protocol specifications. Due to the predictable property, the entire SSB signal can serve as a local replica signal for navigation receivers.

PDCCH dynamically schedules uplink and downlink radio transmission resources. The downlink control information (DCI) within PDCCH randomly varies according to service requirements. Apart from DCI, PDCCH also contains DMRS, which is used for channel estimation. PDCCH's DMRS are predetermined; thus, it can be used for navigation.

PDSCH carries downlink communication data. Its time-frequency resources are flexibly allocated according to service demands. To optimize spectral efficiency, NR allocates most resource grid elements to PDSCH. Beyond random communication data, PDSCH also incorporates fixed DMRS for channel estimation. Once PDSCH is configured, the DMRS is deterministic, allowing it for navigation service.

CSI-RS in NR system is dynamically configured by base station according to communication scenarios. UE measures its power to evaluate the wireless channel quality. This protocol-defined determinism signal can be used for navigation.

5G NR employs an ultra-lean transmission paradigm: Only SSB maintains periodic transmission, while other physical channels are transmitted on demand. SSB, CSI-RS, and DMRS embedded in PDSCH/PDCCH are predictable for the receiver and can be used as a local replica signal in NR navigation receiver. All these predictable navigation signals are referred to as reference signals for brevity in the following paper. Based on the reference signal, time delays between the transmitter and receiver can be estimated through code tracking loops to provide PNT service.

3.2. OFDM Modulation

OFDM combines multiple orthogonal narrowband subcarriers into a single wideband signal and it is a core modulation technology for 5G NR systems. Consider an OFDM signal containing K subcarriers with subcarrier spacing of Δf. For modulation symbol $c_{n,k}$, the useful OFDM symbol $s_u(t)$ is expressed as

$$s_u(t) = \frac{1}{K} \sum_{k=-K/2}^{K/2-1} c_{n,k} e^{j2\pi k \Delta f t} u_{T_u}(t) \tag{2}$$

where $T_u = 1/\Delta f$ represents the useful symbol duration. This paper employs 1024 subcarriers with a subcarrier spacing of 15 kHz. The rectangular window function $u_T(t)$ is defined as

$$u_T = \begin{cases} 1, 0 \le t < T \\ 0, \text{others} \end{cases} \tag{3}$$

The cyclic prefix (CP) is adopted in the NR system to mitigate inter-symbol interference (ISI) caused by the multipath. CP is a copy of OFDM symbol:

$$s_g(t) = \frac{1}{K} \sum_{k=-K/2}^{K/2-1} c_{n,k} e^{j2\pi k \Delta f(t)} u_{T_g}(t) \tag{4}$$

where T_g denotes the CP duration and is determined by maximum multipath delay.

CP is appended at the beginning of the useful OFDM symbol and the complete time-domain representation of an OFDM symbol with CP is given by

$$s_{\text{sym}}(t) = \frac{1}{K} \sum_{k=-K/2}^{K/2-1} c_{n,k} e^{j(2\pi k \Delta f t)} u_{T_{\text{sym}}}(t) \tag{5}$$

where $T_{\text{sym}} = T_u + T_g$ is the duration of total OFDM with CP.

3.3. PSD of NR Reference Signal

The PSD of navigation signals enables the quantitative analysis of tracking accuracy and anti-jamming capability, serving as a critical tool for navigation performance evaluation [74]. According to the NR signal model, reference signals are distributed across subcarriers and are part of the NR signal. ZPC ensures orthogonality between reference signals on different subcarriers while eliminating interference caused by the CP and intercarrier. Therefore, when the correlation integration period is sufficiently long, the PSD of the NR reference signal can be expressed as the summation of the PSDs of all individual subcarrier reference signals:

$$G_{\text{RS}}(f) = \frac{1}{N_{\text{RS}}} \sum_{(n,k)\in\Omega} G_{n,k}(f) \tag{6}$$

where N_{RS} and Ω denotes the total number and location set of the reference signal in an OFDM resource grid, respectively. $G_{n,k}(f)$ represents the PSD of the subcarrier reference signal at the n-th symbol and k-th subcarrier. Its corresponding time-domain signal $s_{n,k}(t)$ comprises the modulation symbol $c_{n,k}$ multiplied by a complex exponential (subcarrier waveform):

$$s_{n,k}(t) = c_{n,k} e^{j2\pi k \Delta f t} u_T(t) = c_{n,k} p_k(t) \tag{7}$$

where $p_k(t)$ denotes the pulse shaping waveform of the k-th subcarrier.

For the subcarrier reference signal, its PSD $G_{n,k}(f)$ can be derived from the PSD $G_{\text{PRN}}(f)$ of the modulation symbol $c_{n,k}$ and the spectrum $P_k(f)$ of the subcarrier waveform [75]:

$$G_k(f) = \frac{1}{T_u} |P_k(f)|^2 G_{\text{PRN}}(f) \tag{8}$$

where $P_k(f)$ is the Fourier transform of the subcarrier's pulse shaping waveform:

$$P_k(f) = FT[p_k(t)] = T_u \, e^{-j\pi(f-f_k)T_u} \operatorname{sinc}[(f - f_k)T_u] \tag{9}$$

where $f_k = k\Delta f$ represents the k-th subcarrier frequency.

The modulation symbol $c_{n,k}$ of the reference signal can be regarded as an ideal pseudo-random sequence (PRN), and its PSD remains constant at unity:

$$G_{\text{PRN}}(f) = 1 \tag{10}$$

This paper uses a default NR and receiver configuration as listed in Table 1. Its PSD curve is illustrated in Figure 2. The PSD of the NR reference signal exhibits multiple spectral peaks at corresponding frequencies. Between these peaks, spectral notches appear due to the absence of a subcarrier reference signal. NR parameters can adjust the number and location of PSD peaks. Hence, the NR parameters determine its tracking performance.

Table 1. Default configurations in LEO NR navigation.

	Parameter	Value
NR Basic	FFT Length	1024
	Subcarrier Spacing (kHz)	15
	Sampling Rate (MHz)	15.36
	Period (ms)	10
	Slot Number Per Period	5
	Symbol Length Per Slot	8
	Used BandWidth (MHz)	1.8
	Resource Block Number	10
	Subcarrier Number	240
PDSCH	Modulation	QPSK
	DMRS Mapping	Type A
	DMRS-TypeA-Position	2
	DMRSLength	1
	DMRS-Additional-Position	0
	DMRS-Configuration-Type	2
	DMRS proportion α	3.50%
Delay-locked loop (DLL)	B_L (Hz)	20
	Damping Coefficient	0.707
	Integration Time (ms)	10
	Early–Late Spacing Δ (Sampling Time)	2
	Transmit Power (W)	10
	Proportion of Reference Signal in NR signal	19.44%
	Receiver Front-end Bandwidth (MHz)	15.36
	Sampling Rate (MHz)	15.36

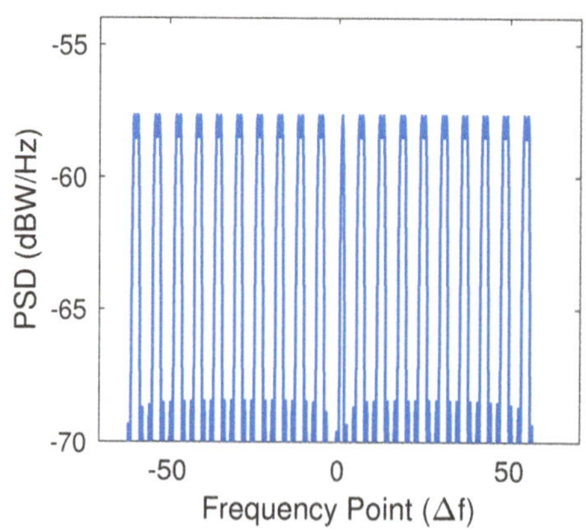

Figure 2. Power spectral density (PSD) of NR. The unit of horizontal axis is Δf.

4. NR Navigation System

This section presents the system block diagram of the NR transmitter and navigation receiver. We give a detailed introduction to the receiver tracking loop. A novel correlation integration scheme is proposed to avoid the interference caused by the CP and inter-carrier. Based on the tracking loop, an extended Kalman filter (EKF) is proposed to realize LEO NR positioning.

4.1. Transmitting and Reception System Block Diagram

As illustrated in Figure 3, an NR opportunity navigation system comprises three components: an NR signal transmitter, wireless channel, and navigation receiver.

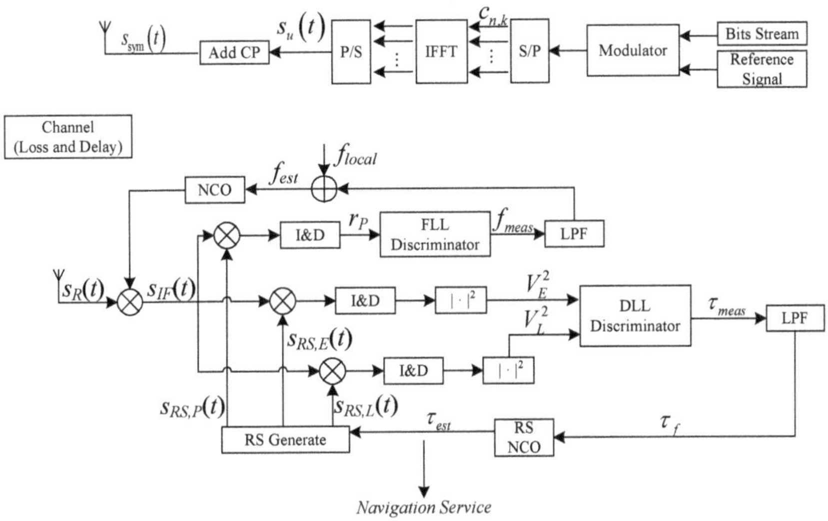

Figure 3. NR navigation system block diagram.

We first introduce the NR transmitter, whose physical layer is similar to the OFDM system. According to NR protocol [31], the higher layer of the base station generates a reference signal and random downlink bit streams. These bit streams undergo digital modulation (e.g., QAM or 16QAM), followed by resource grid mapping. As a result, the serial data streams are transformed to parallel modulation symbols $c_{n,k}$. OFDM modulation is implemented via inverse fast Fourier transform (IFFT), converting parallel signals to serial waveforms to generate the useful modulated signal $s_u(t)$. The cyclic prefix is subsequently appended on useful signal $s_u(t)$ to form the final transmitted waveform $s_{\text{sym}}(t)$ (the frequency up and down conversions are omitted). OFDM signals experience amplitude attenuation and time delay through the wireless propagation channel. For LEO satellite signals, additional Doppler frequency shift occurs due to relative motion between satellites and receivers.

The architecture of the NR navigation receiver has an analogous process flow to conventional GNSS receivers. First, the received NR signal is mixed with a local carrier to generate an intermediate-frequency (IF) signal:

$$s_{\text{IF}}(t) = \text{Re}\left(s_R(t)e^{-j2\pi f_{\text{local}}t}\right) = [s_{\text{RS}}(t) + s_{\text{COM}}(t)]e^{-j(2\pi f_e t+\varphi_e)} \tag{11}$$

where f_{local} denotes the local carrier frequency, and f_e and φ_e represent the frequency and phase offset of the IF signal, respectively. $s_{COM}(t)$ and $s_{RS}(t)$ denote the communication data and navigation-available reference signal, respectively. The $s_{COM}(t)$ may influence the navigation performance, and this influence should be avoided as much as possible for the navigation receiver.

The NR receiver consists of two core processing loops: (1) a delay-locked loop (DLL) for code delay estimation using the non-coherent early-late processor (NELP); and (2) a frequency-locked loop (FLL) dedicated to estimating carrier frequency offset. The DLL and FLL exhibit functional complementarity in signal tracking: the DLL eliminates transmission delays by adjusting the phase of the local replica signal, while the FLL maintains carrier synchronization by compensating frequency offset.

The NELP employs three parallel processing branches with relative delays τ (prompt), $\tau - d$ (early), and $\tau + d$ (late), where d specifies the correlator spacing. In the early branch, the local replica reference signal $s_{\mathrm{RS,E}}(t)$ correlates with the received IF signal. Then, the correlation results are sent to the integrate and dump (I&D) filter and generate correlation amplitude V_E:

$$V_E = \left| \int_0^{T_{\mathrm{coh}}} s_{\mathrm{IF}}(t) s_{\mathrm{RS,E}}^*(t) dt \right| = R_E(\tau) |\mathrm{sinc}(f_e T_{\mathrm{coh}})| + n_{\mathrm{COR}} \tag{12}$$

where T_{coh} represents the duration of correlation integration. n_{COR} quantifies the interference caused by the CP and inter-carrier. Compared with the traditional GNSS receiver, the CP and inter-carrier communication data significantly impact DLL accuracy. An interference elimination method is proposed in the following subsection.

Similarly, the correlation amplitude V_L of the delayed branch can be obtained. The time delay of the received signal can be measured by utilizing V_E and V_L:

$$\tau_{\mathrm{meas}} = K_{\mathrm{RS}} \frac{V_E^2 - V_L^2}{V_E^2 + V_L^2} \tag{13}$$

where K_{RS} denotes the normalization parameter of the time delay discriminator, derived from the S-curve [2].

To avoid the noise interference and achieve stable tracking time delay, a second-order DLL tracking loop is utilized and it contains a first-order filter [2]:

$$\tau_{\mathrm{f}} = \tau_{\mathrm{f,pre}} + \omega_n \cdot 1.414 \left[\tau_{\mathrm{meas}} - \tau_{\mathrm{meas,pre}} \right] + \frac{T_{\mathrm{coh}}}{2} \omega_n^2 \left[\tau_{\mathrm{meas}} + \tau_{\mathrm{meas,pre}} \right] \tag{14}$$

where $\tau_{\mathrm{meas,\ pre}}$ and $\tau_{\mathrm{f,\ pre}}$ represent the measured and filtered delay from the previous integration period, respectively. The natural frequency ω_n critically affects the tracking accuracy and dynamic response of the NELP. ω_n is calculated by the DLL noise bandwidth $B_{L,\mathrm{DLL}}$:

$$\omega_n = \frac{B_{L,\mathrm{DLL}}}{0.53} \tag{15}$$

Next, the time delay estimate τ_{est} is derived according to the previous estimate $\tau_{\mathrm{est,pre}}$ and the filter output τ_f:

$$\tau_{\mathrm{est}} = \tau_{\mathrm{est,pre}} + T_{\mathrm{coh}} \cdot \tau_f \tag{16}$$

FLL operates on the prompt branch, and the frequency offset f_e of the received signal is measured according to the phase change. The phase can be obtained from correlation value r_P, which is integrated correlation between the IF signal $s_{IF}(t)$ and local replica signal:

$$\begin{aligned} r_P &= \int_0^{T_{\mathrm{coh}}} s_{\mathrm{IF}}(t) s_{\mathrm{RS,P}}^*(t) dt \\ &= \int_0^{T_{\mathrm{coh}}} [s_{\mathrm{RS}}(t) + s_{\mathrm{COM}}(t)] e^{-j(2\pi f_I t + \varphi)} s_{\mathrm{RS,P}}^*(t) dt \\ &= R_P(\tau)\,\mathrm{sinc}(f_e T_{\mathrm{coh}}) e^{j\varphi} + n_{\mathrm{COR}} \end{aligned} \tag{17}$$

When the correlation noise n_{COR} is negligible, the phase variation between consecutive integration periods becomes measurable:

$$r_P\overline{r_{P,\mathrm{pre}}} = |r_P||r_{P,\mathrm{pre}}|e^{j[\varphi-\varphi_{\mathrm{pre}}]} \tag{18}$$

where $r_{P,\mathrm{pre}}$ and φ_{pre} denote the correlation value and phase offset in previous integration period, respectively.

Hence, the frequency offset is calculated as follows:

$$f_{\mathrm{meas}} = \frac{\arg\left[r_P\overline{r_{P,\mathrm{pre}}}\right]}{2\pi T_{\mathrm{coh}}} \tag{19}$$

Like DLL, the FLL employs a low-pass filter to eliminate high-frequency noise from f_{meas}. The filtered output updates the local carrier frequency f_{local} to stably track the frequency of the received signal.

Different from the conventional communication receiver, the time-delay configuration of the local replica signal is a new requirement for the NR opportunity navigation receiver. The total delay τ decomposes into integer and fractional components relative to the sampling interval T_s:

$$\tau = (\tau_{\mathrm{int}} + \tau_{\mathrm{frac}})T_s \tag{20}$$

Integer delay τ_{int} is implemented through discrete signal shifting, while fractional delay τ_{frac} is achieved by precise phase adjustment of subcarriers. For example, the k-th subcarrier accomplishes early and late phase shifts τ_{frac} by multiplying modulated symbol $c_{n,k}$ with complex exponential functions:

$$\begin{cases} c_{n,k,E} = c_{n,k}e^{-j2\pi k\tau_{\mathrm{frac}}/K} \\ c_{n,k,L} = c_{n,k}e^{j2\pi k\tau_{\mathrm{frac}}/K} \end{cases} \tag{21}$$

4.2. Interference-Free Correlation: ZPC

First, we introduce the orthogonality condition of OFDM subcarriers: Subcarriers with different frequencies are integrated within a useful symbol period T_u, i.e.,:

$$\int_0^{T_u} c_{n,k}c_{m,l}^* e^{j2\pi k\Delta f t}e^{-j2\pi l\Delta f t}dt = \begin{cases} 0, & k \neq l \\ T_u c_{n,k}c_{m,k}^*, & k = l \end{cases} \tag{22}$$

However, NR base stations append cyclic prefixes after OFDM modulation to prevent ISI. The CP extends the symbol duration to T_{sym}. There exists interference when directly correlating OFDM signals containing CP, as depicted in Figure 4a; in other words, cross-correlation between different subcarriers becomes non-zero:

$$\int_0^{T_{\mathrm{sym}}} c_{n,k}c_{m,l}^* e^{j2\pi k\Delta f t}e^{-j2\pi l\Delta f t}dt \neq 0, \quad k \neq l \tag{23}$$

Using T_{sym} as the integration duration breaks subcarriers' orthogonality and induces mutual interference, degrading NR DLL tracking accuracy.

This paper proposes an interference-free correlation: zero-padding correlation (ZPC). As depicted in Figure 4b, the ZPC process involves two aspects: (1) For every OFDM symbol, confining the integration duration to T_u preserves subcarrier orthogonality, ensuring that integration between different subcarriers equals zero; and (2) zero-padding the cyclic prefix (CP) intervals effectively eliminates intersymbol interference (ISI). The proposed correlation method preserves subcarrier orthogonality while eliminating the influence of CP and communication data.

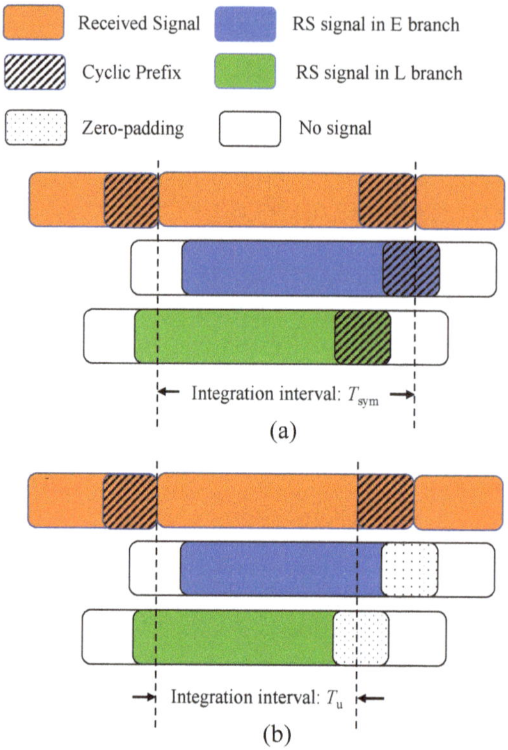

Figure 4. Different correlation method: (**a**) conventional CP-based method with integration interval T_{sym} (replica signal includes CP); (**b**) effective integration interval of proposed ZPC method is T_{u} (replica signal utilizes zero-padding).

4.3. EKF for LEO Navigation

The extended Kalman filter (EKF) is a nonlinear estimation technique and demonstrates particular efficacy in handling the real-time positioning of GNSS navigation receivers [2]. To ensure compatibility with subsequent research, the receiver's state vector incorporates position, velocity, and clock error parameters:

$$\mathbf{x}(k) = \left[\mathbf{r}(k), \dot{\mathbf{r}}(k), \delta t(k), \dot{\delta t}(k)\right]^{\mathrm{T}} \tag{24}$$

where $\mathbf{r}(k) = [x(k), y(k), z(k)]$ denotes the receiver position at epoch k in Earth-centered, Earth-fixed (ECEF) coordinates, and $\dot{\mathbf{r}}(k)$ represents receiver's velocity. δt and $\dot{\delta t}$ are the receiver's clock bias and drift rate, respectively.

The system dynamic model is constructed based on receiver motion characteristics:

$$\mathbf{x}(k+1) = \mathbf{F}\mathbf{x}(k) + \mathbf{w}(k) \tag{25}$$

where $\mathbf{F}$ is the state transition matrix and its definition is as follows:

$$\mathbf{F} = \begin{pmatrix} \mathbf{I}_3 & \mathbf{I}_3 T_s & \mathbf{0}_{3,1} & \mathbf{0}_{3,1} \\ \mathbf{0}_3 & \mathbf{I}_3 & \mathbf{0}_{3,1} & \mathbf{0}_{3,1} \\ \mathbf{0}_{1,3} & \mathbf{0}_{1,3} & 1 & 0 \\ \mathbf{0}_{1,3} & \mathbf{0}_{1,3} & 0 & 1 \end{pmatrix} \tag{26}$$

where τ_s denotes the EKF state propagation interval, equal to the coherent integration duration T_{coh} in this study. $\mathbf{I}_n$ represents an $n \times n$ identity matrix, while $\mathbf{0}_{m\times n}$ indicates an $m \times n$ zero matrix.

$\mathbf{w}(k)$ is the process noise vector and its covariance matrix $\mathbf{Q}$ models system uncertainties:

$$\mathbf{Q} = \begin{pmatrix} \frac{1}{3}\mathbf{S}_a\tau_s^3 & \frac{1}{2}\mathbf{S}_a\tau_s^2 & \mathbf{0}_{3,1} & \mathbf{0}_{3,1} \\ \frac{1}{2}\mathbf{S}_a\tau_s^2 & \mathbf{S}_a\tau_s & \mathbf{0}_{3,1} & \mathbf{0}_{3,1} \\ \mathbf{0}_{1,3} & \mathbf{0}_{1,3} & S_{c\phi}\tau_s + \frac{1}{3}S_{cf}\tau_s^3 & \frac{1}{2}S_{cf}\tau_s^2 \\ \mathbf{0}_{1,3} & \mathbf{0}_{1,3} & \frac{1}{2}S_{cf}\tau_s^2 & S_{cf}\tau_s \end{pmatrix} \tag{27}$$

where $\mathbf{S}_a$ denotes the PSD vector of receiver acceleration in ECEF coordinates, S_{cf} represents clock frequency drift PSD, and $S_{c\phi}$ corresponds to clock phase noise PSD. In this paper, $\mathbf{S}_a$, S_{cf}, and $S_{c\phi}$ are, respectively, set to diag(10, 10, 10), 1, and 1. Hence, $\mathbf{Q}$ remains constant for specified motion states.

The pseudorange and pseudorange rate are used as observations in this paper, and these observations are measured by the DLL and FLL, respectively. In addition to real geometric distance and velocity, these measurements also contain various error components. For i-th LEO satellite, its pseudorange measurement $\rho_i(k)$ can be expressed as

$$\rho_i(k) = d_i(k) + c[\delta t(k) + \delta t_{\text{sat},i}(k) + I_i(k) + T_i(k)] + \varepsilon_i(k) \tag{28}$$

where $d_i(k) = \|\mathbf{r}_{s,i}(k) - \mathbf{r}(k)\|$ denotes the true geometric distance between receiver and satellite, with $\mathbf{r}_{s,i}(k)$ representing the satellite's ECEF position coordinates. c denotes the speed of light while $\delta t(k)$ and $\delta t_{\text{sat},i}(k)$ correspond to receiver and satellite clock offsets, respectively. The ionospheric and tropospheric delays are modeled as $I_i(k)$ and $T_i(k)$, while $\varepsilon_i(k)$ is the ranging measurement error caused by receiver noise.

The pseudorange rate $\dot{\rho}_i(k)$, measured from FLL, incorporates the satellite-receiver relative velocity $\dot{r}_i(k)$ and receiver's clock drift $\dot{\delta t}(k)$:

$$\dot{\rho}_i(k) = \dot{r}_i(k) + c\dot{\delta t}(k) + \varepsilon_{v,i}(k) \tag{29}$$

where $\varepsilon_{v,i}(k)$ represents the velocity measurement error.

Assuming an LEO NR navigation receiver tracking N satellites, the augmented observation vector $\boldsymbol{\rho}(k)$ contains all the pseudorange and pseudorange rate measurements, i.e.,:

$$\boldsymbol{\rho}(k) = [\rho_1(k), \cdots, \rho_N(k), \dot{\rho}_1(k), \cdots, \dot{\rho}_N(k)]^{\mathrm{T}} \tag{30}$$

Therefore, the receiver's tracking loop provides $2N$ observations for estimating eight state parameters, establishing an overdetermined system when $N > 4$.

The nonlinear observation equation is formulated as

$$\boldsymbol{\rho}(k) = h[\mathbf{x}(k)] + \mathbf{v}_k \tag{31}$$

where $h(\cdot)$ denotes the pseudorange and pseudorange rate measurement function, and $\mathbf{v}_k = [\varepsilon_1(k), \varepsilon_2(k), \cdots, \varepsilon_N(k),\ \varepsilon_{v,1}(k), \varepsilon_{v,2}(k), \cdots, \varepsilon_{v,N}(k)]$ represents observation noise. Its covariance matrix $\mathbf{R}_k$ characterizes the measurement noise:

$$\mathbf{R}_k = \mathrm{E}\left[\mathbf{v}_k\mathbf{v}_k^{\mathrm{T}}\right] \tag{32}$$

$\mathbf{R}_k$ has a close relationship with the carrier-to-noise ratio and is acquired according to measured noise.

EKF uses the Jacobian matrix $\mathbf{H}_k$ to linearize the observation model, and it is the partial derivative of the measurement function to the state vector:

$$\mathbf{H}_k = \frac{\partial h(\mathbf{x}(k))}{\partial \mathbf{x}(k)} \tag{33}$$

This matrix quantifies the sensitivity of observations to variations in receiver position, velocity, and clock bias parameters.

Moreover, the covariance matrix $\mathbf{P}_k$ characterizes the estimation accuracy of the state vector, and its definition is as follows:

$$\mathbf{P}_k = \mathrm{E}\left[(\mathbf{x}(k) - \hat{\mathbf{x}}(k))(\mathbf{x}(k) - \hat{\mathbf{x}}(k))^{\mathrm{T}}\right] \tag{34}$$

where $\mathbf{x}(k)$ and $\hat{\mathbf{x}}(k)$ represent the true states and estimated states, respectively.

Figure 5 illustrates the EKF's update process, which contains prediction and correction modules. At the beginning of the EKF update, two initial parameters are required: (1) initial state vector $\mathbf{x}(0)$, obtained via least-squares positioning using first-epoch observations; and (2) initial covariance matrix $\mathbf{P}(0)$, evaluated by the uncertainties in position, velocity, and clock ambiguity.

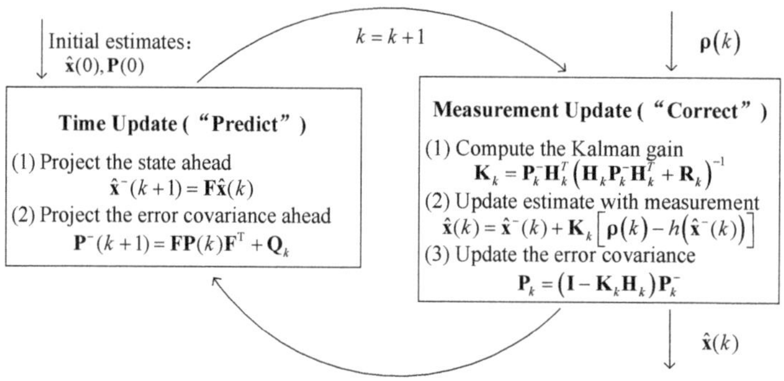

Figure 5. EKF update process.

In EKF, iterations update execute recursive prediction-correction cycles: In the prediction phase, the filter propagates the state estimate $\hat{\mathbf{x}}^-(k+1)$ and covariance $\mathbf{P}^-(k+1)$ from $\hat{\mathbf{x}}(k)$ and $\mathbf{P}(k)$ by using the system dynamics model. Then, the correction phase refines these predictions through observation vector $\rho(k)$ and Kalman gain matrix $\mathbf{K}_k$:

$$\mathbf{K}_k = \mathbf{P}_k^-\mathbf{H}_k^{\mathrm{T}}(\mathbf{H}_k\mathbf{P}_k^-\mathbf{H}_k^{\mathrm{T}} + \mathbf{R}_k)^{-1} \tag{35}$$

The updated state estimate and covariance are as follows:

$$\hat{\mathbf{x}}(k) = \hat{\mathbf{x}}^-(k) + \mathbf{K}_k(\rho(k) - h(\hat{\mathbf{x}}^-(k))) \tag{36}$$

$$\mathbf{P}(k) = (\mathbf{I} - \mathbf{K}_k\mathbf{H}_k)\mathbf{P}_k^- \tag{37}$$

This closed-loop EKF outputs real-time navigation solutions while maintaining an optimal balance between model predictions and sensor measurements.

5. Improvement of ZPC Tracking Performance

This section validates the interference-free property of ZPC and compares ZPC with other works in tracking performance. The experiment results demonstrate the improvement of ZPC tracking accuracy and its advantages.

5.1. Validate the ZPC's Interference-Free Property.

An experiment is proposed to demonstrate the ZPC's interference-free property, which improves the tracking accuracy. There are three configurations in the experiments. (1) The received signal is a conventional NR signal containing both reference signals and unknown communication data. The local replica signal has a CP and its integration duration is T_{sym}. (2) The input signal is the same as (1). Implementing ZPC utilizes a zero-padded local replica signal with a T_u integration duration. (3) The received signal only contains the reference signal without communication data, and the receiver uses ZPC. The tracking accuracy of DLL with different configurations is plotted in Figure 6. Comparative results reveal two critical findings: First, the ZPC configuration (red curve) achieves significantly higher tracking accuracy than conventional CP-based processing with duration T_{sym} (blue curve). Because the CP introduces interference to the subcarriers' orthogonality when the integrated duration is T_{sym}, the subcarrier maintains orthogonality when the ZPC is adopted (the integrated duration is T_u). ZPC significantly eliminates interference caused by the CP and inter-carriers, thereby improving tracking accuracy. Second, configurations (2) and (3) have identical DLL standard deviation (STD) curves, which confirms their same tracking performance. This indicates that the NR communication signal in other carriers has no interference on the DLL when using ZPC. The reason is that ZPC maintains subcarrier orthogonality through optimized integration window selection and zero-padding CP. Overall, ZPC ensures the orthogonality and serves as an interference-free correlation mechanism, effectively eliminating the influence of the CP and inter-carriers while significantly enhancing tracking accuracy.

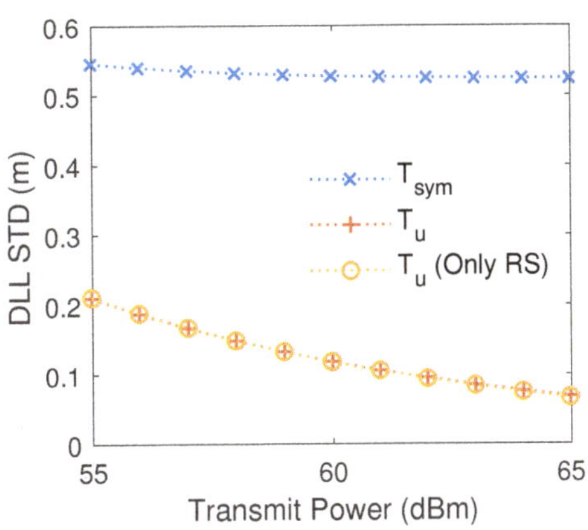

Figure 6. The relationship between delay-locked loop (DLL) tracking accuracy and different correlation methods. DLL standard deviation (STD) represents the DLL tracking accuracy.

5.2. Tracking Performance Comparison of ZPC Against Other Works

This subsection compares our proposed ZPC receiver with other representative NR navigation receivers by tracking accuracy. The ZPC receiver is a cooperative navigation receiver that utilizes all detectable signal components as a local replica signal, including SSB, CSI-RS, and DMRS in PDSCH and PDCCH. Shamaei first proposed an NR receiver exclusively using SSB signals [37]. This reception strategy ignores other available navigation signals and does not consider the LEO navigation scenario. Neinavaie introduces a cognitive (CON) method for LEO OFDM-like signals to identify navigation reference signals without considering cyclic prefix influence [30,38].

For comprehensive comparison, three NR signal configurations are used in the experiment:

- **Signal 1**: using default configuration as baseline (see Table 1).
- **Signal 2**: equivalent bandwidth to Signal 1 with PDSCH spectral center shifted from zero frequency to $260\Delta f$.
- **Signal 3**: PDSCH bandwidth expands (1.8 MHz $\rightarrow$ 10.62 MHz), slot number increase (5 $\rightarrow$ 9), and DMRS-Additional-Position adjustment (0 $\rightarrow$ 1). For every NR frame, the resource elements of the reference signal increase from 830 to 6850, leading to an augmentation in the reference signal's power.

Figure 7 compares the DLL tracking accuracy of different navigation reception strategies for three NR signals. The experimental results reveal four key findings. (1) The SSB navigation receiver (proposed by Shamaei) exhibits identical tracking performance for different NR signals. This receiver only utilizes SSB signals, so the PDSCH changes do not affect its tracking performance. (2) Our proposed ZPC receiver achieves superior tracking accuracy compared to SSB receivers. This enhancement stems from comprehensively utilizing all known reference signals, including SSB, DMRS in PDCCH and PDSCH, and CSI-RS. (3) Compared with CON, the ZPC receiver significantly enhances tracking accuracy for identical NR signals. Because the CON receiver does not consider the CP process, noise caused by CP exists. Moreover, the PBCH changes over time, and the CON receiver cannot recognize the PBCH signal. Notably, this paper assumes the CON receiver operates in an ideal scenario, whereas real-world implementations face additional challenges from dynamic PDSCH bandwidth allocation and an aperiodic reference signal. (4) The tracking accuracy of ZPC can significantly improve by using different NR configurations. because navigation signals demonstrate varying PSD and CNR across different NR configurations. So, the base stations can flexibly configure NR signals according to navigation requirements.

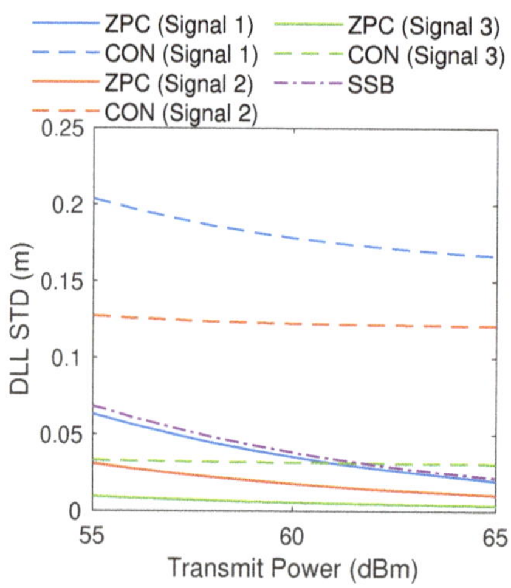

Figure 7. DLL tracking accuracy for different NR signals and reception strategies. For signals 1, 2, and 3, ZPC, respectively, achieves 80.3%, 85.6%, and 82.6% higher tracking accuracy than CON.

6. NR Tracking Performance Analysis for Different NR and Receiver Configuration

This section evaluates the impact of different NR signal and receiver configurations on tracking accuracy. The analysis results offer systematic insights for LEO OFDM tracking loop.

6.1. Impact of NR Signal Configuration on Tracking Accuracy

As described in Section 4, NELP is used as a code tracking loop and its theoretical tracking accuracy is as follows [76]:

$$\sigma_{\mathrm{NELP}} = \sqrt{\frac{B_{L,\mathrm{DLL}}(1-0.5B_L T_{\mathrm{coh}})\int_{-\beta_r/2}^{\beta_r/2} G(f)\sin^2(\pi f\Delta)df}{\left(\frac{C_{\mathrm{nav}}}{N_0}\right)\left(2\pi\int_{-\beta_r/2}^{\beta_r/2} fG(f)\sin(\pi f\Delta)df\right)^2}} \times \sqrt{1+\frac{\int_{-\beta_r/2}^{\beta_r/2} G(f)\cos^2(\pi f\Delta)df}{T_{\mathrm{coh}}\left(\frac{C_{\mathrm{nav}}}{N_0}\right)\left(\int_{-\beta_r/2}^{\beta_r/2} G(f)\cos(\pi f\Delta)df\right)^2}} \tag{38}$$

where $G(f)$ denotes the normalized PSD of the navigation signal, and β_r represents the receiver front-end bandwidth. $\Delta = 2d$ is the time interval between the early and late branches, and $B_{L,\mathrm{DLL}}$ is the bandwidth of DLL. C_{nav}/N_0 is the carrier-to-noise ratio, which can be determined by LEO transmit power (Appendix A) .

Unless otherwise specified, all NR signals follow default configurations listed in Table 1. This section only utilizes PDSCH DMRS as a local replica signal to evaluate the impact of NR configurations on tracking accuracy. The NR's PDSCH contains 10 resource blocks (RBs), each comprising 12 subcarriers with a 15 kHz spacing. Hence, the default NR signal contains 120 subcarriers, and its bandwidth is 1.8 MHz. The center frequency of PDSCH is located at zero frequency.

The DLL tracking accuracy significantly depends on three key configurations: *signal bandwidth*, *spectral distribution*, and the *proportion* of NR reference signal. The base station flexibly configures these parameters according to the communication scenario. First, we discuss the relationship between tracking accuracy and signal bandwidth. The experiment chooses three signal bandwidths: 1.8 MHz, 3.6 MHz, and 5.4 MHz. Their corresponding reference signal PSD and DLL standard deviation (STD) are depicted in Figure 8. The comparison results reveal that expanded NR bandwidth significantly enhances the DLL tracking accuracy. At a fixed transmitting power of 60 dBm, the DLL tracking accuracy improves from 0.1187 m (1.8 MHz) to 0.0605 m (3.2 MHz) and 0.0506 m (5.4 MHz), respectively, by 49.03% and 65.80%. When the NR bandwidth increases, its spectral redistribution moves to higher frequencies. This spectral redistribution leads to higher tracking accuracy. Moreover, the tracking accuracy increases with transmitting power. Each five dBm power increment generates approximately 44% improvement in DLL tracking accuracy.

The spectral location of the NR signal constitutes another critical factor affecting tracking performance. For a fixed bandwidth NR signal, its center frequencies are placed at $0, 130\Delta f$, and $260\Delta f$, respectively. The receiver front-end bandwidth is 15.36 MHz to ensure complete signal reception. The relationship of NR signal spectral location and tracking accuracy is shown in Figure 9. We can find that the tracking accuracy increases with the center frequency of the NR signal. Under 60 dBm transmitting power, the tracking accuracy achieves 86.35% enhancement when the center frequency shifts from zero frequency to $260\Delta f$. Moving the NR spectral to a higher frequency is conducive to improving tracking accuracy. Notably, the NR spectral should be within the receiver front bandwidth to ensure the integrity of the received NR signal.

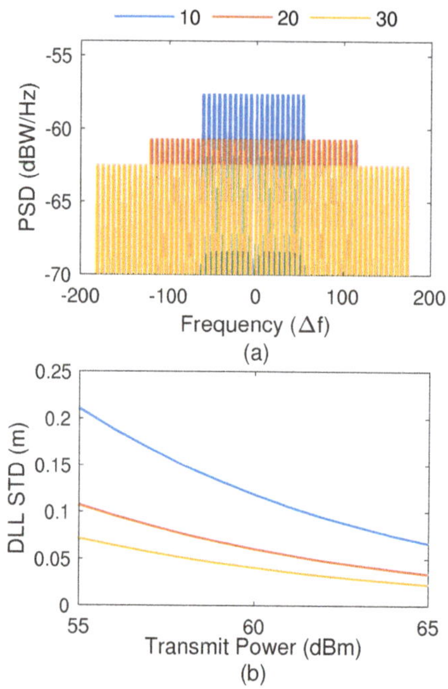

Figure 8. Relationship between NR signal bandwidth and tracking accuracy. (**a**) PSD of NR reference signal; (**b**) DLL STD with different bandwidth NR signal. At 60 dBm transmitting power, the DLL has 65.8% accuracy gain at 5.4 MHz compared to 1.8 MHz.

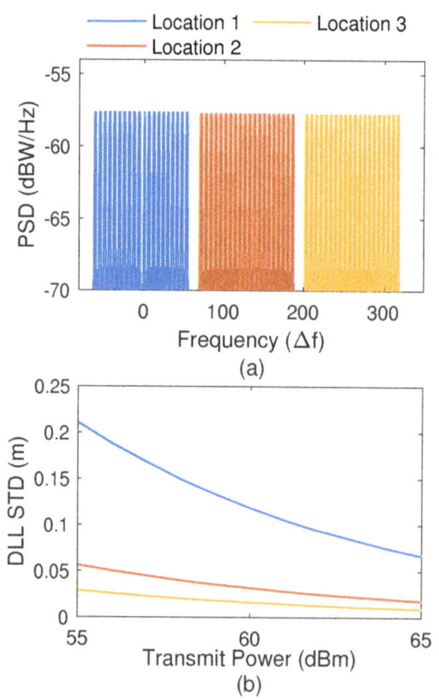

Figure 9. Relationship between signal frequency location and tracking accuracy. (**a**) PSD of NR reference signal; (**b**) DLL STD with different NR frequency location.

The proportion of reference signals in NR signal is dynamically configured according to channel quality [31], significantly affecting DLL tracking accuracy. According to NR protocol, the parameter *DMRS-Additional-Position* adjusts PDSCH DMRS proportions. In the experiments, *DMRS-Additional-Position* is configured to 0, 1, and 2, achieving reference signal proportions of 4.17%, 8.33%, and 12.5%, respectively. Its tracking accuracy is plotted in Figure 10, and we can find that the DLL tracking accuracy increases with the proportion of the reference signal. At 60 dBm transmitting power, elevating the reference signal proportion from 4.17% to 8.33% and 12.5% yields 31.52% and 44.95% accuracy enhancements, respectively. Increasing the reference signal power allocation enhances the CNR of the navigation signal. Notably, increasing the reference proportions reduces the communication capacity, and reference signals occupy the same resource blocks as data transmission. This one-to-one compromise suggests that a 12.5% reference signal allocation would consume 12.5% of the total channel capacity, leaving 87.5% capacity for data. The optimal ratio should be determined by specific application requirements for navigation accuracy versus data throughput.

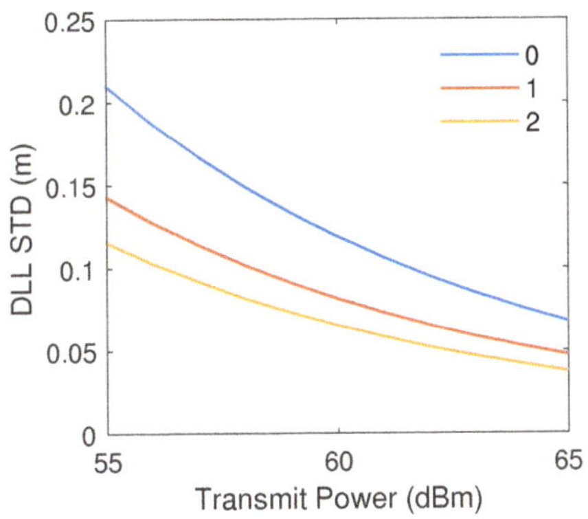

Figure 10. Relationship between portion of NR reference signal and tracking accuracy.

Overall, for fixed transmitting power, there are three NR configurations to enhance DLL tracking accuracy: first, maximizing signal bandwidth; second, the spectral of the NR signal is moved away from zero frequency as far as possible; and third, enhancing the proportion of the reference signal to increase navigation signal power.

6.2. Impact of Receiver Configuration on Tracking Accuracy

DLL tracking accuracy is also determined by NR receiver configurations: (1) coherent integration time (T_{coh}), (2) DLL loop bandwidth (B_L), and (3) different local replica signals. T_{coh} determines the integration duration of the integrate-and-dump filter, while the loop bandwidth B_L governs both DLL tracking accuracy and dynamic response speed. Figure 11 and Table 2 demonstrate DLL tracking accuracy with different coherent integration time T_{coh} and DLL bandwidth B_L, and the results reveal that (1) at 60 dBm transmitting power with fixed $B_L = 20$ Hz, decreasing T_{coh} to 5 ms (relative to the baseline $T_{\text{coh}} = 10$ ms, $B_L = 20$ Hz) reduces DLL tracking accuracy by 2.52%, while elevating T_{coh} to 15 ms enhances accuracy by 2.52%; and (2) at 60 dBm with fixed $T_{\text{coh}} = 10$ ms, reducing B_L from 20 Hz (baseline) to 5 Hz and 1 Hz yields DLL tracking accuracy improvements of 48.10% and 76.58% respectively, indicating the stronger influence of loop bandwidth on DLL tracking accuracy. Generally, extending T_{coh} or reducing B_L enhances DLL tracking accuracy.

Table 2. DLL tracking accuracy summary for different coherent integration time T_{coh} and DLL loop bandwidth B_L at 60 dBm transmit power.

T_{coh} (ms)	B_L (Hz)	DLL STD (m)	DLL STD Improvement
5	20	0.122	−2.52%
10	20	0.119	0% (baseline)
15	20	0.116	2.52%
10	5	0.062	47.9%
10	1	0.028	76.5%

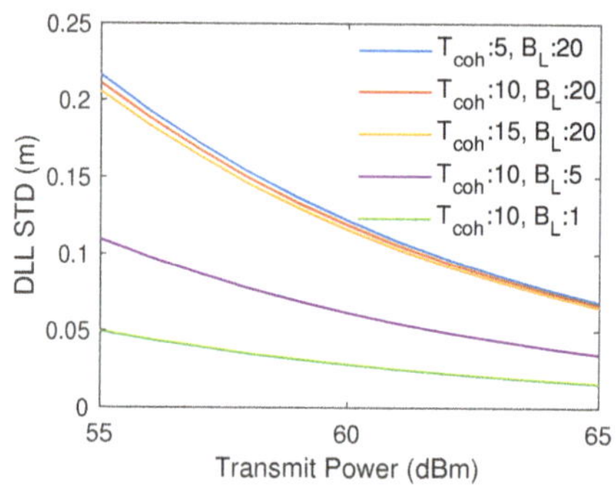

Figure 11. DLL tracking accuracy with different coherent integration time T_{coh} and DLL loop bandwidth B_L.

Multiple reference signals, such as SSB, CSI-RS, and DMRS in PDCCH and PDSCH, can be used as local replica signals for an NR navigation receiver. The selection of local replica signals significantly influences DLL tracking accuracy. Previous experiments only utilize PDSCH DMRS as the local replica signal. This experiment systematically evaluates the impact of four NR replica signals: (1) SSB only; (2) PDSCH DMRS only; (3) combined PDSCH DMRS and SSB; and (4) combined PDSCH DMRS, SSB, and CSI-RS. The DLL tracking accuracy with different replica signals is plotted in Figure 12, and the proportion of the replica reference signal is listed in Table 3. The results show the following: First, exclusive use of PDSCH DMRS yields inferior tracking accuracy compared to other configurations; and second, composite replica signals demonstrate measurable improvements in tracking accuracy. At 60 dBm transmitting power, the combined PDSCH DMRS+SSB achieves a 3.48% accuracy enhancement over SSB only, while the full combination (PDSCH DMRS+SSB+CSI-RS) provides a 5.06% improvement. The proportion of CSI-RS and PDSCH DMRS is lower than 4%, so the improvement in tracking accuracy is not apparent. This enhancement in tracking accuracy is attributed to the increased correlator output power achieved through reference signal superposition. Hence, all predictable NR signals should be used as reference signals to improve navigation performance.

Table 3. Resource elements distribution in an NR frame.

	PBCH	CSI	PDSCH DMRS	PDSCH Data	Sum
Resource Elements	830	80	200	4600	5710
Resource Proportion (Power Proportion)	14.54%	1.40%	3.50%	80.56%	100.00%

In summary, this section analyzes the impact of NR signal and receiver configurations on DLL tracking accuracy, offering systematic insights for LEO OFDM navigation design. Based on the tracking results, the positioning performance is discussed in the following.

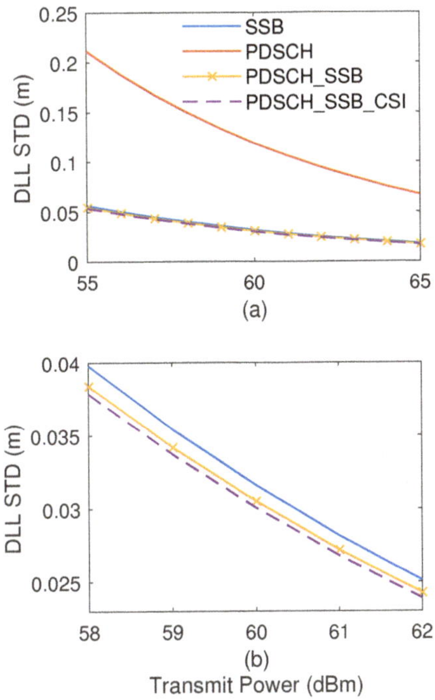

Figure 12. DLL tracking accuracy with different local replica signals. (**a**) Transmitting power range: 55~65 dBm. (**b**) Zoomed-in range: 58~62 dBm.

7. Positioning Performance of LEO NR Signal

This section first simulates the motion trajectory of LEO satellites to generate the wireless channel characteristics. Next, the positioning performance of the ZPC receiver will be discussed, covering static and dynamic scenarios as well as the impacts of various error sources.

7.1. LEO Navigation Scenario

The propagation channel is a key part of the NR navigation system and is generated by LEO scenario simulation. We use Starlink's orbit data to establish the LEO satellite motion scenario [77]. The static receiver is located in Xi'an Jiaotong University, Shaanxi, China (108.9788°E, 34.2481°N), at an altitude of 450 m. The dynamic receiver moves around the campus at a speed of 60 km/h. The simulation starts at 12:03 UTC on 5 November 2024, and lasts 15 s. As shown in Figure 13, the elevation angles of six visible LEO satellites are higher than 10 degrees, and their orbit parameters are shown in Table 4. Therefore, the satellite-receiver kinematic simulation generates their geometric distance and relative velocity. According to the distance and velocity, we obtain three key propagation channel characteristics: (1) path loss, (2) time delay, and (3) Doppler shift. These parameters are sampled at T_{coh} interval to capture rapid channel variations. The received NR signal can be built based on the transmitting signal and propagation channel characteristics. The NR receiver includes multiple received signals from different LEO satellites, which experience various channels. We use DLL and FLL to measure the time delay and relative velocity of the received NR signals. Finally, at each T_{coh} interval, the EKF updates these measured values to estimate the receiver state vector, such as position, velocity, and clock error.

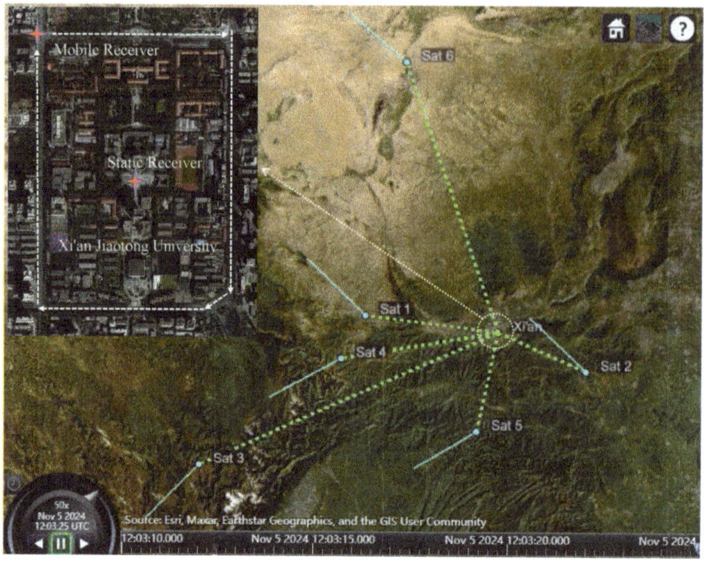

Figure 13. The trajectory of LEO satellites and receiver.

Table 4. Orbital data from Starlink satellites [77].

Parameter	Sat 1	Sat 2	Sat 3	Sat 4	Sat 5	Sat 6
Semi-major Axis a (km)	7136	6980	6970	7050	6820	7130
Eccentricity e	0.0001568	0.0001459	0.0001603	0.0001432	0.0001464	0.0001406
Inclination i (°)	53.0531	53.2146	53.2157	43.0039	43.0019	53.1584
RAAN Ω (°)	189.0089	191.0164	305.9367	291.2627	294.6071	192.7046
Argument of Perigee ω (°)	63.1495	78.6725	89.8188	270.5086	262.0388	61.5981
Mean Anomaly M (°)	82.5489	166.1697	347.1216	212.3624	208.9401	164.3645
Altitude [1] (km)	765	609	599	679	449	759

[1] The radius of Earth is 6371 km.

7.2. Positioning Performance

This section compares the three-dimensional (3D) positioning performance between the proposed ZPC navigation receiver and the CON receiver. As described in Section 4, the simulation of the LEO NR system consists of three core components: the NR transmitter, the LEO satellite channel, and the NR navigation receiver. The NR navigation receiver contains two sequential operational phases: signal acquisition and tracking. The acquisition module initially estimates the pseudo-random noise (PRN) sequence identifier, code phase delay, and carrier frequency offset. Like conventional GNSS receivers, the acquisition process detects peaks generated by correlating local replica and received signals. For brevity, the acquisition module is not expanded upon.

After successful acquisition, the receiver enters tracking mode. The DLL and FLL in the NR receiver estimate the pseudorange and pseudorange rate, respectively. Based on these measurements, the EKF output receiver's state vector is in real time. The simulation experiment includes six viewed satellites, as described in Section 7.1, and their tracking errors are plotted in Figure 14. All the tracking error curves have a similar trend: There are transient responses at the beginning of the tracking (0–0.2 s), then the error curves tend to stabilize. The settling time is determined by the tracking loop bandwidth B_L and is about $4.24/B_L$ [2].

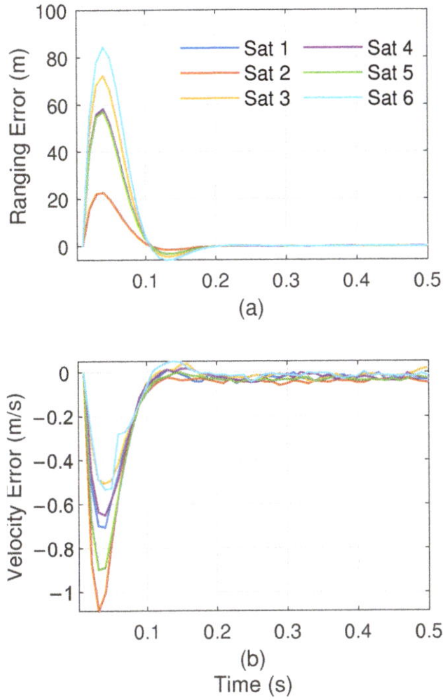

Figure 14. ZPC receiver tracking errors over time: (**a**) ranging error; (**b**) velocity error.

The proposed ZPC and CON receiver's tracking accuracy comparison is plotted in Figure 15. All simulation parameters for the ZPC and CON receivers are identical except for the replica signal and correlation process. This simulation configuration helps to compare the difference of receivers with high repeatability. The tracking accuracy is quantified through the standard deviation (STD) of ranging and velocity errors. The comparison results show that (1) significant differences exist in ranging and velocity error STD for different satellites, with this difference being caused by distinct orbital characteristics, leading to different Doppler shifts and carrier-to-noise ratios; and (2) the CON architecture ranging accuracy ranges from 17.8 to 27.9 cm, and velocity precision ranges from 3.66 to 4.93 cm/s. In contrast, the ZPC receiver demonstrates superior ranging accuracy (3.0~4.5 cm) and velocity accuracies (0.7~1.14 cm/s). ZPC achieves about an 80% improvement in both ranging and velocity accuracy. The CON receiver has two shortages: Omitting the CP process leads to inter-carrier interference (ICI), and it cannot identify PBCH signals decreasing the output power of correlation. The ZPC receiver addresses these shortages, and all known NR reference signals are utilized as local replica signals.

Next, we analyze the positioning performance of the LEO NR system, considering only the receiver noise while excluding other error sources. Four navigation reception strategies are utilized: static ZPC, mobile ZPC, static CON, and mobile CON. Figure 16 and Table 5 present the state estimation results from the EKF: the positioning and velocity error in the north–south, east–west, and altitude dimensions (the ECEF coordinate is generally used in the EKF filter; for clarity, ECEF outputs are converted to the ENU coordinate). The experiment results show the following. (1) The static ZPC receiver exhibits positioning error fluctuation during initial operation, stabilizing after 3 s of continuous operation. The means of positioning errors in the north, east, and horizontal directions measure

−1.84 cm, 2.51 cm, and 4.54 cm, respectively. Moreover, the mean of the three-dimensional positioning accuracy is approximately 5.61 cm. (2) For static and mobile ZPC receivers, the difference of position and velocity errors is lower than 4%. Hence, they have similar position performances, and the ZPC receiver is robust in mobile scenarios. Moreover, the CON receiver demonstrates a similar position performance in both static and mobile scenarios. (3) The CON receiver demonstrates lower positioning accuracies: 5.85 cm, −17.33 cm, and −44.38 cm in the north, east, and horizontal directions. Compared to the CON receiver, our proposed ZPC receiver achieves 88.3% improvement in positioning accuracy. (4) The ZPC receiver also demonstrates a significantly enhanced velocity measurement compared to the CON receiver. The 3D velocity error mean of ZPC is 0.59 cm/s, while CON exhibits more significant errors of 2.42 cm/s. This corresponds to a 75.62% improvement in velocity accuracy.

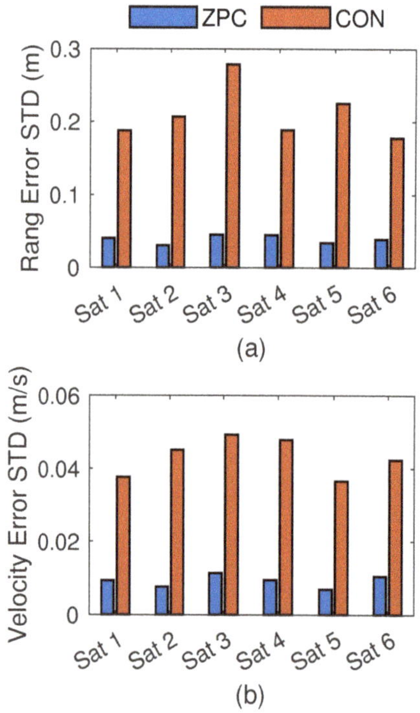

Figure 15. Stable tracking accuracy for different LEO satellites in static scenario: (**a**) ranging accuracy; (**b**) velocity accuracy.

Table 5. Position and velocity error statistics: ZPC and CON receiver.

		Mean				Std			
		North	East	Down	3D	North	East	Down	3D
Position error (m)	ZPC (Static)	−0.0184	0.0251	−0.0454	0.0561	0.0055	0.0050	0.0096	0.0059
	ZPC (Mobile)	−0.0152	0.0184	−0.0507	0.0585	0.0087	0.0097	0.0139	0.0091
	CON (Static)	0.0585	−0.1733	−0.4438	0.4807	0.0184	0.0172	0.0249	0.0242
	CON (Mobile)	0.0570	−0.1765	−0.4327	0.4717	0.0199	0.0270	0.0231	0.0288
Velocity error (m/s)	ZPC (Static)	0.0005	−0.0039	0.0069	0.0487	0.0169	0.0267	0.0459	0.0281
	ZPC (Mobile)	−0.0012	−0.0009	0.0115	0.0484	0.0160	0.0265	0.0465	0.0301
	CON (Static)	0.0189	−0.0126	0.0351	0.2157	0.0760	0.1361	0.1876	0.1212
	CON (Mobile)	0.0171	−0.0099	0.0414	0.2200	0.0762	−0.1383	0.1918	0.1240

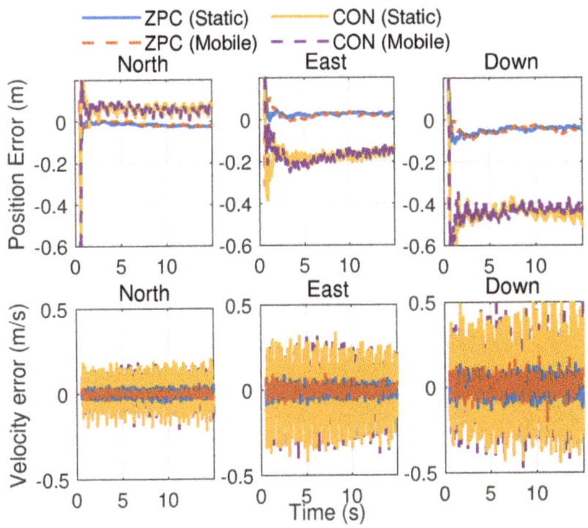

Figure 16. Position and velocity error comparison: ZPC and CON receiver.

7.3. Positioning Performance with Error Components

The previous simulation experiment only considered receiver thermal noise, but actual LEO NR systems necessitate comprehensive error modeling. These error sources include (1) the clock error of receiver and satellite; and (2) the propagation delay uncertainties induced by ionospheric and tropospheric effects. These errors are empirically configured as follows [78]:

- Receiver clock phase offset: 0.1 m;
- Receiver clock frequency drift: 0.01 m/s;
- Satellite clock phase offset STD: 0.5 m;
- Ionospheric error STD: 0.1 m;
- Tropospheric error STD: 0.2 m.

Over the simulation duration, these error components are stable and can be viewed as constant. For the i-th satellite, its satellite clock offset $\delta t_{\text{sat},i}$, ionospheric delay I_i, and tropospheric delay T_i can be modeled as an equivalent time-delay error $\delta t_{eq,i}$, e.g.,:

$$\delta t_{eq,i} = \delta t_{\text{sat},i} + I_i + T_i \tag{39}$$

Moreover, the satellite-receiver geometry impacts the ionosphere and troposphere delay. In the simulation, the six satellites' equivalent range delays $c\delta t_{eq,i}$ are 0.238, 0.506, −0.747, 0.258, −0.466, and −0.062 m, respectively. Figure 17 and Table 6 demonstrate the effect of different error components on the positioning error. The experiment sets four error combinations: (1) white noise only; (2) white noise adds receiver clock offset; (3) composition of white noise, clock error, and ionospheric delay; and (4) composition of white noise, clock error, and ionospheric and tropospheric delays. The comparison results reveal that (1) the first and second configuration have identical positioning accuracies, indicating that the clock error has negligible positioning accuracy degradation due to receiver clock estimation in the state vector; and (2) the third and fourth configurations generate substantial error output, resulting in a notable EKF positioning bias. When comparing the results of the third and fourth configurations, it is evident that incorporating tropospheric effects exacerbates the positioning outcomes. Since the receiver is unable to estimate these errors, alternative error-elimination techniques should be adopted to eliminate them.

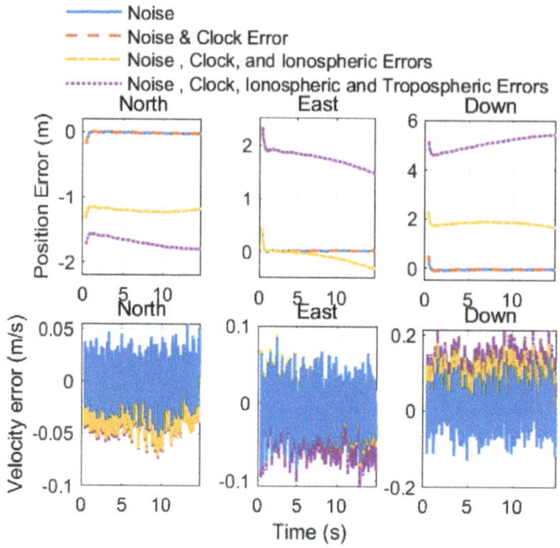

Figure 17. Position and velocity error with different error components.

Table 6. Position and velocity error statistics for different error components.

		Mean				Std			
		North	East	Down	3D	North	East	Down	3D
Position error (m)	Noise	−0.0184	0.0252	−0.0454	0.0561	0.0055	0.0050	0.0096	0.0059
	Noise and clock	−0.0184	0.0252	−0.0454	0.0561	0.0055	0.0050	0.0096	0.0059
	Noise, clock, and ION [1]	−1.2169	−0.1098	1.8210	2.1951	0.0135	0.0928	0.0537	0.0445
	Noise, clock, ION, and TRO [2]	−1.7307	1.7201	5.1643	5.7143	0.0540	0.1142	0.1938	0.1591
Velocity error (m/s)	Noise	0.0005	−0.0039	0.0069	0.0487	0.0169	0.0267	0.0459	0.0281
	Noise and clock	0.0005	−0.0039	0.0069	0.0487	0.0169	0.0267	0.0459	0.0281
	Noise, clock, and ION	−0.0184	−0.0055	0.0502	0.0696	0.0170	0.0268	0.0460	0.0341
	Noise, clock, ION, and TRO	−0.0199	−0.0279	0.0857	0.1024	0.0172	0.0274	0.0461	0.0347

[1] ION denotes inonospheric; [2] TRO denotes tropospheric.

It is obvious that the simulation is essential for the LEO NR navigation system. The simulation can estimate positioning performance and analyze the influence of error components. The simulation method is cost-effective and repeatable in the LEO navigation system design.

8. Conclusions

This paper investigates the navigation potential of OFDM-based LEO broadband constellations and demonstrates that it can achieve a 5.61 cm 3D positioning accuracy. We assume the satellites transmit a 5G NR signal and introduce the structure of NR, focusing on reference signals applicable for navigation service such as SSB, DMRS, and CSI-RS, along with CP OFDM modulation. The PSD of NR reference signals is derived and used for the theoretical analysis of tracking performance.

Subsequently, the workflow of the LEO NR navigation system is outlined, encompassing the transmitter, channel, and navigation receiver. A ZPC is proposed for the tracking loop to eliminate CP and inter-carrier interference, ensuring that navigation performance remains unaffected by CP and communication data. Based on the tracking loop, an EKF positioning algorithm is developed.

We validate the interference-free property of ZPC and compare ZPC with conventional SSB and CON receivers in tracking performance. The experiment results demonstrate that the ZPC receiver has a significant improvement in tracking accuracy.

The comprehensive analysis evaluates navigation performance impacted by NR signal bandwidth, spectral allocation, reference signal proportion, and replica signal combinations. This parametric investigation provides theoretical support for tracking accuracy enhancement of the LEO NR system.

A LEO satellite motion scenario is simulated to characterize channel properties. Based on the EKF, the positioning performance of the LEO NR system and the impacts of error sources are systematically evaluated. The simulation results provide configuration recommendations for implementing opportunity navigation in LEO broadband constellations. This study proposes a low-cost, repeatable, and efficient methodology for analyzing LEO navigation performance through software simulation.

As LEO broadband constellations are primarily designed for communication services, future research should focus on developing integrated communication and navigation receivers.

Author Contributions: Conceptualization and original draft preparation, L.D. and X.Q.; theory and methodology, Y.Y. and H.L.; validation and data analysis, L.D. and J.M.; data curation and supervision, L.D. and T.W.; project administration, Y.Y. All authors have read and agreed to the published version of the manuscript.

Funding: This research was funded by "Key Laboratory of Intelligent Space TTC&O (Space Engineering University), Ministry of Education, NO. CYK2024-01-01", and Grant "50903020401".

Data Availability Statement: Data are contained within the article.

Conflicts of Interest: The authors declare no conflicts of interest.

Appendix A. Carrier-to-Noise Ratio of LEO NR Signal

This paper focuses on the navigation performance analysis of LEO NR signals. For navigation researchers, the carrier-to-noise ratio (CNR) is often used in navigation performance analysis. Next, we derive the relationship between LEO satellite transmitting power and CNR at the receiver. Assuming the transmitting power of the LEO satellite is P_{TX}, the received power P_{RX} after path loss (PL) attenuation is

$$P_{\mathrm{RX}} = P_{\mathrm{TX}} - PL \tag{A1}$$

where the unit of P_{TX} and P_{RX} is dBm.

Only the reference signal is used for navigation service in opportunity navigation scenarios. The reference signal includes system information broadcasting and channel estimation signals, occupying a subset of NR signals. Let α denote the proportion of the reference signal in the OFDM's resource grid. Therefore, the navigation signal power C_{nav} in the receiver becomes

$$C_{\mathrm{nav}} = 10\log_{10}(\alpha) + P_{\mathrm{RX}} \tag{A2}$$

The resulting carrier-to-noise-density ratio is

$$\frac{C_{\mathrm{nav}}}{N_0} = P_{\mathrm{TX}} - PL + 10\log_{10}(\alpha) - N_0 \tag{A3}$$

where N_0 denotes the noise power spectral density ant its representative value is -203.98 dBW/Hz [2].

For a LEO satellite at 600 km altitude with reference signal ratio $\alpha = 19.44\%$, Equation (A3) simplifies to

$$\frac{C_{\text{nav}}}{N_0} \approx P_{\text{TX}} + 10.87 \tag{A4}$$

The transmitting power of LEO satellites ranges 40~57 dBm (10 W~500 W) [79,80]. Considering antenna gains (0~40 dBi), the equivalent isotropic radiated power (EIRP) exhibits significant variation. For brevity, this paper assumes $P_{\text{TX}} = 55$~65 dBm, yielding a navigation CNR of 65.87~75.87 $\text{dB} \cdot \text{Hz}$.

References

1. Langley, R.B.; Teunissen, P.J.G.; Montenbruck, O., Introduction to GNSS. In *Springer Handbook of Global Navigation Satellite Systems;* Springer International Publishing: Cham, Switzerland, 2017; pp. 3–23. [CrossRef]
2. Xie, G. *Principles of GPS and Receiver Design*; Electronic Industry Press: Beijing, China, 2017.
3. Schmidt, D.; Radke, K.; Camtepe, S.; Foo, E.; Ren, M. A Survey and Analysis of the GNSS Spoofing Threat and Countermeasures. *ACM Comput. Surv.* **2016**, *48*, 1–31. [CrossRef]
4. Reid, T.G.; Neish, A.M.; Walter, T.; Enge, P.K. Broadband LEO Constellations for Navigation. *NAVIGATION J. Inst. Navig.* **2018**, *65*, 205–220. [CrossRef]
5. Reid, T.G.; Neish, A.M.; Walter, T.F.; Enge, P.K. Leveraging commercial broadband leo constellations for navigating. In Proceedings of the 29th International Technical Meeting of the Satellite Division of the Institute of Navigation (Ion Gnss+ 2016), Portland, OR, USA, 12–16 September 2016; Volume 12, p. 2016.
6. Chen, X.; Wei, Q.; Zhan, Y.; Kuang, L. Performance analysis of the packet-based PNT service in NGSO broadband satellite communication systems. *China Commun.* **2023**, *20*, 247–259. [CrossRef]
7. Yang, Y.; Mao, Y.; Ren, X.; Jia, X.; Sun, B. Demand and key technology for a LEO constellation as augmentation of satellite navigation systems. *Satell. Navig.* **2024**, *5*, 11. [CrossRef]
8. Iannucci, P.A.; Humphreys, T.E. Economical Fused LEO GNSS. In Proceedings of the 2020 IEEE/ION Position, Location and Navigation Symposium, PLANS 2020, Portland, OR, USA, 20–23 April 2020; pp. 426–443. [CrossRef]
9. Reid, T.G.R.; Chan, B.; Goel, A.; Gunning, K.; Manning, B.; Martin, J.; Neish, A.; Perkins, A.; Tarantino, P. Satellite navigation for the age of autonomy. In Proceedings of the 2020 IEEE/ION Position, Location and Navigation Symposium (PLANS), Portland, OR, USA, 20–23 April 2020; pp. 342–352.
10. Lei, W.; Deren, L.; Ruizhi, C.; Wenju, F.; Xin, S.; Hao, J. Low Earth Orbiter (LEO) Navigation Augmentation: Opportunities and Challenges. *Chin. J. Eng. Sci.* **2020**, *22*, 144. [CrossRef]
11. Danchik, R.J. An overview of transit development. *Johns Hopkins APL Tech. Dig.* **1998**, *19*, 18–26.
12. Wang, L.; Chen, R.; Li, D.; Zhang, G.; Shen, X. Initial Assessment of the LEO Based Navigation Signal Augmentation System from Luojia-1A Satellite. *Sensors* **2018**, *18*, 3919. [CrossRef]
13. Li, W.; Yang, Q.; Du, X.; Li, M.; Zhao, Q.; Yang, L.; Qin, Y.; Chang, C.; Wang, Y.; Qin, G. LEO augmented precise point positioning using real observations from two CENTISPACE™ experimental satellites. *GPS Solut.* **2024**, *28*, 1–13. [CrossRef]
14. Miller, N.S.; Koza, J.T.; Morgan, S.C.; Martin, S.M.; Neish, A.; Grayson, R.; Reid, T. SNAP: A Xona Space Systems and GPS Software-Defined Receiver. In Proceedings of the 2023 IEEE/ION Position, Location and Navigation Symposium (PLANS), Monterey, CA, USA, 24–27 April 2023; pp. 897–904. [CrossRef]
15. Xiao, L.; Lei, Z. Analysis of iridium-augmented GPS positioning performance. *J. Eng.* **2019**, *2019*, 7139–7143. [CrossRef]
16. Prol, F.S.; Bhuiyan, M.Z.H.; Kaasalainen, S.; Lohan, E.S.; Praks, J.; Celikbilek, K.; Kuusniemi, H. Simulations of Dedicated LEO-PNT Systems for Precise Point Positioning: Methodology, Parameter Analysis, and Accuracy Evaluation. *IEEE Trans. Aerosp. Electron. Syst.* **2024**, *60*, 6499–6516. [CrossRef]
17. Stock, W.; Schwarz, R.T.; Hofmann, C.A.; Knopp, A. Survey On Opportunistic PNT With Signals From LEO Communication Satellites. *IEEE Commun. Surv. Tutor.* **2024**, *27*, 77–107. [CrossRef]
18. Wang, L.; Lü, Z.; Tang, X.; Zhang, K.; Wang, F. LEO-augmented GNSS based on communication navigation integrated signal. *Sensors* **2019**, *19*, 4700. [CrossRef] [PubMed]
19. Gutt, G.; Lawrence, D.; Cobb, S.; O'Connor, M. Recent PNT improvements and test results based on low earth orbit satellites. In Proceedings of the Annual Precise Time and Time Interval Systems and Applications Meeting, PTTI, Reston, VA, USA, 28 January–1 February 2018; pp. 72–79. [CrossRef]
20. Wei, Q.; Chen, X.; Jiang, C.; Huang, Z. Time-of-Arrival Estimation for Integrated Satellite Navigation and Communication Signals. *IEEE Trans. Wirel. Commun.* **2023**, *22*, 9867–9880. [CrossRef]
21. Xue, L.; Li, X.; Wu, W.; Dong, J. Multifunctional Signal Design for Measurement, Navigation and Communication Based on BOC and BPSK Modulation. *Remote Sens.* **2022**, *14*, 1653. [CrossRef]

22. Huang, X.; Zhao, X.; Zhu, X.; Ou, G. *MC-BOC: A New Interoperable Modulation and Performance Analysis for BeiDou B1 Signal*; Springer: Singapore, 2018.
23. Li, X.; Rao, Z.; Xue, L. A Channel Compensation Technique Based on Frequency-Hopping Binary Offset Carrier Modulated Signal. *Remote Sens.* **2023**, *15*, 1849. [CrossRef]
24. Li, X.; Zeng, X.; Xue, L. Integrated Communication and Measurement System with BOC-Assisted OFDM. *Drones* **2022**, *7*, 14. [CrossRef]
25. Reid, T.G.R.; Walter, T.; Enge, P.K.; Lawrence, D.; Cobb, H.S.; Gutt, G.; O'Connor, M.; Whelan, D. Navigation from Low Earth Orbit: Part 1: Concept, Current Capability, and Future Promise. In *Position, Navigation, and Timing Technologies in the 21st Century*; John Wiley & Sons, Ltd.: Hoboken, NJ, USA, 2020; Chapter 43, pp. 1359–1379. [CrossRef]
26. Kassas, Z.Z.M. Navigation from Low-Earth Orbit: Part 2: Models, Implementation, and Performance. In *Position, Navigation, and Timing Technologies in the 21st Century*; John Wiley & Sons, Ltd.: Hoboken, NJ, USA, 2020; Chapter 43, pp. 1381–1412. [CrossRef]
27. Singh, U.K.; Shankar, M.R.B.; Ottersten, B. Opportunistic localization using LEO signals. In Proceedings of the 2022 56th Asilomar Conference on Signals, Systems, and Computers, Pacific Grove, CA, USA, 31 October–2 November 2022; pp. 894–899.
28. Cho, Y.S. *MIMO-OFDM Wireless Communications with MATLAB*; Wiley Publishing: Hoboken, NJ, USA, 2010.
29. Humphreys, T.E.; Iannucci, P.A.; Komodromos, Z.M.; Graff, A.M. Signal Structure of the Starlink Ku-Band Downlink. *IEEE Trans. Aerosp. Electron. Syst.* **2023**, *59*, 6016–6030. [CrossRef]
30. Neinavaie, M.; Kassas, Z.M. Unveiling Starlink LEO Satellite OFDM-Like Signal Structure Enabling Precise Positioning. *IEEE Trans. Aerosp. Electron. Syst.* **2024**, *60*, 2486–2489. [CrossRef]
31. ETSI. 5G NR Physical Channels and Modulation (3GPP TS 38.211 Version 15.2.0 Release 15). 2018. Available online: https://shop.standards.ie/en-ie/standards/etsi-ts-138-211-v15-2-0-2018-07--1399445_saig_etsi_etsi_3585366/ (accessed on 18 June 2025).
32. Tataria, H.; Shafi, M.; Molisch, A.F.; Dohler, M.; Sjoland, H.; Tufvesson, F. 6G Wireless Systems: Vision, Requirements, Challenges, Insights, and Opportunities. *Proc. IEEE* **2021**, *109*, 1166–1199. [CrossRef]
33. Dogra, A.; Jha, R.K.; Jain, S. A Survey on beyond 5G Network with the Advent of 6G: Architecture and Emerging Technologies. *IEEE Access* **2021**, *9*, 67512–67547. [CrossRef]
34. Shahid, H.; Amatetti, C.; Campana, R.; Tong, S.; Panaitopol, D.; Vanelli-Coralli, A.; Mohamed, A.; Zhang, C.; Khalifa, E.; Medeiros, E.; et al. Emerging Advancements in 6G NTN Radio Access Technologies: An Overview. In Proceedings of the 2024 Joint European Conference on Networks and Communications and 6G Summit, EuCNC/6G Summit 2024, Antwerp, Belgium, 3–6 June 2024; pp. 593–598. [CrossRef]
35. Hu, B.; Li, X.; Xue, L. A Pilot-Based Integration Method of Ranging and LS Channel Estimation for OFDM Systems. *Drones* **2022**, *6*, 400. [CrossRef]
36. Abdallah, A.; Khalife, J.; Kassas, Z.M. Exploiting On-Demand 5G Downlink Signals for Opportunistic Navigation. *IEEE Signal Process. Lett.* **2023**, *30*, 389–393. [CrossRef]
37. Shamaei, K.; Kassas, Z.M. Receiver Design and Time of Arrival Estimation for Opportunistic Localization with 5G Signals. *IEEE Trans. Wirel. Commun.* **2021**, *20*, 4716–4731. [CrossRef]
38. Neinavaie, M.; Kassas, Z.M. Cognitive Sensing and Navigation With Unknown OFDM Signals With Application to Terrestrial 5G and Starlink LEO Satellites. *IEEE J. Sel. Areas Commun.* **2024**, *42*, 146–160. [CrossRef]
39. Kassas, Z.M.; Kozhaya, S.; Kanj, H.; Saroufim, J.; Hayek, S.W.; Neinavaie, M.; Khairallah, N.; Khalife, J. Navigation with Multi-Constellation LEO Satellite Signals of Opportunity: Starlink, OneWeb, Orbcomm, and Iridium. In Proceedings of the 2023 IEEE/ION Position, Location and Navigation Symposium, PLANS 2023, Monterey, CA, USA, 24–27 April 2023; pp. 338–343. [CrossRef]
40. Neinavaie, M.; Khalife, J.; Kassas, Z.M. Acquisition, Doppler Tracking, and Positioning with Starlink LEO Satellites: First Results. *IEEE Trans. Aerosp. Electron. Syst.* **2022**, *58*, 2606–2610. [CrossRef]
41. Khalife, J.; Neinavaie, M.; Kassas, Z.Z. The First Carrier Phase Tracking and Positioning Results With Starlink LEO Satellite Signals. *IEEE Trans. Aerosp. Electron. Syst.* **2022**, *58*, 1487–1491. [CrossRef]
42. Saroufim, J.; Hayek, S.W.; Kassas, Z.M. Simultaneous LEO satellite tracking and differential LEO-aided IMU navigation. In Proceedings of the 2023 IEEE/ION Position, Location and Navigation Symposium (PLANS), Monterey, CA, USA, 24–27 April 2023; pp. 179–188.
43. McLemore, B.; Psiaki, M.L. Navigation using Doppler shift from LEO constellations and INS data. *IEEE Trans. Aerosp. Electron. Syst.* **2022**, *58*, 4295–4314. [CrossRef]
44. Yang, C.; Zang, B.; Gu, B.; Zhang, L.; Dai, C.; Long, L.; Zhang, Z.; Ding, L.; Ji, H. Doppler positioning of dynamic targets with unknown LEO satellite signals. *Electronics* **2023**, *12*, 2392. [CrossRef]
45. Kassas, Z.M. Navigation with cellular signals of opportunity. In *Position, Navigation, and Timing Technologies in the 21st Century*; Wiley: Hoboken, NJ, USA, 2020.
46. Kassas, Z.M.; Khairallah, N.; Kozhaya, S. Ad Astra: Simultaneous Tracking and Navigation With Megaconstellation LEO Satellites. *IEEE Aerosp. Electron. Syst. Mag.* **2024**, *39*, 46–71. [CrossRef]

47. Neinavaie, M.; Khalife, J.; Kassas, Z.M. Cognitive Opportunistic Navigation in Private Networks with 5G Signals and beyond. *IEEE J. Sel. Top. Signal Process.* **2022**, *16*, 129–143. [CrossRef]
48. Jardak, N.; Jault, Q. The potential of LEO satellite-based opportunistic navigation for high dynamic applications. *Sensors* **2022**, *22*, 2541. [CrossRef] [PubMed]
49. Khairallah, N.; Kassas, Z.M. Ephemeris Tracking and Error Propagation Analysis of LEO Satellites With Application to Opportunistic Navigation. *IEEE Trans. Aerosp. Electron. Syst.* **2024**, *60*, 1242–1259. [CrossRef]
50. Wang, J.; Zhang, Y.; Zhao, S.; He, J.; Shen, Y.; Jiang, X. A Survey on Authentication in Satellite Internet. *J. Netw. Netw. Appl.* **2022**, *2*, 183–194. [CrossRef]
51. Iannucci, P.A.; Humphreys, T.E. Fused Low-Earth-Orbit GNSS. *IEEE Trans. Aerosp. Electron. Syst.* **2024**, *60*, 3730–3749. [CrossRef]
52. Shamaei, K.; Kassas, Z.M. A Joint TOA and DOA Acquisition and Tracking Approach for Positioning with LTE Signals. *IEEE Trans. Signal Process.* **2021**, *69*, 2689–2705. [CrossRef]
53. Yang, C.; Arizabaleta-Diez, M.; Weitkemper, P.; Pany, T. An Experimental Analysis of Cyclic and Reference Signals of 4G LTE for TOA Estimation and Positioning in Mobile Fading Environments. *IEEE Aerosp. Electron. Syst. Mag.* **2022**, *37*, 16–41. [CrossRef]
54. Yang, C.; Pany, T.; Weitkemper, P. Effect of Antenna Ports on TOA Estimation with 4G LTE Signals in Urban Mobile Environments. In Proceedings of the 33rd International Technical Meeting of the Satellite Division of the Institute of Navigation, ION GNSS+ 2020, Online, 22–25 September 2020; pp. 2166–2181. [CrossRef]
55. Wang, P.; Morton, Y.J. Impact Analysis of Intercell Interference in Cellular Networks for Navigation Applications. *IEEE Trans. Aerosp. Electron. Syst.* **2023**, *59*, 685–694. [CrossRef]
56. Liu, Z.; Chen, L.; Zhou, X.; Shen, N.; Chen, R. Multipath tracking with LTE signals for accurate TOA estimation in the application of indoor positioning. *Geo-Spat. Inf. Sci.* **2023**, *26*, 31–43. [CrossRef]
57. Wang, P.; Morton, Y.J. Multipath Estimating Delay Lock Loop for LTE Signal TOA Estimation in Indoor and Urban Environments. *IEEE Trans. Wirel. Commun.* **2020**, *19*, 5518–5530. [CrossRef]
58. Abdallah, A.; Kassas, Z. Multipath Mitigation via Synthetic Aperture Beamforming for Indoor and Deep Urban Navigation. *IEEE Trans. Veh. Technol.* **2021**, *70*, 8838–8853. [CrossRef]
59. Whiton, R.; Chen, J.; Johansson, T.; Tufvesson, F. Urban Navigation with LTE using a Large Antenna Array and Machine Learning. In Proceedings of the IEEE Vehicular Technology Conference, Helsinki, Finland, 19–22 June 2022; pp. 1–5. https://doi.org/10.1109/VTC2022-Spring54318.2022.9860844.
60. Khalife, J.; Kassas, Z.M. Differential Framework for Submeter-Accurate Vehicular Navigation With Cellular Signals. *IEEE Trans. Intell. Veh.* **2023**, *8*, 732–744. [CrossRef]
61. Abdallah, A.A.; Kassas, Z.M. UAV navigation with 5G carrier phase measurements. In Proceedings of the 34th International Technical Meeting of the Satellite Division of the Institute of Navigation, ION GNSS+ 2021, Online, 20–24 September 2021; pp. 3294–3306. [CrossRef]
62. Xhafa, A.; Del Peral-Rosado, J.A.; López-Salcedo, J.A.; Seco-Granados, G. Evaluation of 5g positioning performance based on utdoa, aoa and base-station selective exclusion. *Sensors* **2022**, *22*, 101. [CrossRef]
63. Koivisto, M.; Talvitie, J.; Rastorgueva-Foi, E.; Lu, Y.; Valkama, M. Channel Parameter Estimation and TX Positioning With Multi-Beam Fusion in 5G mmWave Networks. *IEEE Trans. Wirel. Commun.* **2022**, *21*, 3192–3207. [CrossRef]
64. Lapin, I.; Seco-Granados, G.; Renaudin, O.; Zanier, F.; Ries, L. Joint Delay and Phase Discriminator Based on ESPRIT for 5G NR Positioning. *IEEE Access* **2021**, *9*, 126550–126563. [CrossRef]
65. Wen, F.; Kulmer, J.; Witrisal, K.; Wymeersch, H. 5G Positioning and Mapping with Diffuse Multipath. *IEEE Trans. Wirel. Commun.* **2021**, *20*, 1164–1174. [CrossRef]
66. Gante, J.; Sousa, L.; Falcao, G. Dethroning GPS: Low-Power Accurate 5G Positioning Systems Using Machine Learning. *IEEE J. Emerg. Sel. Top. Circuits Syst.* **2020**, *10*, 240–252. [CrossRef]
67. Liu, Z.; Chen, L.; Zhou, X.; Jiao, Z.; Guo, G.; Chen, R. Machine Learning for Time-of-Arrival Estimation With 5G Signals in Indoor Positioning. *IEEE Internet Things J.* **2023**, *10*, 9782–9795. [CrossRef]
68. Liu, Z.; Chen, L.; Jiao, Z.; Lu, X.; Ruan, Y.; Chen, R. MLLoc: Machine Learning Location System Based on RAIM in Mobile Networks. *IEEE Sens. J.* **2024**, *24*, 16953–16960. [CrossRef]
69. Khalife, J.; Neinavaie, M.; Kassas, Z.M. Navigation with differential carrier phase measurements from megaconstellation LEO satellites. In Proceedings of the 2020 IEEE/ION Position, Location and Navigation Symposium (PLANS), Portland, OR, USA, 20–23 April 2020; pp. 1393–1404.
70. Kozhaya, S.; Kassas, Z.M. On the Fundamental Tracking Performance and Design Considerations of Radio Navigation. *IEEE J. Sel. Areas Commun.* **2024**, *42*, 2395–2409. [CrossRef]
71. Zhao, C.; Qin, H.; Li, Z. Doppler Measurements from Multiconstellations in Opportunistic Navigation. *IEEE Trans. Instrum. Meas.* **2022**, *71*, 1–9. [CrossRef]

72. Cassel, R.S.; Scherer, D.R.; Wilburne, D.R.; Hirschauer, J.E.; Burke, J.H. Impact of improved oscillator stability on LEO-based satellite navigation. In Proceedings of the 2022 International Technical Meeting of the Institute of Navigation, Long Beach, CA, USA, 25–27 January 2022; pp. 893–905.
73. Zhao, C.; Qin, H.; Wu, N.; Wang, D. Analysis of Baseline Impact on Differential Doppler Positioning and Performance Improvement Method for LEO Opportunistic Navigation. *IEEE Trans. Instrum. Meas.* **2023**, *72*, 1–10. [CrossRef]
74. Yao, Z.; Lu, M. *Next-Generation GNSS Signal Design: Theories, Principles and Technologies*; Springer: Berlin/Heidelberg, Germany, 2021.
75. Xie, G. *Principles of GNSS: GPS, GLONASS, and Galileo*; Publishing House of Electronics Industry: Beijing, China, 2019; p. 443.
76. Betz, J.W.; Kolodziejski, K.R. Generalized theory of code tracking with an early-late discriminator part II: Noncoherent Processing and Numerical Results. *IEEE Trans. Aerosp. Electron. Syst.* **2009**, *45*, 1538. [CrossRef]
77. Celestrak. Starlink TLE Data. Available online: https://celestrak.org/NORAD/elements/gp.php?GROUP=starlink (accessed on 5 November 2024).
78. Groves, P. *Principles of GNSS, Inertial, and Multisensor Integrated Navigation Systems, 7nd ed.*; Artech: London, UK, 2013.
79. Ippolito, L.J., Jr. *Satellite Communications Systems Engineering: Atmospheric Effects, Satellite Link Design and System Performance*; John Wiley & Sons: Hoboken, NJ, USA, 2017.
80. 3rd Generation Partnership Project; Technical Specification Group Radio Access Network; Study on Narrow-Band Internet of Things (NB-IoT)/enhanced Machine Type Communication (eMTC) support for Non-Terrestrial Networks (NTN) (Release 17). 3GPP Technical Report TR 36.763 V0.2.0, 3rd Generation Partnership Project (3GPP), 2021. Available online: https://portal.3gpp.org/desktopmodules/Specifications/SpecificationDetails.aspx?specificationId=3747 (accessed on 5 November 2024).

remote sensing

MDPI

Article

Robust and Adaptive Ambiguity Resolution Strategy in Continuous Time and Frequency Transfer

Kun Wu [1,2,3], Weijin Qin [1,2,3,*], Daqian Lv [4,5], Wenjun Wu [1,2,3], Pei Wei [1,2] and Xuhai Yang [1,2,3]

1 National Time Service Center, Chinese Academy of Sciences, Xi'an 710600, China; wukun@ntsc.ac.cn (K.W.); wuwj@ntsc.ac.cn (W.W.); weipei@ntsc.ac.cn (P.W.); yyang@ntsc.ac.cn (X.Y.)
2 Key Laboratory of Time Reference and Applications, Chinese Academy of Sciences, Xi'an 710600, China
3 University of Chinese Academy of Sciences, Beijing 100049, China
4 College of Electronic Engineering, National University of Defense Technology, Hefei 230037, China; daqian_lv@nudt.edu.cn
5 National Key Laboratory of Electromagnetic Space Security, Jiaxing 314033, China
* Correspondence: qwj@ntsc.ac.cn; Tel.: +86-029-83893326

Abstract: The integer precise point positioning (IPPP) technique significantly improves the accuracy of positioning and time and frequency transfer by restoring the integer nature of carrier-phase ambiguities. However, in practical applications, IPPP performance is often degraded by day-boundary discontinuities and instances of incorrect ambiguity resolution, which can compromise the reliability of time transfer. To address these challenges and enable continuous, robust, and stable IPPP time transfer, this study proposes an effective approach that utilizes narrow-lane ambiguities to absorb receiver clock jumps, combined with a robust sliding-window weighting strategy that fully exploits multi-epoch information. This method effectively mitigates day-boundary discontinuities and employs adaptive thresholding to enhance error detection and mitigate the impact of incorrect ambiguity resolution. Experimental results show that at an averaging time of 76,800 s, the frequency stabilities of GPS, Galileo, and BDS IPPP reach 4.838×10^{-16}, 4.707×10^{-16}, and 5.403×10^{-16}, respectively. In the simulation scenario, the carrier-phase residual under the IGIII scheme is 6.7 cm, whereas the robust sliding-window weighting method yields a lower residual of 5.2 cm, demonstrating improved performance. In the zero-baseline time link, GPS IPPP achieves stability at the 10^{-17} level. Compared to optical fiber time transfer, the GPS IPPP solution demonstrates superior long-term performance in differential analysis. For both short- and long-baseline links, IPPP consistently outperforms the PPP float solution and IGS final products. Specifically, at an averaging time of 307,200 s, IPPP improves average frequency stability by approximately 29.3% over PPP and 32.6% over the IGS final products.

Keywords: IPPP; time and frequency transfer; ambiguity resolution; day-boundary discontinuity; sliding window

1. Introduction

Time and frequency transfer technology plays a critical role in sustaining the global time scale based on International Atomic Time (TAI). Various techniques have been developed to achieve precise time and frequency transfer, including two-way satellite time and frequency transfer (TWSTFT), common-view (CV) and precise point positioning (PPP) techniques based on Global Navigation Satellite System (GNSS) satellites, and optical fiber link time and frequency transfer. Among these, PPP has emerged as a dominant method for time and frequency transfer due to its cost-effectiveness and capability to simultaneously determine parameters such as station coordinates and receiver clock offsets using both carrier-phase and pseudorange observations. However, in the standard PPP mathematical model, the ambiguity parameters absorb both uncalibrated satellite hardware delays and receiver hardware delays. Consequently, these ambiguities are estimated as float values. This can introduce additional errors into the estimated receiver clock offset, potentially degrading the accuracy of time and frequency transfer. To restore the integer nature of carrier-phase ambiguities, several methods of server-side products to support this process have been proposed, primarily categorized as the Uncalibrated Phase Delay (UPD) method (theoretically equivalent to the Observable-Specific Signal Bias (OSB) method [1–3]), the Integer Recovery Clock (IRC) or phase clock method, and the decoupled clocks method. In this study, we utilize post-processed OSB products provided by the International GNSS Service (IGS) analysis center to recover the integer properties of the ambiguities. This enables high-precision positioning and time and frequency transfer with integer ambiguity resolution, known as integer PPP (IPPP). Experiments employing IPPP time and frequency transfer techniques with post-processed IPPP products have demonstrated the achievement of frequency transfer stability at the level of 1×10^{-16} at averaging times of 5 days [4,5]. While time and frequency transfer utilizing integer ambiguity resolution already attains high accuracy, aspects concerning its continuity (especially during signal outages or high ionospheric activity) and robustness (resistance to biases and errors under diverse conditions) warrant further consideration and improvement.

Currently, the IGS analysis center generates post-processed precise products based on single-day processing. During this process, day-boundary discontinuities manifest in the estimated receiver clock offsets due to variations in pseudorange biases [6]. To mitigate these discontinuities, Yao and Levine (2013) [7] proposed the shift-stacking (Rinex-Shift) method. Similarly, Dach et al. (2006b) [8] and Guo et al. (2025) [9] employed a multi-day batch processing strategy to attenuate the impact of day-boundary jumps, thereby improving the continuity and stability of PPP time and frequency transfer. Petit et al. (2015b) [10] implemented a combination of bridging and extrapolation methods to achieve continuous IPPP time and frequency transfer. However, these processing strategies tend to complicate the data processing workflow to a certain extent. Subsequently, Guo et al. (2025) [9] employed IPPP using multi-day batch-processed precise products to further reduce day-boundary discontinuities. This approach relies on specific server-side auxiliary products, which significantly limits its applicability. Nevertheless, no universally accepted solution exists to completely eliminate day-boundary discontinuity effects.

Within single-epoch processing, ambiguity parameters exhibit strong correlation with the receiver clock offset parameter. Incorrect integer ambiguity resolution directly propagates into the estimated receiver clock offset. Consequently, the time and frequency transfer link may exhibit some jumps or outliers, leading to degraded stability [11]. Furthermore, post-processed precise products are inevitably subject to anomalies or periods of degraded accuracy. If these products contain biases due to system failures or anomalies [12], these errors can be transferred into the estimated ambiguities or receiver clock offsets. Therefore, an IPPP time and frequency transfer solution necessitates a robust IPPP processing strategy capable of detecting and mitigating false ambiguity fixes. This is essential to maintain the continuity and stability of the time and frequency transfer. In conventional approaches, a float narrow-lane ambiguity is considered successfully fixed when its residual is smaller than 0.25 cycles [13]. Nonetheless, abnormal values may still occur within this fixed threshold. To address this issue, the IGIII scheme was proposed, in which the constraint variance is appropriately inflated based on the magnitude of the residuals [14]. However, it relies solely on single-epoch information, which may reduce its ability to detect localized anomalies. Moreover, in most IPPP processing strategies, the threshold used to determine the correctness of ambiguity fixing, such as the ratio test, is typically chosen empirically and cannot readily adapt to dynamic environmental changes [15,16]. Thus, static thresholds are not necessarily optimal for all processing scenarios, and adaptive thresholding may provide enhanced effectiveness in detecting and mitigating errors.

In this study, we implement IPPP time and frequency transfer utilizing OSB products, which provide greater flexibility in handling multi-frequency signal combinations. To address key challenges in operational IPPP links, we propose a novel strategy that effectively mitigates the classical day-boundary discontinuity problem in time and frequency transfer links. This strategy enables mitigation of day-boundary discontinuities through relatively simple yet effective narrow-lane ambiguity compensation using products from different IGS analysis centers. It thereby eliminates the dependency on specific post-processed products and avoids complex data processing procedures. In addition, we design a robust ambiguity-resolution constraint scheme to enhance the stability of time and frequency transfer results. This method employs robust error detection metrics for anomaly identification. By integrating a sliding-window mechanism that incorporates historical data, it achieves enhanced sensitivity to localized incorrect ambiguity fixes. The manuscript is structured as follows: First, we detail the methodology for IPPP time and frequency transfer using OSB products. This is followed by a comprehensive description of the processing strategy for day-boundary discontinuity mitigation and the implementation of the robust constraint scheme. Subsequently, we present the experimental design and evaluate the performance of time and frequency transfer links across multiple stations. Finally, we present the principal conclusions of this work.

2. Method

2.1. Materials and IPPP Approach

Multi-frequency GNSS Observation Model

For GNSS satellite s and receiver r, the multi-frequency pseudorange and carrier-phase observation equations can be expressed as

$$\begin{cases} P_{r,n}^{s} = \rho_r^s + cdt_r - cdt^s + \gamma_n \cdot I_{r,1}^{s} + m_{r,w}^{s} Z_{r,w} + (b_{r,n} - b_n^s) + e_{r,n}^{s} \\ L_{r,n}^{s} = \rho_r^s + cdt_r - cdt^s - \gamma_n \cdot I_{r,1}^{s} + m_{r,w}^{s} Z_{r,w} + \lambda_n (N_{r,n}^{s} + B_{r,n} - B_n^s) + \varepsilon_{r,n}^{s} \end{cases} \tag{1}$$

where P and L denote raw pseudorange and carrier-phase observations, respectively; n represents the frequency index; ρ_r^s is the true geometric range between receiver and satellite; c is the speed of light in a vacuum; dt_r and dt^s are receiver and satellite clock offsets; $Z_{r,w}$ is the zenith wet tropospheric delay; $m_{r,w}^{s}$ is the mapping function for zenith wet delay tropospheric delay; $I_{r,1}^{s}$ is the slant ionospheric delay at the first frequency; $\gamma_n = \lambda_n^2/\lambda_1^2$ is the ionospheric scaling factor; λ_n is the signal wavelength (m); $N_{r,n}^{s}$ denotes the integer carrier-phase ambiguity in units of cycles; $b_{r,n}/b_n^s$ denote receiver/satellite pseudorange hardware delays; likely, $B_{r,n}/B_n^s$ denote receiver/satellite phase hardware delays; $e_{r,n}^{s}/\varepsilon_{r,n}^{s}$ represent pseudorange/carrier-phase observation noise, including multipath effects.

In this study, the following error sources are corrected using established models and thus omitted from the equations: satellite/receiver antenna phase center offsets (PCOs) and variations (PCVs), relativistic effects, solid earth tides, ocean tidal loading, and phase wind-up. For first-order ionospheric delay elimination in dual-frequency scenarios (frequencies 1 and 2), the ionosphere-free combination is applied:

$$\begin{cases} P_{\mathrm{IF}}^{s} = \alpha_{\mathrm{IF}} P_1^s + \beta_{\mathrm{IF}} P_2^s = \rho_r^s + cdt_r - cdt^s + m_{r,w}^{s} Z_{r,w} + (b_{r,\mathrm{IF}} - b_{\mathrm{IF}}^s) + e_{r,\mathrm{IF}}^{s} \\ L_{\mathrm{IF}}^{s} = \alpha_{\mathrm{IF}} L_1^s + \beta_{\mathrm{IF}} L_2^s = \rho_r^s + cdt_r - cdt^s + m_{r,w}^{s} Z_{r,w} + \lambda_{\mathrm{IF}} (N_{\mathrm{IF}}^{s} + B_{r,\mathrm{IF}} - B_{\mathrm{IF}}^s) + \varepsilon_{r,\mathrm{IF}}^{s} \end{cases} \tag{2}$$

$$\begin{cases} \alpha_{\mathrm{IF}} = \frac{f_1^2}{f_1^2 - f_2^2}, \beta_{\mathrm{IF}} = -\frac{f_2^2}{f_1^2 - f_2^2} \\ b_{r,\mathrm{IF}} = \alpha_{\mathrm{IF}} b_{r,1} + \beta_{\mathrm{IF}} b_{r,2}, b_{\mathrm{IF}}^s = \alpha_{\mathrm{IF}} b_1^s + \beta_{\mathrm{IF}} b_2^s \\ B_{r,\mathrm{IF}} = (\alpha_{\mathrm{IF}} \lambda_1 B_{r,1} + \beta_{\mathrm{IF}} \lambda_2 B_{r,2}) / \lambda_{\mathrm{IF}}, B_{\mathrm{IF}}^s = (\alpha_{\mathrm{IF}} \lambda_i B_i^s + \beta_{\mathrm{IF}} \lambda_2 B_2^s) / \lambda_{\mathrm{IF}} \\ N_{r,\mathrm{IF}}^s = \left(\alpha_{\mathrm{IF}} \lambda_1 N_{r,1}^s + \beta_{\mathrm{IF}} \lambda_2 N_{r,2}^s\right) / \lambda_{\mathrm{IF}} \end{cases} \tag{3}$$

where IF denotes the ionosphere-free combination; α_{IF} and β_{IF} are the corresponding combination coefficients. In the pseudorange observation model, $b_{r,\mathrm{IF}}$ and b_{IF}^s represent the combined hardware delays at the receiver and satellite ends, respectively. Similarly, in the carrier-phase observation model, $B_{r,\mathrm{IF}}$ and B_{IF}^s $\varepsilon_{r,\mathrm{IF}}^s$ refer to the receiver and satellite phase hardware delays, respectively. The terms $e_{r,\mathrm{IF}}^s$ and $\varepsilon_{r,\mathrm{IF}}^s$ represent the measurement noise and multipath effects in the ionosphere-free pseudorange and phase observations. λ_{IF} is the wavelength of the ionosphere-free combination, and N_{IF}^s is the corresponding carrier-phase ambiguity.

To implement PPP with ambiguity resolution, we adopt a cascaded ambiguity fixing strategy, introducing the wide-lane and narrow-lane ambiguities N_{WL} and N_{NL} (where N_1 is typically regarded as N_{NL} for simplicity). These ambiguities are jointly estimated using the Melbourne–Wübbena (MW) combination [17] and the ionosphere-free combination, as defined by the following equations:

$$\begin{aligned} L_{r,\mathrm{MW}}^s &= \lambda_{\mathrm{WL}}(N_{\mathrm{WL}} + \Delta_{\mathrm{WL}}) \\ &= \left(\frac{L_{r,1}^s + z_{r,1}^s}{\lambda_1} - \frac{L_{r,2}^s + z_{r,2}^s}{\lambda_2}\right) - \frac{f_1\left(P_{r,1}^s + z_{r,1}^s\right) + f_2\left(P_{r,2}^s + z_{r,2}^s\right)}{f_1 + f_2} \end{aligned} \tag{4}$$

$$\begin{aligned} &L_{\mathrm{IF}}^s = \rho^s + cdt_r - cdt^s + + m_{r,w}^s Z_{r,w} + \lambda_{\mathrm{NL}}\left(N_1^s + \frac{\lambda_{\mathrm{WL}}}{\lambda_2} N_{\mathrm{WL}}^s\right) + \lambda_{\mathrm{IF}}\left(B_{r,\mathrm{IF}} - B_{\mathrm{IF}}^s\right) + \varepsilon_{r,\mathrm{IF}}^s \\ &\lambda_{\mathrm{WL}} = \frac{c}{f_1 - f_2}, \ \lambda_{\mathrm{NL}} = \frac{c}{f_1 + f_2} \end{aligned} \tag{5}$$

where λ_{WL} and λ_{NL} represent the wavelengths of the wide-lane and narrow-lane combinations, respectively. Δ_{WL} denotes the fractional component of the wide-lane ambiguity, corresponding to the non-integer portion less than one full cycle. Equation (5) describes the transformation relationship between the ionosphere-free carrier-phase observation model and the wide-lane and narrow-lane ambiguity parameters. $z_{r,1}^s$ and $z_{r,2}^s$ denote the antenna phase center offsets (PCOs) for different frequencies, as defined in [18]. It is particularly important to note that PCO errors can be frequency-dependent. When the Z-component of dual-frequency PCOs is inconsistent, the discrepancy may reach several decimeters, potentially leading to incorrect ambiguity fixes [19].

$$N_{r,\mathrm{NL}}^s = \frac{f_1 + f_2}{f_1} N_{r,\mathrm{IF}}^s - \frac{f_2}{f_1 - f_2} N_{r,\mathrm{WL}}^s \tag{6}$$

Due to the relatively long wavelength of the wide-lane combination (up to several tens of centimeters), it is less sensitive to errors and can be reliably fixed by rounding. Subsequently, based on the transformation relationship among N_{IF}, N_{WL}, and N_{NL}, the narrow-lane ambiguity can be resolved using the LAMBDA method.

IPPP Implementation Utilizing OSB Products

From the above equations, it is evident that the critical step in restoring float ambiguity parameters to integer solutions lies in separating the integer component of ambiguities from the fractional hardware biases originating from satellites and receivers. To achieve this, various analysis centers provide distinct products to eliminate satellite hardware delays from integer ambiguity estimates. When combined with between-satellite single differencing, these products further mitigate receiver hardware delays. Consequently, the fractional components can be isolated from ambiguity parameters, enabling integer ambiguity resolution.

In this study, we utilize the post-processed OSB products provided by the Centre National d'Études Spatiales (CNES) analysis center (note that this differs from the traditional IRC-based fixing method). The IPPP fixing procedure presented herein is generalizable and can also be applied to clock and OSB products from other IGS analysis centers. The fundamental workflow for implementing OSB products alongside other post-processed precise products is illustrated in Figure 1.

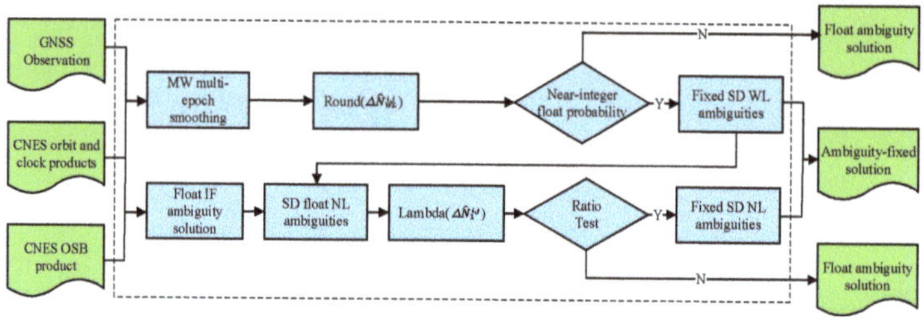

Figure 1. Flowchart of the ambiguity resolution strategy based on CNES OSB products using a dual-frequency ionosphere-free combination. Green boxes denote data or products while blue boxes denote processes.

To recover the integer nature of the ambiguities, a cascaded ambiguity resolution strategy is adopted, as proposed by Ge et al. [20]. In this approach, the ionosphere-free ambiguity is decomposed into wide-lane and narrow-lane components, and all three ambiguity terms are subsequently estimated. The relationship among these components is described by Equation (6).

$$\Delta\tilde{N}^{i,j}_{\mathrm{IF}} = \tilde{N}^{i}_{\mathrm{IF}} - \tilde{N}^{j}_{\mathrm{IF}} \tag{7}$$

The first step is to estimate the float ionosphere-free ambiguities, which forms the foundation for implementing IPPP. In this study, the satellite with the highest elevation angle is selected as the reference satellite. Between-satellite single differences (SDs) are then formed between the reference satellite and the remaining satellites involved in the ambiguity resolution. Using an extended Kalman filter (EKF), the float single-differenced ionosphere-free ambiguities are denoted as $\Delta\tilde{N}^{i,j}_{\mathrm{IF}}$, where i refers to the satellite under consideration and j denotes the reference satellite.

The next step is to fix the wide-lane ambiguities. The float wide-lane ambiguities can be derived from the MW combination. The MW combined observation equation is given as follows:

$$\Delta\hat{N}^{i,j}_{\mathrm{WL}} = \lambda_{WL}\Delta N^{i,j}_{\mathrm{WL}} + \lambda_{\mathrm{WL}}\Delta u^{i,j}_{\mathrm{WL}} + \lambda_{\mathrm{WL}}\Delta O^{i,j}_{\mathrm{r,WL}} + \lambda_{\mathrm{WL}}\Delta\varepsilon^{i,j}_{\mathrm{WL}} \tag{8}$$

$\Delta O^{i,j}_{\mathrm{r,WL}}$ represents the phase center offset (PCO) correction. When the PCO corrections for the first and second frequencies are equal, $\Delta O^{i,j}_{\mathrm{r,WL}}$ becomes zero. $\Delta u^{i,j}_{\mathrm{WL}}$ denotes satellite and receiver hardware bias. After applying the OSB corrections and compensating for the vertical PCO error, the satellite-side hardware bias and PCO-related bias in the wide-lane ambiguity are effectively removed—this is referred to as the Wide-lane Satellite Bias (WSB). Subsequently, by forming between-satellite single differences, the receiver-side hardware biases—known as the Wide-lane Receiver Bias (WRB)—can also be eliminated, thereby restoring the integer nature of the wide-lane ambiguity.

$$\Delta\hat{N}^{i,j}_{\mathrm{WL}} = \left(\hat{N}^{i}_{1} - \hat{N}^{i}_{2}\right) - \left(\hat{N}^{j}_{1} - \hat{N}^{j}_{2}\right) \tag{9}$$

The MW combination has a relatively long wavelength (approximately 86 cm for GPS), and Equation (9) possesses inherent integer characteristics. After multi-epoch smoothing, the wide-lane ambiguities can be directly fixed by applying rounding.

Finally, the narrow-lane ambiguities are fixed. Due to the short wavelength of $\hat{N}_1^i$ (approximately 19 cm for GPS), measurement noise and multipath effects can cause significant deviations between $\hat{N}_1$ and the true ambiguity N_1. Moreover, the strong correlation among ambiguities associated with different satellites must be taken into account. Therefore, the LAMBDA method is employed to reliably fix N_1. Once fixing N_1, it is implemented as a virtual observation to constrain the Kalman filter state, yielding the ambiguity-fixed solution.

2.2. Fundamentals of Time and Frequency Transfer

The receiver clock offset dt_r estimated in IPPP represents the time difference between the local receiver clock and the time scale t_{ref} of the precise satellite products. It is important to note that the receiver clock in this context refers specifically to the phase clock, as it is estimated solely from carrier-phase observations. This approach avoids contamination from the pseudorange-derived clock estimates during the ambiguity resolution process [10].

Since the time scale t_{ref} provided by the precise satellite products is consistent across stations, the local time difference between any two stations A and B can be derived by a simple differencing of their respective clock offsets. The resulting time difference between the two local receiver clocks is given by

$$\Delta t = dt_A - dt_B = (t_{\text{A}} - t_{\text{ref}}) - (t_{\text{B}} - t_{\text{ref}}) \tag{10}$$

where t_{A} and t_{B} are the local time at stations A and B, respectively, and t_{ref} is the time scale of the satellite products.

2.3. Day-Boundary Discontinuity Compensation

Currently, the majority of precise post-processed products provided by the IGS analysis center are generated in daily batches. On the one hand, satellite time scale discontinuities may occur at daily boundaries in the precise satellite products. On the other hand, during IPPP processing, strong correlations between parameters may cause the receiver clock estimates to be affected by hardware delays embedded in the ambiguity parameters—particularly wide-lane and narrow-lane delays—resulting in polluted clock solutions. Considering that these uncertainties may vary across time and location, their propagation into time and frequency transfer links between stations can lead to discontinuities at day boundaries.

According to Equation (5), a quantitative relationship between the receiver clock offset parameter and the narrow-lane (NL) ambiguity can be derived [21]. For example, a single cycle jump in the NL ambiguity (assuming GPS L1 and L2 frequencies of 1575.42 MHz and 1227.60 MHz, respectively, yielding an NL wavelength of approximately 0.19 m) would result in a corresponding jump in the clock offset of approximately 0.353 ns. Therefore, NL ambiguity cycles can be exploited to absorb clock discontinuities and repair the day-boundary discontinuities in the time link.

To minimize the impact of such day-boundary discontinuities, the correction can be performed either at the PPP solution level or during time link formation [21]. In this study, we propose a convenient and effective correction strategy at the link level to mitigate day-boundary jumps as much as possible. The proposed method consists of the following steps:

(I) Estimating the clock jump across the day boundary.

Since NL ambiguities may change between daily batches, such changes will inevitably introduce clock jumps equivalent to integer multiples of the NL wavelength. Based on this, standard processing using single-day products is first performed. After detecting a day-boundary discontinuity and verifying it over multiple epochs, the float-level clock jump can be estimated directly.

(II) Bridging the discontinuity in the time link.

Once the float jump in the clock offset has been determined, it is rounded to the nearest integer multiple of the NL cycle, based on the known conversion factor. Since we are dealing with between-satellite single-difference ambiguities, the compensation can be implemented in one of two ways: either by keeping the ambiguity of the reference satellite fixed and adjusting the ambiguity of the other satellite, or by fixing the ambiguity of the secondary satellite and applying the correction to the reference satellite. This procedure effectively bridges the cycle-induced clock jump and restores the continuity of the time link across the day boundary.

2.4. Robust Sliding-Window Weighted Strategy

During single-epoch integer ambiguity fixing, degraded signal quality, excessive multipath effects, or imperfections in post-processed precise products may introduce anomalies into the observations. These deviations are often absorbed by the fractional part of the float ambiguity estimates. When the fractional residual approaches 0.5 cycles, it may result in incorrect ambiguity fixes to a neighboring integer value. Due to the strong correlation between ambiguities and receiver clock parameters, such erroneous integer constraints can directly compromise the receiver clock estimation, potentially leading to instability in the time and frequency transfer link—manifesting as abrupt jumps or isolated outliers.

In conventional residual analysis for ambiguity fixing, the float residuals of narrow-lane ambiguities serve as an indicator of fixing quality. A commonly adopted threshold is 0.25 cycles, below which the fixed solution is considered reliable. However, under abnormal conditions, relying solely on single-epoch residuals may not provide sufficient sensitivity for detecting outliers.

To address this issue, we propose a robust ambiguity fixing constraint method that incorporates multi-epoch narrow-lane residual information, as shown in Figure 2. The key steps are as follows:

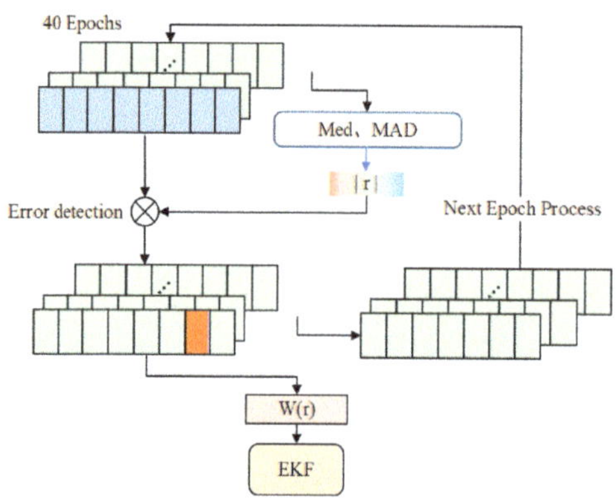

Figure 2. Structure of the robust sliding-window weighting algorithm. The blue blocks represent the narrow-lane residuals currently being processed; the green blocks indicate residuals classified as normal, while the orange blocks denote residuals identified as outliers.

(I) Construction of Detection Metric

To enhance sensitivity to local anomalies, we employ a sliding window to statistically evaluate the float residuals of narrow-lane ambiguities across multiple consecutive epochs for all satellites involved in the ambiguity resolution process. The selection of an appropriate sliding window size must strike a balance between estimation accuracy and sensitivity to errors [22], as well as between real-time responsiveness and the use of historical information. In the case of sudden ambiguity fixing errors, a small window is more responsive to transient anomalies and can effectively capture such events [23,24]. However, an overly small window may result in misidentification of normal residuals as outliers. Previous studies have shown that one hour of epoch data provides a sufficient number of points for anomaly detection [21]. In this study, the primary motivation for using a small window is to preserve finer details, which is particularly suitable for detecting abrupt changes while maintaining low latency and enabling timely identification of ambiguity fixing errors. Based on extensive experimental evaluations of different window lengths, a 40-epoch window was selected by considering both positioning accuracy and computational efficiency. From the residual set, the median and median absolute deviation (MAD) are computed to construct a robust error detection metric. Compared with the standard deviation, the MAD provides better resistance to outliers and is thus more suitable for robust detection in noisy datasets.

$$|r| = \frac{|res - med|}{MAD} \tag{11}$$

where *res* denotes the fractional part of the float narrow-lane ambiguity; *med* represents the median of the residual dataset; *MAD* is the median absolute deviation of the parameter dataset; and r is the test statistic used for error detection.

(II) Error Detection

An error detection procedure is applied to the set of narrow-lane ambiguity residuals. A dynamic threshold is used for the test, defined as $k = 1.345 \times \text{MAD}$, which corresponds to 95% Gaussian efficiency. If an outlier is detected, it will be down-weighted in the subsequent step, and the residual from the current epoch will be removed from the sliding window used to maintain the residual set. Otherwise, the PPP processing proceeds to the next epoch as usual.

(III) Adaptive Down-Weighting

For satellites identified with abnormal narrow-lane ambiguity behavior, a Huber-based down-weighting strategy is employed. The Huber function is used to compute the weight factor of each affected satellite, which is then applied to adjust the observation weight accordingly. This procedure reduces the influence of incorrect ambiguities on the estimation of receiver clock offset and ensures robust PPP ambiguity resolution.

$$w(r) = \begin{cases} 1, |r| \leq k \\ \frac{k}{|r| * \alpha}, |r| > k \end{cases} \tag{12}$$

Here, $w(r)$ denotes the weighting factor constructed based on the error detection metric, k is the dynamic threshold, and α represents the variance inflation factor.

3. Data and Strategy

3.1. Dataset

To validate the effectiveness of the proposed method and evaluate the performance of IPPP time and frequency transfer, we collected 30 days of multi-frequency GNSS observations at a 30-s interval from six globally distributed IGS reference stations, covering day of year (DOY) 72–101, 2025 (corresponding to MJD 60746–60776). The corresponding post-processed precise products, including satellite orbits, clock offsets, and OSB corrections, were obtained from the CNES analysis center. All stations are equipped with high-performance atomic clocks. The geographical distribution of the stations is illustrated in Figure 3, and the hardware configuration of each station is summarized in Table 1. The baseline distances for the selected time links are listed in Table 2.

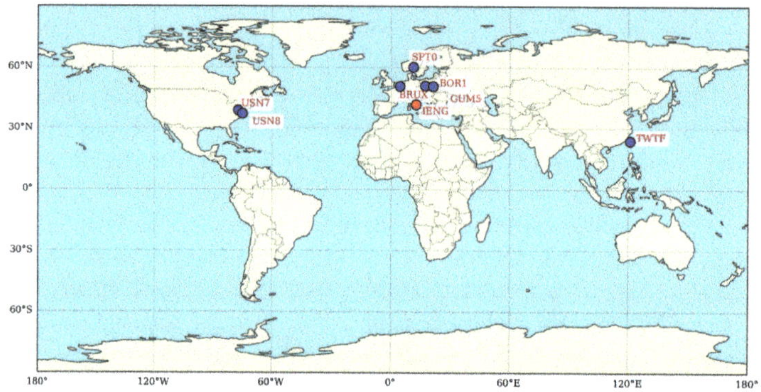

Figure 3. Spatial distribution of the experimental stations. IENG (red dot) serves as the center, linked with the other stations (blue dots) to establish time links. An optical fiber link is used between BOR1 and GUM5.

Table 1. The hardware details of the stations.

Station	Receiver Type	Antenna Type	Atomic Clock Type
IENG	SEPT POLARX5TR-5.4.0	SEPCHOKE_MC-NONE	EXTERNAL H-MASER
BRUX	SEPT POLARX5TR-5.5.0	JAVRINGANT_DM-SCIS	EXTERNAL IMASER 3000
SPT0	SEPT POLARX5TR-5.5.0	TRM59800.00-OSOD	EXTERNAL H-MASER
TWTF	SEPT POLARX4TR-2.9.6	SEPCHOKE_B3E6-SPKE	EXTERNAL H-MASER
BOR1	TRIMBLE NETR9-5.45	TRM59800.00-NONE	EXTERNAL H-MASER
GUM5	TTS-5	NAX3G + C NONE	EXTERNAL H-MASER

Table 2. The distance of the time links.

Link	Type	Distance
BRUX-IENG	Short-baseline link	688 km
SPT0-IENG	Short-baseline link	1457 km
TWTF-IENG	Long-baseline link	9736 km
USN7-USN8	Zero-baseline link	0 km
BOR1-GUM5	Optical fiber link	267 km

For the zero-baseline experiment, two co-located stations in the United States, USN7 and USN8, were selected as the test sites. For the medium- and long-baseline experiments, IENG was chosen as the central reference station, forming three time and frequency transfer links. The IGS final precise products were used as the reference. Additionally, two domestic Polish stations, BOR1 and GUM5, were selected. Although the two stations collect only GPS observations, an optical fiber time transfer link is available between them. Owing to its superior frequency stability compared to GNSS time links and the absence of day-boundary discontinuities [9], this optical link can serve as a benchmark for independently validating the performance of IPPP time and frequency transfer.

3.2. Processing Strategies

The dual-frequency ionosphere-free combination is formed based on GPS L1C and L2W observations, and all parameters are solved using single-epoch processing. IPPP employs a stepwise ambiguity fixing strategy, in which the wide-lane (WL) satellite-differenced ambiguities are sequentially rounded to integers and fixed. Once fixed, these ambiguities are used to generate virtual observations to constrain the PPP equations. Subsequently, the narrow-lane (NL) satellite-differenced ambiguities are fixed using the LAMBDA method and validated by the ratio test. The reference satellite is selected as the one with the highest elevation angle. A partial ambiguity resolution (PAR) strategy is applied by selecting ambiguities in descending order of satellite elevation. Upon successful NL ambiguity fixing, the receiver clock offset is estimated under the combined constraints of WL and NL ambiguities. Notably, to further reduce the probability of incorrect ambiguity fixes, a more rigorous fixing strategy is adopted: after obtaining a continuous and stable PPP solution, the Kalman filter states are updated only when the ambiguities are fixed correctly for eight consecutive epochs [1,25]. In all processing schemes, all datasets are processed in static mode, antenna phase center offsets and variations are corrected using ATX files, the receiver clock offset is modeled as white noise, and the receiver coordinates are estimated as constants. Further details on the error handling and modeling strategies can be found in Table 3.

Table 3. Details of the PPP processing strategies.

Item	Strategy
Observations	1 day pseudorange and carrier phase
Sampling rate of solutions	30s
Elevation cut off	7°
PCV and PCO	igs20_2350.atx
Tropospheric delay	Dry delay: Corrected by Saastamoinen model; Wet delay: Estimated as random walk
Ionospheric delay	Dual-frequency ionosphere-free combination correction
Relative effect	Applied
Phase-windup effect	Corrected
Positions	Estimated as daily constants
Receiver clock	Estimated as white noise
Tidal effects	Ocean tide file
Estimator	EKF

4. Results and Analysis

In this section, the availability of the IPPP algorithm based on OSB products is first validated. Subsequently, the results of the day-boundary discontinuity repair experiments and related simulations are presented. Finally, a comprehensive performance evaluation of the proposed IPPP time and frequency transfer scheme is conducted using multiple transfer links across various baselines. In the tests, the performance of GPS IPPP is compared with the GPS PPP float solution (to avoid contamination of phase clocks by pseudorange-related biases, carrier phase-only processing was employed in both the IPPP and PPP experimental schemes), optical fiber links, and the IGS final products employing the modified Allan deviation (MDEV) as the primary stability metric. Notably, to ensure consistency with the IGS final products, the experimental results were resampled to obtain a uniform interval of 300 s. Observations from all selected stations span from MJD 60,746 to 60,776.

4.1. Distribution of the WL and NL Ambiguity Fractional-Cycle Residuals

It is generally accepted that the NL residuals represent the quality of ambiguity fixing, and effective integer ambiguity resolution can only be achieved when these residuals are sufficiently small. In both WL and NL ambiguity resolution, the subset of satellites used for ambiguity fixing is initially selected based on residuals, with a threshold set at 0.25 cycles. Figures 4 and 5 illustrate the probability distribution of float WL and NL ambiguity residuals during the 30-day short baseline link (SPT0-IENG) experiment, showing a large concentration of residuals near zero to ensure reliable ambiguity resolution. Table 4 summarizes the ratios of WL and NL fractional residuals within 0.25 cycles. The statistical average was computed based on the measurements from both stations. Specifically, 97.90% of WL fractional residuals are within 0.25 cycles, and 87.84% are within 0.15 cycles; for NL fractional residuals, 99.70% fall within 0.25 cycles, and 97.83% within 0.10 cycles. Moreover, the standard deviations of both WL and NL residuals are within picosecond-level precision, indicating good accuracy. These results demonstrate that the OSB products from the CNES satisfy the requirements for IPPP and enable precise time and frequency transfer.

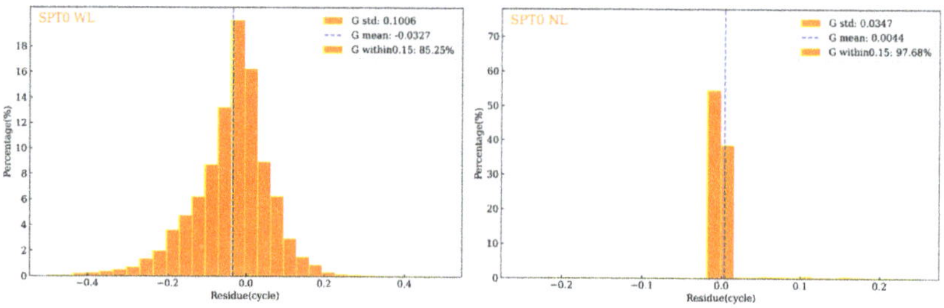

Figure 4. Statistics of WL and NL fractional residuals over 30 days (SPT0).

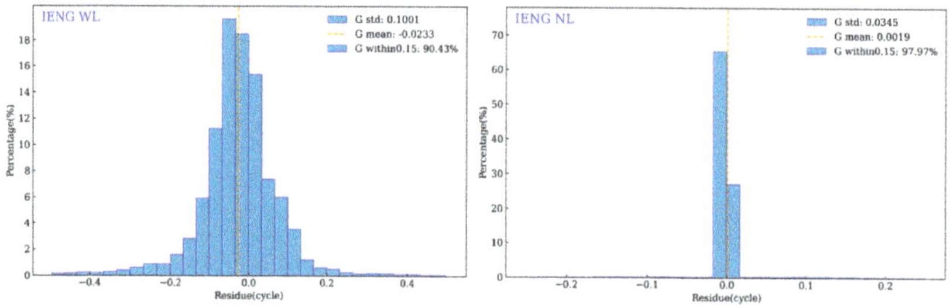

Figure 5. Statistics of WL and NL fractional residuals over 30 days (IENG).

Table 4. Percentages of WL and NL fractional residuals within 30 days (%).

Station	WL (Within 0.25)	WL (Within 0.15)	NL (Within 0.25)	NL (Within 0.10)
SPT0	97.54	85.25	99.61	97.68
IENG	98.26	90.43	99.79	97.97

4.2. Repair of Day-Boundary Discontinuities

In this section, GPS observation data from the SPT0-IENG link between day of year (DOY) 72 and 78 in 2025 were collected to conduct continuity tests of IPPP time and frequency transfer. The experiments are divided into two groups: before and after discontinuity repair. Note that a small vertical offset is applied to the time transfer results in the figure to facilitate better comparison. As shown by the blue curve in Figure 6, when performing standard single-day batch processing for time and frequency transfer, a discontinuous clock jump occurs at the day boundary (between DOY 74–75, 75–76 and 76–77) with a magnitude of approximately 0.353 ns. Using the compensation algorithm proposed in this study, the clock jump at the day boundary is actively absorbed by the satellite-differenced NL ambiguities, thereby repairing the discontinuity. The orange curve in the figure indicates that after applying the NL ambiguity-based clock jump absorption, the discontinuities in the time link between DOY 74–75, 75–76 and 76–77 have been bridged, and the discontinuities are minimized.

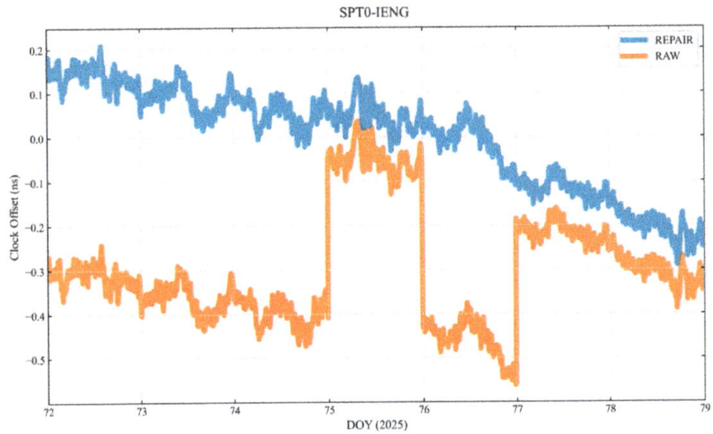

Figure 6. Comparison of day-boundary discontinuity repair for the SPT0–IENG (about 1457 km) link over 7 days. The blue curve shows the time link before repair, and the orange curve shows the result after repair.

4.3. Measured-Data and Simulation Time Link Experiments

4.3.1. IPPP Time Links of GPS, BDS, and GAL

To evaluate the general applicability of the proposed algorithm, IPPP time transfer experiments were conducted using GPS, Galileo (GAL), and BeiDou (BDS) signals. For this purpose, 30-day time transfer links were established between the BRUX and IENG stations using GPS L1C and L2W signals, GAL L1C and L5Q signals, and BDS B1C and B2a signals, respectively. The goal was to assess the frequency stability performance across different GNSS constellations.

As shown in Figure 7, continuous time transfer was successfully achieved using IPPP for all three systems—GPS, GAL, and BDS. Among them, BDS exhibited slightly inferior frequency stability compared to GPS and GAL. Tables 5 and 6 present the standard deviation (STD) and the frequency stability of the time link. In terms of the STD of the time link, GPS IPPP achieved 0.744 ns, GAL IPPP 0.727 ns, and BDS IPPP 0.859 ns. At an averaging time of 1200 s, the frequency stability reached 1.258×10^{-14} for GPS, 1.038×10^{-14} for GAL, and 1.411×10^{-14} for BDS. When the averaging time increased to 9600 s, the corresponding values were 2.061×10^{-15}, 2.394×10^{-15}, and 2.893×10^{-15}. At 76,800 s, the stability further improved to 4.838×10^{-16} (GPS), 4.707×10^{-16} (GAL), and 5.403×10^{-16} (BDS). Compared with GPS and GAL, the frequency transfer stability of BDS time links demonstrates relatively inferior performance. Possible contributing factors include the relatively lower performance of onboard clocks; the limited number of visible satellites when using the newly introduced BDS-3 signals (B1C/B2a) alone; lower global availability of BDS ground stations, which may lead to inconsistencies in orbit and clock product precision; and higher pseudorange multipath noise levels.

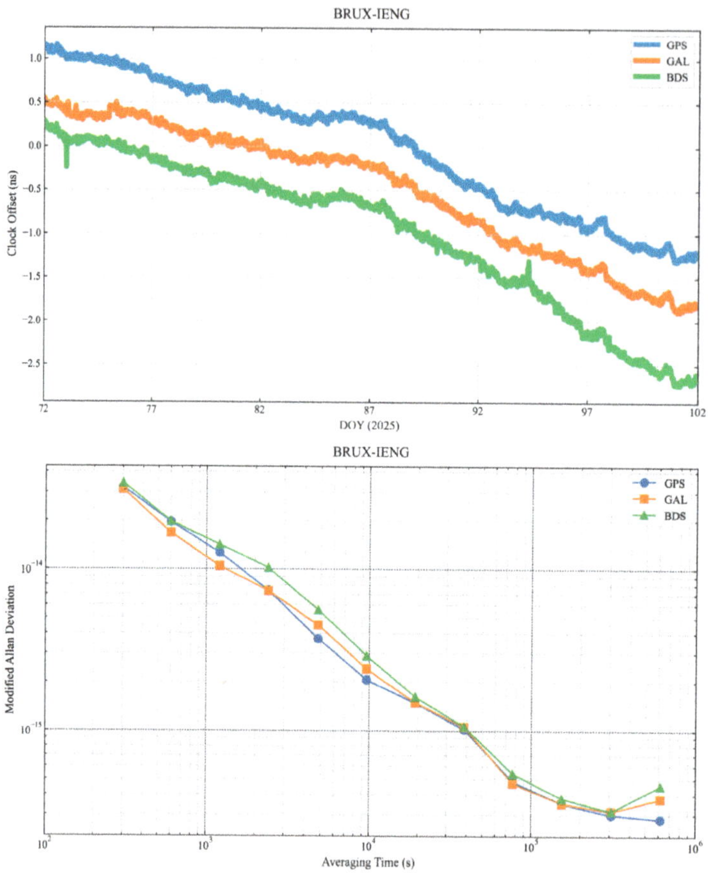

Figure 7. Time transfer results and MDEV frequency stability comparison of GPS, Galileo, and BDS IPPP solutions over the BRUX-IENG (about 688km) link (DOY 72–101).

Table 5. STD of GPS, GAL, and BDS IPPP time transfer links.

System	GPS	GAL	BDS
STD/ns	0.744	0.727	0.859

Table 6. Frequency stability of the BRUX-IENG test time links from DOY 72 to 101.

Averaging Time/s	**Solution**		
	GPS	**GAL**	**BDS**
1200	1.258×10^{-14}	1.038×10^{-14}	1.411×10^{-14}
9600	2.061×10^{-15}	2.394×10^{-15}	2.893×10^{-15}
76800	4.838×10^{-16}	4.707×10^{-16}	5.403×10^{-16}

4.3.2. Simulation of Incorrect Ambiguity Fixes

To compare the performance of the IGIII and the proposed robust sliding-window weighting strategies under worst-case conditions, a simulated experiment involving incorrect ambiguity fixes was conducted over the 7-day time link between SPT0 and IENG (DOY 72–78). Specifically, incorrect ambiguity fixes at epochs 3000, 4000, and 5000 were introduced at station SPT0. As shown in Figure 8, such incorrect ambiguity fixes cause significant fluctuations in the time transfer link between BRUX and SPT0, with the carrier-phase residual increasing sharply to 9.6 cm, ultimately leading to a degradation in time and frequency stability.

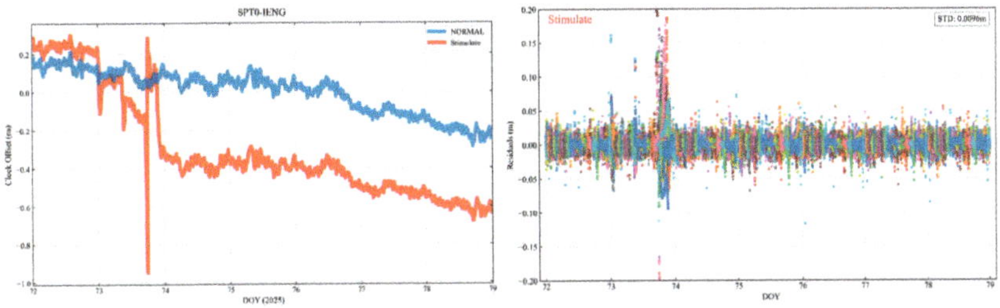

Figure 8. Time transfer link comparison and carrier-phase residuals for the SPT0–IENG baseline (about 1457 km) after simulating incorrect ambiguity fixes at station SPT0 (DOY 72–78). An arbitrary vertical offset was added to the time link results for visual comparison. In the right figure, each color corresponds to a different satellite.

As shown in Figures 9 and 10, the IGIII triple-segment down-weighting strategy (parameter 1 is 0.25 cycles and parameter 2 is 0.15 cycles) [11,26] fails to fully eliminate the anomalies caused by incorrect ambiguity resolution. Table 7 presents the STD of carrier-phase residuals under different strategies. Both the time transfer link and the carrier-phase residuals exhibit abnormal fluctuations, with the STD of the residuals increasing to 6.7 cm. In contrast, the proposed robust sliding-window weighting strategy, which employs dynamic thresholding, effectively reduces the impact of incorrect ambiguity fixes by suppressing abrupt changes in the time transfer link. As a result, the STD of the carrier-phase residuals is reduced to 5.2 cm. These findings indicate that the dynamic thresholds derived from the robust sliding-window weighting algorithm can more comprehensively and accurately detect incorrect ambiguity fixes, thereby improving the reliability and stability of time transfer links.

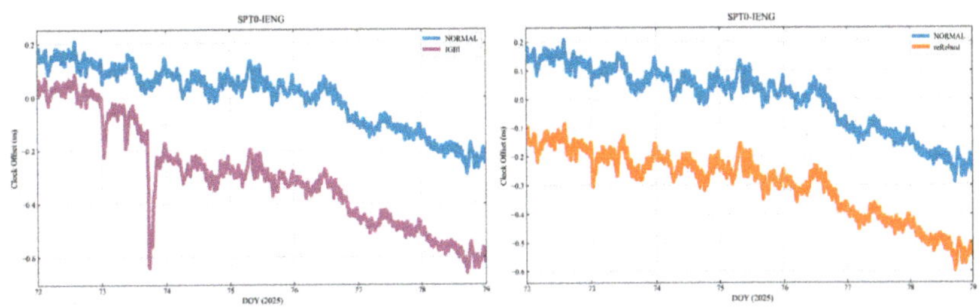

Figure 9. Time transfer link comparison at station SPT0 after applying the IGIII scheme and the robust sliding-window weighting strategy (DOY 72–78).

Table 7. STD of carrier-phase residuals under different strategies.

Strategy	Stimulate	IGIII	Robust
STD/cm	9.6	6.7	5.2

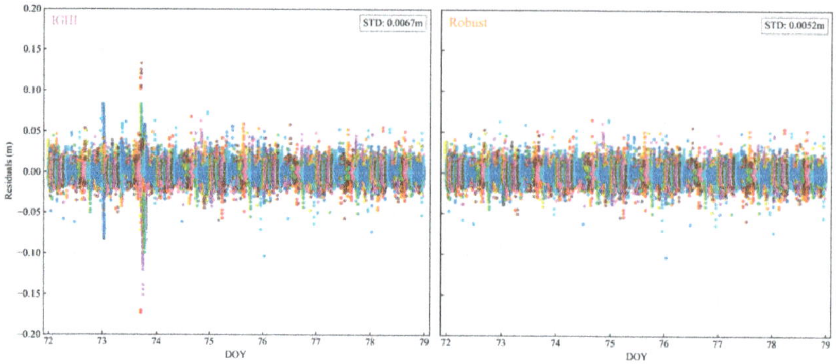

Figure 10. Carrier-phase residuals at SPT0 after processing with the IGIII scheme and the proposed robust sliding-window weighting algorithm. Each color corresponds to a different satellite.

4.4. Zero-Baseline Experiment

In a zero-baseline configuration, the signal propagation paths from the satellites to the two stations are nearly identical, thereby minimizing geometry-related errors. Additionally, the use of a shared high-performance external clock significantly reduces the impact of receiver-related noise. For these reasons, zero-baseline setups are frequently employed to assess the noise level and performance of time and frequency transfer techniques. In this study, USN7 and USN8—equipped with identical high-stability hydrogen masers and sharing a common antenna—were selected to validate the time transfer performance under a common-clock, zero-baseline condition. Results were also compared against the IGS final products.

Figure 11 presents the zero-baseline time transfer results between USN7 and USN8, using both PPP float solutions and IPPP with ambiguity resolution. The float solution was obtained from a standard processing procedure, while the IPPP solution was derived by constraining the float solution with resolved integer ambiguities. As expected, due to the shared antenna and high-quality external clock, the time transfer results exhibit minimal fluctuations. For visualization purposes, the curves were shifted to align the baseline levels. Compared to the PPP float solution, the IPPP results show a noticeable improvement in smoothness and stability. Furthermore, as shown in red circles, the time transfer link between the USN7 and USN8 stations exhibits significant fluctuations when ambiguities are incorrectly fixed, resulting in a noticeable degradation of frequency stability. Traditional triple-segment down-weighting schemes, such as the IGIII method (parameter 1 is 0.25 cycles and parameter 2 is 0.15 cycles) [11,26], are not fully effective in suppressing these anomalies. In contrast, the proposed sliding window-based robust weighting strategy dynamically adjusts the weights and helps recover correct ambiguity fixing. This adaptive mechanism successfully mitigates the outliers in the time transfer link, resulting in fewer isolated points and enhanced robustness against abnormal data.

The STD of the clock differences over a 30-day period was calculated. The GPS IPPP Robust, GPS IPPP IGIII, and GPS PPP solutions yielded STDs of approximately 9 ps, 10 ps, and 29 ps, respectively, all indicating high time transfer precision. In terms of frequency stability analysis, GPS IPPP exhibits increasingly improved stability over longer averaging intervals, as shown in Table 8. For zero-baseline time transfer using GPS IPPP Robust, frequency stabilities at averaging times of 1200 s, 9600 s, and 38,400 s are denoted as 3.699×10^{-15}, 5.235×10^{-16}, and 1.087×10^{-16}, respectively—representing improvements of 15.1%, 59.8%, and 82.0% over the corresponding float PPP solutions. When the averaging time extended to 76,800 s, the frequency stability of GPS IPPP Robust reached 3.332 $\times$ 10^{-17}. The reliability of the proposed robust sliding-window weighting strategy was thus confirmed, and it was subsequently adopted as a default component in all IPPP experimental schemes involving baselines of different lengths. For simplicity of notation, all GPS IPPP experimental schemes are assumed to incorporate both the day-boundary discontinuity repair algorithm and the robust sliding-window weighted strategy by default.

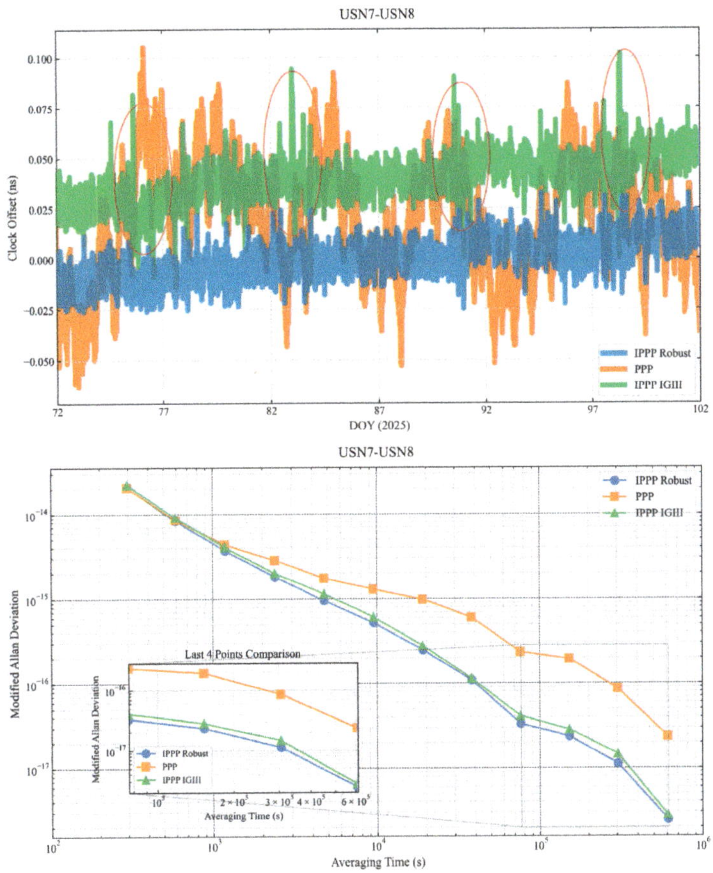

Figure 11. Frequency stability analysis of IPPP IGIII, IPPP Robust, and IPPP time transfer on USN7-USN8 zero-baseline common-clock link (DOY 72–101). Outliers in the time transfer link are marked by red circles in the upper panel of the figure.

Table 8. Frequency stability of the USN7-USN8 test time links from DOY 72 to 101.

Averaging Time/s	Solution		
	GPS IPPP Robust	**GPS PPP**	**GPS IPPP IGIII**
1200	3.699 $\times$ 10^{-15}	4.368 $\times$ 10^{-15}	4.061 $\times$ 10^{-15}
9600	5.235 $\times$ 10^{-16}	1.304 $\times$ 10^{-15}	6.161 $\times$ 10^{-16}
38,400	1.087 $\times$ 10^{-16}	6.055 $\times$ 10^{-16}	1.411 $\times$ 10^{-16}

4.5. Additional Baseline Experiments

4.5.1. Optical Fiber Link Experiment

To benchmark IPPP time transfer performance, a comparative analysis was conducted against a high-precision optical fiber link between BOR1 and GUM5.

As shown in Figure 12, the performance of the optical fiber link surpasses that of the GPS IPPP and PPP differential schemes for averaging times shorter than 9600 s. This is primarily because the transmission medium in optical fiber links is more stable—largely due to consistent refractive index and signal power loss characteristics. These results also indicate that the IPPP- and PPP-based time transfer links are subject to higher short-term frequency noise.

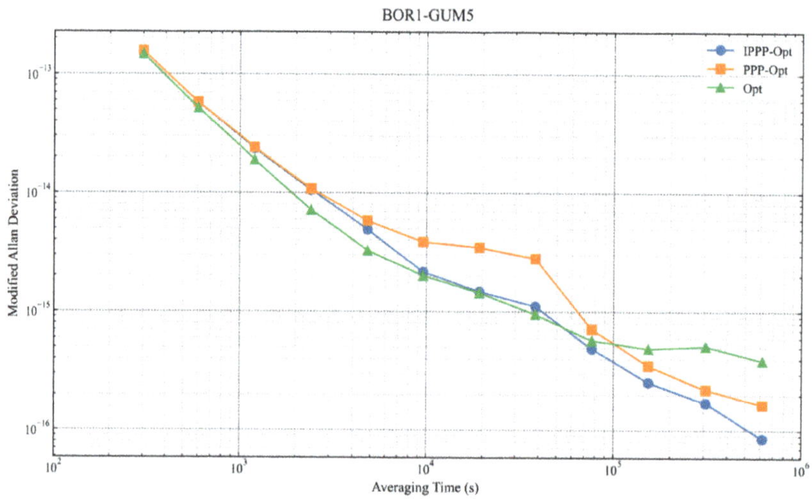

Figure 12. MDEV of time transfer links obtained by differencing the GPS IPPP, GPS PPP, and optical fiber link solutions and the corresponding optical fiber link (about 267 km).

For longer averaging times, however, the IPPP differential scheme outperforms the optical fiber link in terms of time transfer precision, which may be attributed to clock instability or other limiting factors on the optical fiber link.

4.5.2. Short-Baseline and Long-Baseline Experiment

Using IENG as the reference station, two short-baseline links (BRUX-IENG and SPT0-IENG) and one long-baseline link (TWTF–IENG) were tested. To evaluate the performance of the IPPP-based time transfer solution over different baseline lengths, the results were compared with time links derived from the IGS final products.

Figures 13 and 14 present the comparisons and MDEV analyses of the GPS IPPP, GPS PPP, and IGS final product time links for the tested baselines. It is evident from the time series comparison that the IGS-derived links exhibit day-boundary discontinuities, which is consistent with the fact that they are based on single-day batch processing. In contrast, the GPS IPPP and PPP links—enhanced with a compensation algorithm—show significantly reduced discontinuities at the day boundaries.

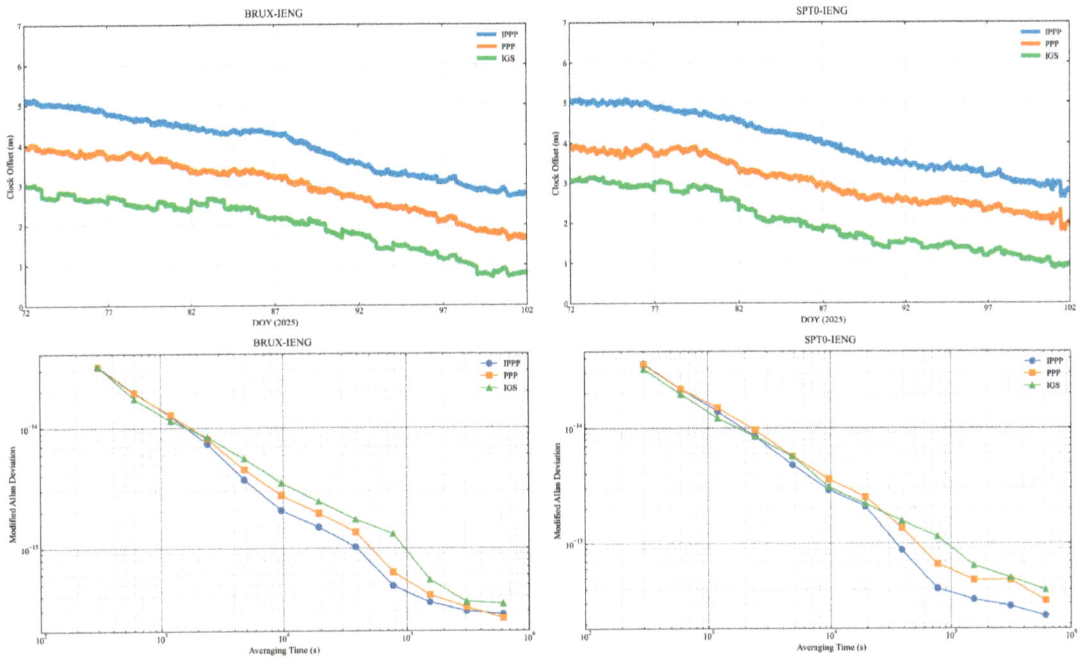

Figure 13. Time transfer links over short baselines (BRUX–IENG, about 688 km, and SPT0–IENG, about 1457 km) from DOY 72–101 are shown in the upper panel, with the corresponding MDEV stability analysis presented in the lower panel. The blue line represents the GPS IPPP solution, the orange line denotes the GPS PPP float solution, and the green line indicates the solution based on IGS final products. The time differences in the top panel remove a linear drift and add an arbitrary offset.

Tables 9–11 present the frequency stability of the three solutions. When the averaging time is relatively short, the frequency stability of the three solutions remains comparable. This is expected, as the IPPP solution is derived by applying integer ambiguity constraints to the PPP float solution, and thus inherits its short-term stability characteristics [13]. Interestingly, for long baselines, the IGS solution sometimes exhibits slightly better stability, likely due to the fact that the path-consistent errors are not effectively mitigated by differencing over long distances. At an averaging time of 76,800 s, all three schemes exhibit similar levels of frequency stability. As the averaging time exceeds 153,600 s for all tested links, both GPS IPPP and PPP demonstrate improved frequency stability compared to the IGS final products, with IPPP consistently delivering the best performance. In the comparison with IGS final products over both short- and long-baseline links, for the averaging time of 307,200 s, the GPS IPPP solution achieves an average improvement of approximately 29.3% in frequency stability over the PPP float solution, and about 32.6% over the IGS final product.

Table 9. Frequency stability of the BRUX-IENG test time links from DOY 72 to 101.

Averaging Time/s	Solution		
	GPS IPPP	GPS PPP	IGS
1200	1.258×10^{-14}	1.279×10^{-14}	1.146×10^{-14}
9600	2.061×10^{-15}	2.735×10^{-15}	3.463×10^{-15}
38,400	1.021×10^{-15}	1.356×10^{-15}	1.723×10^{-15}
153,600	3.530×10^{-16}	4.067×10^{-16}	5.427×10^{-16}

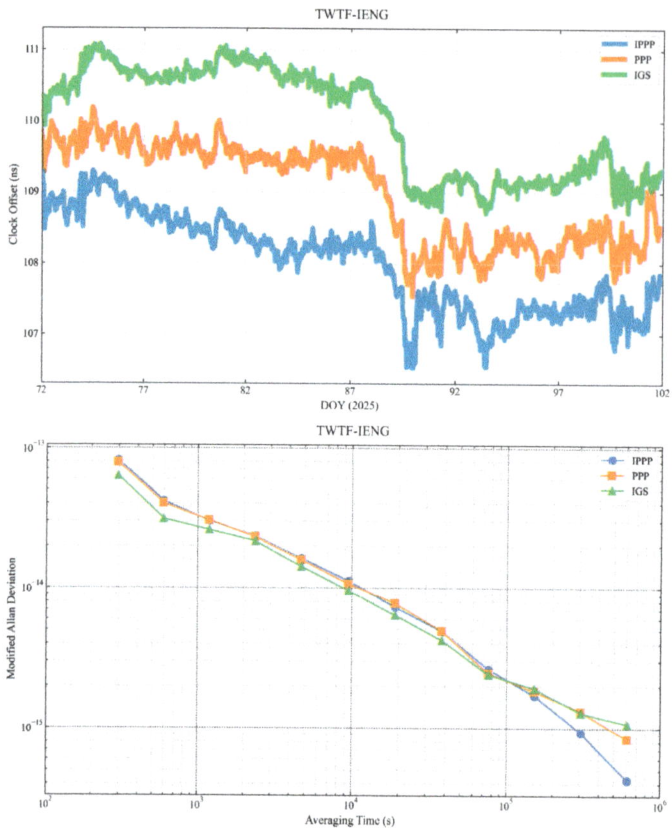

Figure 14. Time transfer links over long baselines (TWTF–IENG) from DOY 72–101 are shown in the upper panel, with the corresponding MDEV stability analysis presented in the lower panel. The blue line represents the GPS IPPP solution, the orange line denotes the GPS PPP float solution, and the green line indicates the solution based on IGS final products. The time differences in the top panel remove a linear drift and add an arbitrary offset.

Table 10. Frequency stability of the SPT0-IENG test time links from DOY 72 to 101.

Averaging Time/s	Solution		
	GPS IPPP	GPS PPP	IGS
1200	1.393×10^{-14}	1.492×10^{-14}	1.212×10^{-14}
9600	2.908×10^{-15}	3.572×10^{-15}	3.067×10^{-15}
38,400	8.725×10^{-16}	1.350×10^{-15}	1.561×10^{-15}
153,600	3.243×10^{-16}	4.795×10^{-16}	6.405×10^{-16}

Table 11. Frequency stability of the TWTF-IENG test time links from DOY 72 to 101.

Averaging Time/s	Solution		
	GPS IPPP	GPS PPP	IGS
1200	3.020×10^{-14}	3.050×10^{-14}	2.603×10^{-14}
9600	1.126×10^{-14}	1.598×10^{-14}	1.427×10^{-14}
38,400	4.929×10^{-15}	4.961×10^{-15}	4.250×10^{-15}
153,600	1.721×10^{-15}	1.835×10^{-15}	1.934×10^{-15}

5. Conclusions

The IPPP technique enhances the accuracy of time and frequency transfer by restoring the integer nature of carrier-phase ambiguities, particularly improving long-term frequency stability. In this study, integer ambiguity resolution was performed using OSB products. Additionally, a day-boundary discontinuity correction algorithm and a robust sliding-window weighting strategy were proposed and employed to assess the effectiveness of IPPP in time and frequency comparison. The following conclusions can be drawn from the analysis:

(1) The proposed day-boundary correction method effectively mitigates clock jumps at daily boundaries. The robust sliding-window weighting approach fully utilizes the narrow-lane residuals across multiple adjacent epochs, enabling more reliable detection and suppression of erroneous ambiguity fixes.

(2) Over a 7-day dataset, the IPPP time transfer technique achieved frequency stabilities of 4.838×10^{-16} (GPS), 4.707×10^{-16} (Galileo), and 5.403×10^{-16} (BDS) at an averaging time of 76,800 s. In the simulation experiments, the proposed robust sliding-window weighting algorithm demonstrated its ability to derive adaptive thresholds, enabling more comprehensive detection of anomalies compared to the IGIII method. Specifically, when applying the IGIII scheme, the carrier-phase residual reached 6.7 cm, whereas with the robust sliding-window approach, the residual was reduced to 5.2 cm.

(3) Across both zero-baseline and longer baseline links, the IPPP time transfer technique consistently outperformed the traditional float PPP solution in terms of accuracy and stability—especially for long-term averaging. Especially in differential comparisons with optical fiber time transfer, the IPPP solution showed better long-term stability than the optical fiber link itself. In the zero-baseline experiment, the frequency stability of GPS IPPP showed improvements of 15.1%, 59.8%, and 82.0% over the float PPP solution at averaging times of 1200 s, 9600 s, and 38,400 s, respectively. At 76,800 s, GPS IPPP achieved a frequency stability of 3.332×10^{-17}. For both short- and long-baseline links, when compared to the IGS final products at an averaging time of 307,200 s, the IPPP solution showed an average improvement of approximately 29.3% over the PPP float solution and 32.6% over the IGS products.

Since this study is based on dual-frequency ionosphere-free combinations, it inherently requires that the user equipment support at least two frequencies. In the future, we aim to extend this work to single-frequency and multi-GNSS scenarios and further investigate ionospheric error correction, with the goal of enabling more flexible and reliable time transfer under constrained environments.

Author Contributions: Conceptualization, methodology, and validation: K.W. and W.Q.; data supply: W.W.; discussion—D.L., P.W. and X.Y. All authors have read and agreed to the published version of the manuscript.

Funding: Youth Innovation Promotion Association of the Chinese Academy of Science (Grant No. Y2023108), CAS "Light of West China—Western Interdisciplinary Team" Program [xbzg-zdsys-202308], the National Natural Science Foundation of China (Grant No. 12273047).

Data Availability Statement: The original contributions presented in this study are included in the article. Further inquiries can be directed to the author at wukun@ntsc.ac.cn.

Conflicts of Interest: The authors declare no conflict of interest.

References

1. Collins, P.; Bisnath, S.; Lahaye, F.; Héroux, P. Undifferenced GPS Ambiguity Resolution Using the Decoupled Clock Model and Ambiguity Datum Fixing. *Navigation* **2010**, *57*, 123–135. [CrossRef]

2. Laurichesse, D.; Mercier, F.; Berthias, J.-P.; Broca, P.; Cerri, L. Integer Ambiguity Resolution on Undifferenced GPS Phase Measurements and Its Application to PPP and Satellite Precise Orbit Determination. *Navigation* **2009**, *56*, 135–149. [CrossRef]
3. Geng, J.; Meng, X.; Dodson, A.H.; Teferle, F.N. Integer Ambiguity Resolution in Precise Point Positioning: Method Comparison. *J. Geod.* **2010**, *84*, 569–581. [CrossRef]
4. Petit, G.; Leute, J.; Loyer, S.; Perosanz, F. Sub 10−16 Frequency Transfer with IPPP: Recent Results. In Proceedings of the 2017 Joint Conference of the European Frequency and Time Forum and IEEE International Frequency Control Symposium (EFTF/IFCS), BesanÃ§on, France, 9–13 July 2017; IEEE: New York, NY, USA, 2017; pp. 784–787.
5. Fujieda, M.; Yang, S.H.; Gotoh, T.; Hwang, S.W.; Hachisu, H.; Kim, H.; Lee, Y.K.; Tabuchi, R.; Ido, T.; Lee, W.K. Advanced Satellite-Based Frequency Transfer at the 10^{-16} Level. *IEEE Trans. Ultrason. Ferroelectr. Freq. Control* **2018**, *65*, 973–978. [CrossRef]
6. Loyer, S.; Perosanz, F.; Mercier, F.; Capdeville, H.; Marty, J.-C. Zero-Difference GPS Ambiguity Resolution at CNES–CLS IGS Analysis Center. *J. Geod.* **2012**, *86*, 991–1003. [CrossRef]
7. Yao, J.; Levine, J. A New Algorithm to Eliminate GPS Carrier-Phase Time Transfer Boundary Discontinuity. In Proceedings of the 45th Annual Precise Time and Time Interval Systems and Applications Meeting, Bellevue, WA, USA, 2–5 December 2013; Institute of Navigation: Manassas, VA, USA, 2013.
8. Dach, R.; Schildknecht, T.; Hugentobler, U.; Bernier, L.G.; Dudle, G. *Continuous Geodetic Time-Transfer Analysis Methods*; IEEE Transactions on Ultrasonics Ferroelectrics & Frequency Control: Piscataway, NJ, USA, 2006.
9. Guo, J.; Defraigne, P.; Pinat, E. Continuous Time and Frequency Transfer Achieved by Multi-Day GPS/Galileo/BDS Integer Precise Point Positioning. *GPS Solut.* **2025**, *29*, 95. [CrossRef]
10. Gerard, P. Sub-10^{-16} Accuracy GNSS Frequency Transfer with IPPP. *GPS Solut.* **2021**, *25*, 22.
11. Ren, Z.; Lyu, D.; Gong, H.; Peng, J.; Huang, X.; Sun, G. Continuous Time and Frequency Transfer Using Robust GPS PPP Integer Ambiguity Resolution Method. *GPS Solut.* **2023**, *27*, 82. [CrossRef]
12. Li, X.; Liang, D.; Li, X.; Huang, J.; Wu, J.; Gou, H. Quality Monitoring of Real-Time PPP Service Using Isolation Forest-Based Residual Anomaly Detection. *GPS Solut.* **2024**, *28*, 118. [CrossRef]
13. Xu, W.; Shen, W.; Cai, C.; Li, L.; Wang, L.; Ning, A.; Shen, Z. Comparison and Evaluation of Carrier Phase PPP and Single Difference Time Transfer with Multi-GNSS Ambiguity Resolution. *GPS Solut.* **2022**, *26*, 58. [CrossRef]
14. Yang, Y.; Song, L.; Xu, T. Robust Estimator for Correlated Observations Based on Bifactor Equivalent Weights. *J. Geod.* **2002**, *76*, 353–358. [CrossRef]
15. Teunissen, P.J.G.; Verhagen, S. The GNSS Ambiguity Ratio-Test Revisited: A Better Way of Using It. *Surv. Rev.* **2009**, *41*, 138–151. [CrossRef]
16. Guo, J.; Geng, J. Integer Ambiguity Validation through Machine Learning for Precise Point Positioning. *Satell. Navig.* **2025**, *6*, 14. [CrossRef]
17. Chen, C.; Xiao, G.; Chang, G.; Xu, T.; Yang, L. Assessment of GPS/Galileo/BDS Precise Point Positioning with Ambiguity Resolution Using Products from Different Analysis Centers. *Remote Sens.* **2021**, *13*, 3266. [CrossRef]
18. Geng, J.; Yang, S.; Guo, J. Assessing IGS GPS/Galileo/BDS-2/BDS-3 Phase Bias Products with PRIDE PPP-AR. *Satell. Navig.* **2021**, *2*, 17. [CrossRef]
19. Cao, X.; Yu, X.; Ge, Y.; Liu, T.; Shen, F. BDS-3/GNSS Multi-Frequency Precise Point Positioning Ambiguity Resolution Using Observable-Specific Signal Bias. *Measurement* **2022**, *195*, 111134. [CrossRef]
20. Geng, J.; Bock, Y. Triple-Frequency GPS Precise Point Positioning with Rapid Ambiguity Resolution. *J. Geod.* **2013**, *87*, 449–460. [CrossRef]
21. Petit, G.; Kanj, A.; Loyer, S.; Delporte, J.; Mercier, F.; Perosanz, F. 1 × 10−16frequency Transfer by GPS PPP with Integer Ambiguity Resolution. *Metrologia* **2015**, *2*, 301. [CrossRef]
22. Song, W.; Zhang, R.; Yao, Y.; Liu, Y.; Hu, Y. PPP Sliding Window Algorithm and Its Application in Deformation Monitoring. *Sci. Rep.* **2016**, *6*, 26497. [CrossRef]
23. Zheng, X.; Dong, Y.; Zhao, Y.; Zhang, B.; Li, M. TSF-GINS: Based on Time-Fixed Sliding Window with Factor Graph a Global Navigation Satellite System and Inertial Measurement Unit Tightly Coupled Localization System. *Measurement* **2025**, *239*, 115421. [CrossRef]
24. Ge, Y.; Zhou, F.; Liu, T.; Qin, W.; Wang, S.; Yang, X. Enhancing Real-Time Precise Point Positioning Time and Frequency Transfer with Receiver Clock Modeling. *GPS Solut.* **2019**, *23*, 20. [CrossRef]
25. Zhang, X.; Li, P.; Guo, F. Ambiguity Resolution in Precise Point Positioning with Hourly Data for Global Single Receiver. *Adv. Space Res.* **2013**, *51*, 153–161. [CrossRef]
26. Yang, Y. Robust estimation for dependent observation. *Manuscripta Geod.* **1994**, *19*, 10–17.

remote sensing

MDPI

Article

Variance Component Estimation (VCE)-Based Adaptive Stochastic Modeling for Enhanced Convergence and Robustness in GNSS Precise Point Positioning (PPP)

Yanning Zheng [1,2], Yongfu Sun [1], Yubin Zhou [1,*], Shengli Wang [3] and Yixu Liu [1,2]

1 National Deep Sea Center (NDSC), Qingdao 266237, China; zheng_yn@ndsc.org.cn (Y.Z.); sunyongfu@ndsc.org.cn (Y.S.)
2 College of Environmental Science and Engineering, Ocean University of China, Qingdao 266100, China
3 College of Ocean Science and Engineering, Shandong University of Science and Technology, Qingdao 266590, China; shlwang@sdust.edu.cn
* Correspondence: zyb@ndsc.org.cn

Abstract: The stochastic model in Precise Point Positioning (PPP) defines the statistical properties of observations and the dynamic behavior of parameters. An inaccurate stochastic model can degrade positioning accuracy, ambiguity resolution, and other aspects of performance. However, due to the influence of multiple factors, the stochastic model in PPP cannot be precisely predetermined, necessitating the development of an Adaptive Stochastic Model (ASM) based on Variance Component Estimation (VCE). While the benefits of ASMs for PPP float solutions are well documented, their contributions to other performance aspects remain insufficiently explored. This paper presents a comprehensive assessment of an ASM's impact on PPP. First, the implementation of an ASM using VCE is described in detail. Then, experimental results demonstrate that the ASM effectively captures observational conditions through the estimated variance component factors. It enhances both PPP float and fixed solutions when the predefined stochastic model is inadequate, improves cycle-slip detection by tightening the stochastic model (reducing the missed detection rate from 19% to 8%), and accelerates both direct reconvergence and re-initialization after data gaps, with reconvergence times improved by 18% and 55%, respectively.

Keywords: stochastic model; Precise Point Positioning (PPP); Variance Component Estimation (VCE); Ambiguity Resolution (AR); Detection, Identification, Adaptation (DIA); interruption fixing

1. Introduction

GNSS Precise Point Positioning (PPP) is a global high-precision positioning method [1]. The PPP model comprises a functional model and a stochastic model. The latter describes the statistical properties of observational noise and process noise, serving as a critical component for data processing in PPP. A well-defined stochastic model is essential for optimal parameter estimation, reliable precision metrics, and effective hypothesis testing. Conversely, an improper stochastic model can degrade positioning accuracy [2], impact the detection of mismodeling errors with Detection, Identification, Adaptation (DIA) [3–7], and even result in incorrect ambiguity resolution [8–10]. The properties of stochastic noise are influenced by numerous factors, such as hardware design, observational environment, and even the time of day, making it challenging to capture these variations using a predefined stochastic model. Compared to other high-precision GNSS positioning methods like Real-Time Kinematic (RTK), PPP faces more complex stochastic noise characteristics due to its lack of support from regional GNSS networks. Thus, a real-time adaptive stochastic model is vital for refining the PPP model.

The PPP model is typically solved using the Kalman filter (KF), and the Sage–Husa adaptive KF is a classic algorithm for real-time stochastic model adjustment. The Sage–Husa algorithm is based on the principle that the a priori variance of the innovation vector can be expressed as a combination of observational noise variance and process noise variance. If one is accurately known, the other can be derived from the estimated innovation vector [11,12]. This algorithm is used extensively for dynamic positioning and navigation, such as GNSS integrated with Inertial Measurement Units (IMUs) [13,14], dead reckoning [15], and celestial navigation systems [16]. However, the Sage–Husa algorithm has notable limitations: it cannot simultaneously estimate both observational noise and process noise variances, and it requires estimating a full variance matrix, which often has more entries than just the innovation vector and cannot be estimated stably.

Variance Component Estimation (VCE) is another well-established algorithm that uses redundant information to estimate the variance matrix. Unlike the Sage–Husa algorithm, VCE decomposes the variance matrix into a linear combination of predefined matrices and estimates their coefficients, termed variance factors. This approach avoids the challenges associated with estimating the full variance matrix and enables the simultaneous refinement of observational and process noise models. VCE has been extensively used in post-processing applications, such as refining stochastic models with inter-frequency and temporal correlations [17–19] and stochastic models for multi-frequency observations [19,20]. Recent studies have explored integrating VCE with the Kalman filter for real-time stochastic model refinement. Some of them focused on the enhancement of stochastic models, such as applying Helmert's simplified VCE to GPS Standard Point Positioning (SPP) [21], refining the stochastic model of PPP with Online-VCE [22], and the joint estimation of the variances of the dual-frequency GNSS observational noise and the process noise of the receiver code bias [23]. These studies are valuable to the analysis of stochastic models but did not present the benefits to the performance of refined stochastic models. Regarding the benefits of VCE to the performance of GNSS data processing, several different topics have been covered. Regarding the positioning performance, Zhang et al. combined modified VCE and Helmert VCE to improve observation weighting for PPP [24], while Zhang and Zhao adjusted the unit weight variance factor of the pseudo-range and the carrier-phase observations [25]; these results showed improvements in positioning accuracy, stability, and convergence time. For the estimation of non-positioning parameters, Yang et al. developed an adaptive KF using VCE to adjust the process noise variance for tropospheric delay of PPP, significantly improving the root-mean-square (RMS) error of tropospheric delay estimates [26]. For the quality control of data processing, Chang et al. used VCE to refine the stochastic model for ionospheric process noise in a cycle-slip repair algorithm [27], and the results showed that all manually introduced nontrivial cycle slips were correctly repaired.

These studies highlight the benefits of an Adaptive Stochastic Model (ASM) in various aspects of GNSS applications, including improved observational stochastic modeling, enhanced positioning precision, and refined tropospheric delay estimates; meanwhile, many of them are based on a PPP model which retains parameters for systematic errors and needs precise modeling of these parameters [22,24–26]. However, many aspects of GNSS PPP that depend on a proper stochastic model remain underexplored. For instance, ambiguity resolution (AR) typically involves two steps: decorrelation using Z-transformation and ambiguity fixing [28,29]. The Z-transformation operates on the variance matrix of ambiguities [30] while the success rate of fixing is calculated using the variance matrix [31]. Similarly, mismodeling detection algorithms such as DIA rely on hypothesis testing, whose effectiveness is determined by the quality of the stochastic model [32]. Furthermore, when satellite signals are interrupted, a dynamic model with an appropriate process noise model can better link epochs before and after the interruption, accelerating reconvergence and improving performance after data gaps [33–35]. However, it is incomplete to study only the benefits of an ASM for applicational problems; the value of an ASM to PPP lies in reflecting observational conditions through refined stochastic modeling, which should be assessed first. For a better understanding of the value of the ASM algorithm to PPP, an ASM algorithm is designed based on VCE and its influences on several topics of PPP are analyzed, including reflecting observational conditions through the estimation of time-variant variance factors, PPP float solution, PPP-AR, cycle-slip detection with DIA, reconvergence after data interruption, and interruption repair. The remaining sections are as follows: Section 2 proposes the ASM algorithm for the Kalman filter and the mathematical model of PPP; Section 3 evaluates the benefits of ASM across various PPP applications; and Section 4 summarizes the conclusions.

2. Materials and Methods

2.1. ASM for the Kalman Filter

In this section, the ASM for the Kalman filter is proposed, which is implemented with VCE. There are many VCE methods, including the Minimum Norm Quadratic Unbiased Estimator (MINQUE) [36], the best invariant quadratic unbiased estimator (BIQUE) [37], and the Least-Squares-VCE (LS-VCE) [38,39]. The LS-VCE is based on the well-known least-squared principle, which is more versatile and will be used to construct our ASM method. The LS-VCE will be introduced first; then, the implementation of the LS-VCE for the Kalman filter model will be derived; finally, the implementation of the ASM for the Kalman filter will be provided.

2.1.1. LS-VCE

Assume a linear observational model as

$$y = Ax + \varepsilon, \qquad D(\varepsilon) = Q = Q_0 + \sum_{i=1}^{n} \sigma_i Q_i \tag{1}$$

where y is the observational vector, A is the design matrix, x is the estimable parameter, ε is the observational noise, Q is the variance matrix of ε, Q_0 is the known part of variance matrix, and σ_i is the variance factor with corresponding matrix Q_i. A normal equation for the estimation of $\sigma = [\sigma_1, \cdots, \sigma_n]^T$ can be formed as follows:

$$N\hat{\sigma} = L \tag{2}$$

where $\hat{\sigma}$ is the estimated value of σ, and the entries of N and L are calculated with

$$N_{i,j} = \frac{1}{2} tr(Q_i Q^{-1} P_A^{\perp} Q_j Q^{-1} P_A^{\perp}) \tag{3}$$

$$L_i = \frac{1}{2} y^T P_A^{\perp} Q^{-1} Q_i Q^{-1} P_A^{\perp} y - \frac{1}{2} tr(Q_i Q^{-1} P_A^{\perp} Q_0 Q^{-1} P_A^{\perp}) \tag{4}$$

For $i,j \in [1,n]$, where $P_A^{\perp} = E - A(A^T Q^{-1} A)^{-1} A^T Q^{-1}$ is the projector of the orthogonal complement of $R(A)$, and the column space of A [40]; $tr(*)$ obtains the trace of a matrix. The estimated variance factor is $\hat{\sigma} = N^{-1} L$ and the variance of $\hat{\sigma}$ is $Q_{\hat{\sigma}} = N^{-1}$.

2.1.2. LS-VCE for the Kalman Filter

The dynamic model and observational model of the Kalman filter read

$$\boldsymbol{x}_k = \boldsymbol{F}_k \boldsymbol{x}_{k-1} + \boldsymbol{\varepsilon}_{p,k}, \quad D(\boldsymbol{\varepsilon}_{p,k}) = \boldsymbol{Q}_{p,k} \tag{5}$$

$$\boldsymbol{y}_k = \boldsymbol{A}_k \boldsymbol{x}_k + \boldsymbol{\varepsilon}_{m,k}, \quad D(\boldsymbol{\varepsilon}_{m,k}) = \boldsymbol{Q}_{m,k} \tag{6}$$

where $\boldsymbol{x}_{k-1}$ and $\boldsymbol{x}_k$ are the system states at epoch $k-1$ and k, respectively; $\boldsymbol{F}_k$ is the state transition matrix; $\boldsymbol{\varepsilon}_{p,k}$ is the process noise with the corresponding variance matrix $\boldsymbol{Q}_{p,k}$; $\boldsymbol{y}_k$ is the observational vector; $\boldsymbol{A}_k$ is the design matrix; and $\boldsymbol{\varepsilon}_{m,k}$ is the measurement noise with the corresponding variance matrix $\boldsymbol{Q}_{m,k}$. Assume the estimated system state at epoch $k-1$ as

$$\hat{\boldsymbol{x}}_{k-1} = \boldsymbol{x}_{k-1} + \boldsymbol{\varepsilon}_{\hat{x}_{k-1}}, \quad D\left(\boldsymbol{\varepsilon}_{\hat{x}_{k-1}}\right) = \boldsymbol{Q}_{\hat{x}_{k-1}} \tag{7}$$

Then, the predicted system state at epoch k can be obtained from the time update Equations (8) and (9), while the estimated system state can be obtained from the measurement update Equations (10) and (11).

$$\bar{\boldsymbol{x}}_k = \boldsymbol{F}_k \hat{\boldsymbol{x}}_{k-1} \tag{8}$$

$$\boldsymbol{Q}_{\bar{x}_k} = \boldsymbol{F}_k \boldsymbol{Q}_{\hat{x}_{k-1}} \boldsymbol{F}_k^T + \boldsymbol{Q}_{p,k} \tag{9}$$

$$\hat{\boldsymbol{x}}_k = \bar{\boldsymbol{x}}_k + \boldsymbol{K}_k \boldsymbol{v}_k \tag{10}$$

$$\boldsymbol{Q}_{\hat{x}_k} = (\boldsymbol{I} - \boldsymbol{K}_k \boldsymbol{A}_k) \boldsymbol{Q}_{\bar{x}_k} \tag{11}$$

where $\boldsymbol{v}_k = \boldsymbol{y}_k - \boldsymbol{A}_k \bar{\boldsymbol{x}}_k$ is the innovation vector, $\boldsymbol{K}_k = \boldsymbol{Q}_{\bar{x}_k} \boldsymbol{A}_k^T \boldsymbol{Q}_{v_k}^{-1}$ is the Kalman gain matrix, and $\boldsymbol{Q}_{v_k} = \boldsymbol{A}_k \boldsymbol{Q}_{\bar{x}_k} \boldsymbol{A}_k^T + \boldsymbol{Q}_{m,k}$ is the variance matrix of $\boldsymbol{v}_k$. $\boldsymbol{v}_k$ holds all the redundant information of the measurement update and can be used for determining the variance components of $\boldsymbol{Q}_{p,k}$ and $\boldsymbol{Q}_{m,k}$. Assume that $\boldsymbol{Q}_{p,k}$ is divided as

$$\boldsymbol{Q}_{p,k} = \boldsymbol{Q}_{0,p,k} + \sum_{i=1}^{n_p} \sigma_{i,p,k} \boldsymbol{Q}_{i,p,k} \tag{12}$$

while $\boldsymbol{Q}_{m,k}$ is divided as

$$\boldsymbol{Q}_{m,k} = \boldsymbol{Q}_{0,m,k} + \sum_{i=1}^{n_m} \sigma_{i,m,k} \boldsymbol{Q}_{i,m,k} \tag{13}$$

The expression of $\boldsymbol{v}_k$ can be rewritten in the form of (1) as

$$\begin{aligned} \boldsymbol{v}_k = \boldsymbol{y}_k - \boldsymbol{A}_k \bar{\boldsymbol{x}}_k &= \boldsymbol{A}_k \boldsymbol{x}_k + \boldsymbol{\varepsilon}_{m,k} - \boldsymbol{A}_k \boldsymbol{F}_k \hat{\boldsymbol{x}}_{k-1} \\ &= \boldsymbol{A}_k \boldsymbol{x}_k + \boldsymbol{\varepsilon}_{m,k} - \boldsymbol{A}_k \boldsymbol{F}_k \left(\boldsymbol{x}_{k-1} + \boldsymbol{\varepsilon}_{\hat{x}_{k-1}}\right) \\ &= \boldsymbol{\varepsilon}_{m,k} + \boldsymbol{A}_k \boldsymbol{\varepsilon}_{p,k} - \boldsymbol{A}_k \boldsymbol{F}_k \boldsymbol{\varepsilon}_{\hat{x}_{k-1}} \end{aligned} \tag{14}$$

where the symbols in (1) are $\boldsymbol{y} = \boldsymbol{y}_k - \boldsymbol{A}_k \bar{\boldsymbol{x}}_k$, $\boldsymbol{A} = \boldsymbol{0}$, $\boldsymbol{\varepsilon} = \boldsymbol{\varepsilon}_{m,k} + \boldsymbol{A}_k \boldsymbol{\varepsilon}_{p,k} - \boldsymbol{A}_k \boldsymbol{F}_k \boldsymbol{\varepsilon}_{\hat{x}_{k-1}}$, $\boldsymbol{Q}_0 = \boldsymbol{A}_k \boldsymbol{F}_k \boldsymbol{Q}_{\hat{x}_{k-1}} \boldsymbol{F}_k^T \boldsymbol{A}_k^T + \boldsymbol{A}_k \boldsymbol{Q}_{0,p,k} \boldsymbol{A}_k^T + \boldsymbol{Q}_{0,m,k}$, $\sum_{i=1}^{n} \sigma_i \boldsymbol{Q}_i = \sum_{i=1}^{n_p} \sigma_{i,p,k} \boldsymbol{A}_k \boldsymbol{Q}_{i,p,k} \boldsymbol{A}_k^T + \sum_{i=1}^{n_m} \sigma_{i,m,k} \boldsymbol{Q}_{i,m,k}$, $\boldsymbol{\sigma} = \left[\sigma_{1,p,k}, \cdots, \sigma_{n_p,p,k}, \sigma_{1,m,k}, \cdots, \sigma_{n_m,m,k}\right]^T$. Furthermore, since $\boldsymbol{A} = \boldsymbol{0}$, $\boldsymbol{P}_A^{\perp}$ is simplified as $\boldsymbol{P}_A^{\perp} = \boldsymbol{E}$ and Equations (3) and (4) are simplified accordingly as

$$N_{i,j} = \frac{1}{2} tr\left(\boldsymbol{Q}_i \boldsymbol{Q}^{-1} \boldsymbol{Q}_j \boldsymbol{Q}^{-1}\right) \tag{15}$$

$$L_i = \frac{1}{2} \boldsymbol{y}^T \boldsymbol{Q}^{-1} \boldsymbol{Q}_i \boldsymbol{Q}^{-1} \boldsymbol{y} - \frac{1}{2} tr\left(\boldsymbol{Q}_i \boldsymbol{Q}^{-1} \boldsymbol{Q}_0 \boldsymbol{Q}^{-1}\right) \tag{16}$$

Then $\sigma_{i,p,k}$ and $\sigma_{i,m,k}$ can be estimated with the LS-VCE. The normal equation of the kth epoch is noted as

$$N_k\hat{\sigma}_k = L_k \tag{17}$$

2.1.3. Implementation of ASM

In most articles about refining stochastic models with VCE, data of several minutes or even hours are used for the estimation of one set of variance factors. However, the estimation in Equation (17) uses single-epoch data, which could be unstable due to the lack of redundant information. Meanwhile, the subtraction in Equation (16) may yield negative results, leading to negative estimated values of variance factors. To compensate the instability of single-epoch VCE, the single-epoch normal equations can be accumulated as Equation (18), where $N_{\Sigma,k}$ and $W_{\Sigma,k}$ form the accumulated normal equation of epoch k. $N_{\Sigma,0}$ and $W_{\Sigma,0}$ for epoch 0 are obtained from the initial value $\hat{\sigma}_0$ and variance matrix $Q_{\hat{\sigma}_0}$ as $N_{\Sigma,0} = Q_{\hat{\sigma}_0}^{-1}$ and $W_{\Sigma,0} = N_{\Sigma,0}\hat{\sigma}_0$. The solution of $N_{\Sigma,k}$ and $W_{\Sigma,k}$ is $\hat{\sigma}_{\Sigma,k} = Q_{\hat{\sigma}_{\Sigma},k}W_{\Sigma,k}$ with variance matrix $Q_{\hat{\sigma}_{\Sigma},k} = N_{\Sigma,k}^{-1}$.

$$\begin{cases} N_{\Sigma,k} = N_{\Sigma,k-1} + N_k \\ W_{\Sigma,k} = W_{\Sigma,k-1} + W_k \end{cases} \tag{18}$$

However, when the estimated variance factors are time-variant, which happens when the observational environment changes, the accumulated estimations of the variance factors cannot reflect the temporal variation of the factors. To preserve the temporal characteristics of the estimated factors, a fading process is carried out as Equation (19) for $\hat{\sigma}_{\Sigma,k-1}$, the estimation of last epoch, and its corresponding variance matrix $Q_{\hat{\sigma}_{\Sigma},k-1}$ before updating with Equation (18), where $Q_{f,k}$ is a variance matrix for fading. $Q_{f,k}$ is diagonal and its ith diagonal entry reads $Q_{f,k,i} = \alpha_i Q_{\hat{\sigma}_{\Sigma},k-1,i}$, where α_i is the fading factor for the ith variance factor and $Q_{\hat{\sigma}_{\Sigma},k-1,i}$ the ith diagonal entry of $Q_{\hat{\sigma}_{\Sigma},k-1}$.

$$\begin{cases} N_{\Sigma,k-1} = \left(Q_{\hat{\sigma}_{\Sigma},k-1} + Q_{f,k}\right)^{-1} + N_k \\ W_{\Sigma,k-1} = \left(Q_{\hat{\sigma}_{\Sigma},k-1} + Q_{f,k}\right)^{-1}\hat{\sigma}_{\Sigma,k-1} + W_k \end{cases} \tag{19}$$

Figure 1 shows the estimations of the variance factor for pseudo-range observation. The single-epoch solution is noisy; with the normal matrices accumulated, the solution becomes so flat that it does not preserve any temporal variations; finally, with the fading process, the solution is both time-varying and smooth.

The flow chart in Figure 2 shows the update of the Kalman filter with ASM. The update starts with system state and variance factor of the last epoch and observation of the current epoch; the green arrows show how the Kalman filter obtains the estimated system state of the current epoch, while the red ones show the process of updating the variance factor. Since the update of the variance factor is carried out after every update of the filter, it can be used for real-time applications.

2.2. Data Processing Model of GNSS PPP

In this section, the multi-GNSS dual-frequency PPP model will be introduced. For the sake of simplicity, here we provide directly the full-rank observational equations. Equations (20) and (21) are the observational equations of pseudo-range and carrier-phase observations for GPS; Equations (22) and (23) are those of Galileo; and Equations (24) and (25) are those of BDS.

$$P_{j,k}^s - \rho_k^s = G_{p,k}^s x_{p,k} + \Delta t_k + \eta_k^s T_k + \mu_j I_k^s + \varepsilon_{P,j,k}^s \tag{20}$$

$$\Phi_{j,k}^{s} - \rho_{k}^{s} = G_{p,k}^{s}x_{p,k} + \Delta t_{k} + \eta_{k}^{s}T_{k} + \mu_{j}I_{k}^{s} + \lambda_{j}N_{j}^{s} + \varepsilon_{\Phi,j,k}^{s} \tag{21}$$

$$P_{j,k}^{s} - \rho_{k}^{s} = G_{p,k}^{s}x_{p,k} + \Delta t_{k} + ISB_{k}^{E} + \eta_{k}^{s}T_{k} + \mu_{j}I_{k}^{s} + \varepsilon_{P,j,k}^{s} \tag{22}$$

$$\Phi_{j,k}^{s} - \rho_{k}^{s} = G_{p,k}^{s}x_{p,k} + \Delta t_{k} + ISB_{k}^{E} + \eta_{k}^{s}T_{k} + \mu_{j}I_{k}^{s} + \lambda_{j}N_{j}^{s} + \varepsilon_{\Phi,j,k}^{s} \tag{23}$$

$$P_{j,k}^{s} - \rho_{k}^{s} = G_{p,k}^{s}x_{p,k} + \Delta t_{k} + ISB_{k}^{C} + \eta_{k}^{s}T_{k} + \mu_{j}I_{k}^{s} + \varepsilon_{P,j,k}^{s} \tag{24}$$

$$\Phi_{j,k}^{s} - \rho_{k}^{s} = G_{p,k}^{s}x_{p,k} + \Delta t_{k} + ISB_{k}^{C} + \eta_{k}^{s}T_{k} + \mu_{j}I_{k}^{s} + \lambda_{j}N_{j}^{s} + \varepsilon_{\Phi,j,k}^{s} \tag{25}$$

Superscript s indicates the index of satellite, while subscripts j and k indicate the indices of frequency and epoch; $P_{j,k}^{s}$ and $\Phi_{j,k}^{s}$ are pseudo-range and carrier-phase observations, respectively; ρ_{k}^{s} is the approximated distance between satellite and receiver; $x_{p,k}$ is the positioning error with coefficient $G_{p,k}^{s}$; Δt_{k} is the clock error of receiver; ISB_{k}^{E} and ISB_{k}^{C} are Inter-System Bias (ISB) for Galileo and BDS, respectively; T_{k} is the zenith tropospheric wet delay with mapping function η_{k}^{s}; I_{k}^{s} is the total electron content (TEC) in the unit of TECU, with $\mu_{j} = 40.3 \times 10^{16}/f_{j}^{2}$ that converts I_{k}^{s} to signal delay on the jth frequency band in meter; N_{j}^{s} is the estimable ambiguity with wavelength λ_{j}; and $\varepsilon_{P,j,k}^{s}$ and $\varepsilon_{\Phi,j,k}^{s}$ are observational noise for pseudo-range and carrier-phase, respectively. Other systematic errors are corrected and omitted. $x_{p,k}$, Δt_{k}, ISB_{k}^{E}, ISB_{k}^{C}, T_{k}, I_{k}^{s}, and N_{j}^{s} are set as parameters.

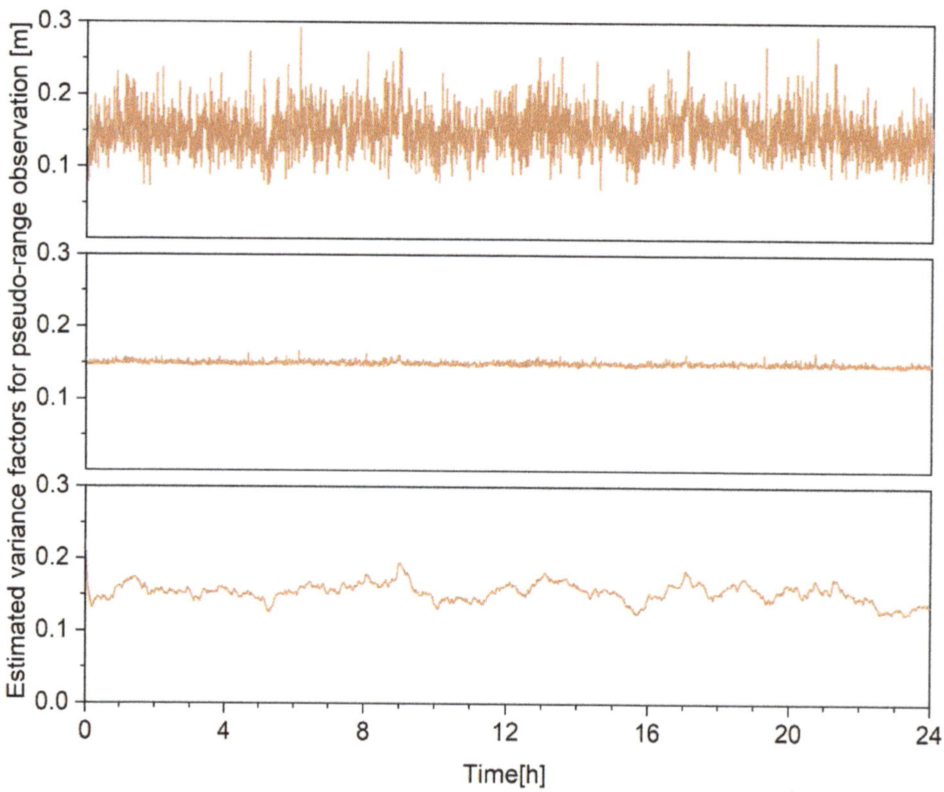

Figure 1. The estimations of the variance factor for pseudo-range observation, single epoch (**top**), accumulating normal matrices of multiple epochs (**middle**), and with fading process (**bottom**).

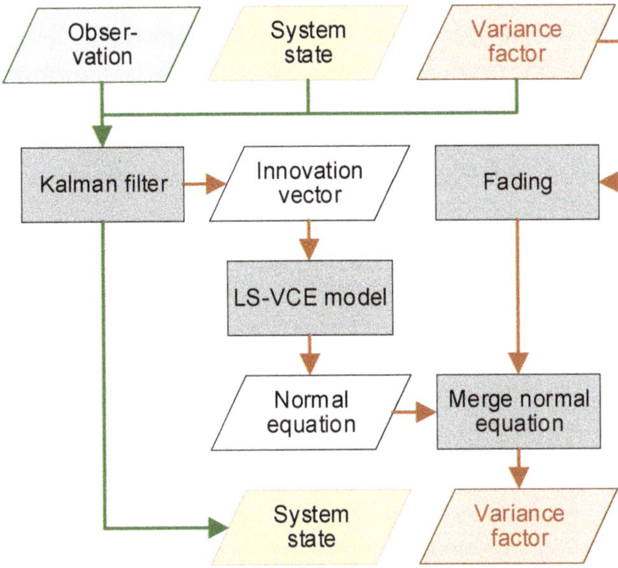

Figure 2. Flow chart for the update of Kalman filter with ASM.

The dynamic model and observational model can be formed as Equations (5) and (6), where $\boldsymbol{x}_k = \left[\boldsymbol{x}_{R,k}^T, \boldsymbol{x}_{S,k}^T\right]^T$, $\boldsymbol{x}_{R,k} = \left[\boldsymbol{x}_{p,k}^T, T_k, \Delta t_k, ISB_k^E, ISB_k^C\right]^T$, $\boldsymbol{x}_{S,k} = \left[\boldsymbol{x}_{S1,k}^T, \cdots, \boldsymbol{x}_{Sn,k}^T\right]^T$, $\boldsymbol{x}_{Ss,k} = \left[I_k^s, N_1^s, N_2^s\right]^T$ for $s = 1, \cdots, n$, n for number of satellites; $\boldsymbol{F_k} = \boldsymbol{E}$; $\boldsymbol{y}_k = \left[\boldsymbol{y}_{1,k}^T, \cdots, \boldsymbol{y}_{n,k}^T\right]^T$, $\boldsymbol{y}_{s,k} = \left[P_{1,k}^s, P_{2,k}^s, \Phi_{1,k}^s, \Phi_{2,k}^s\right]^T$; $\boldsymbol{A}_k = [\boldsymbol{A}_R, \boldsymbol{A}_S]$, $\boldsymbol{A}_R = \left[\boldsymbol{A}_{R,1}^T, \ldots, \boldsymbol{A}_{R,n}^T\right]^T$, $\boldsymbol{A}_{R,s} = \left[\boldsymbol{e}_4 \otimes \boldsymbol{G}_{p,k}^s, \eta_k^s \boldsymbol{e}_4, \boldsymbol{e}_4, \boldsymbol{0}, \boldsymbol{0}\right]$ for GPS, $\boldsymbol{A}_{R,s} = \left[\boldsymbol{e}_4 \otimes \boldsymbol{G}_{p,k}^s, \eta_k^s \boldsymbol{e}_4, \boldsymbol{e}_4, \boldsymbol{e}_4, \boldsymbol{0}\right]$ for Galileo, $\boldsymbol{A}_{R,s} = \left[\boldsymbol{e}_4 \otimes \boldsymbol{G}_{p,k}^s, \eta_k^s \boldsymbol{e}_4, \boldsymbol{e}_4, \boldsymbol{0}, \boldsymbol{e}_4\right]$ for BDS, $\boldsymbol{A}_S = \boldsymbol{E}_n \otimes \boldsymbol{S}$ with

$$\boldsymbol{S} = \begin{bmatrix} \mu_1 & & \\ \mu_2 & & \\ -\mu_1 & 1 & \\ -\mu_2 & & 1 \end{bmatrix} \tag{26}$$

Meanwhile, the variance matrices $\boldsymbol{Q}_{p,k}$ and $\boldsymbol{Q}_{m,k}$ are divided as

$$\boldsymbol{Q}_{p,k} = blkdiag\left(\sigma_{x_p,k}\boldsymbol{E}_3, \sigma_{T,k}, \sigma_{\Delta t,k}, \sigma_{ISB^E,k}, \sigma_{ISB^C,k}, \boldsymbol{E}_n \otimes diag(\alpha_{prc}^s \sigma_{I,k}, \sigma_{N^s,k}, \sigma_{N^s,k})\right) \tag{27}$$

$$\boldsymbol{Q}_{m,k} = diag(\alpha_{ele}^1, \cdots, \alpha_{ele}^n) \otimes diag(\sigma_{P,k}, \sigma_{P,k}, \sigma_{\Phi,k}, \sigma_{\Phi,k}) \tag{28}$$

where α_{ele}^s is the elevational factor calculated with $\alpha_{ele}^s = \frac{1}{\sin(ele^s)}$ for the elevation ele^s of satellite s; $\alpha_{prc}^s = \frac{|P_{prc}|\cdot|P_{sat} - P_{rec}|}{P_{prc}\cdot(P_{sat} - P_{rec})}$ is the ionosphere mapping factor; P_{sat} and P_{rec} are the positions of the satellite and receiver; $P_{prc} = P_{rec} + kD$ is the position of the ionospheric piercing point of the single-layer ionospheric model, where $D = \frac{P_{sat} - P_{rec}}{|P_{sat} - P_{rec}|}$ is the unitary vector from receiver to satellite and k is calculated by solving the equation $|P_{rec} + kD| = R + r$; and $R = 6,378,137$ m and $r = 450,000$ m are the length of the semi-major axis of Earth and height of the single-layer ionospheric model, respectively.

3. Results

In this section, the benefits of the ASM to several topics of PPP will be analyzed, including the estimation of variance factors, PPP float solution, PPP-AR, cycle-slip detection, reconvergence after data interruption, and interruption repair, where PPP tests are carried out for the model of Section 2.2 with either the predefined stochastic model or the ASM. Please note that although these tests are carried out in post-processing, the ASM can be used for real-time applications.

3.1. ASM Reflecting the Observational Conditions

To assess the ability of the ASM of reflecting the observational conditions, PPP tests with the ASM will be carried out for the model in Section 2.2 for global stations to obtain the estimated variance factors. Multi-GNSS observational data from 106 global stations on DOY 1~3 of 2023 is collected and processed with the configuration in Table 1. The precise products come from the Center for Orbit Determination in Europe (CODE), an analysis center of International GNSS Service (IGS), and include orbit, clock error, code bias, and phase bias [41,42].

Table 1. Processing scheme for the analyses of variance factors.

Items		Values
Data input	Observations	GPS: L1/L2; BDS: B1I/B3I; Galileo: E1/E5a
	Sampling rate	30 s
	Satellite orbit and clock	CODE final MGEX product
Parameter estimation	Estimator	Kalman filter
	PPP models	Uncombined PPP
Measurement noise	Pseudo range	0.2 m
	Carrier phase	2 mm
Dynamic model	Position	White noise
	Tropospheric wet delay	Random walk
	Receiver clock error	White noise
	Code bias	Random walk
	Phase bias	Random walk
	Ambiguity	Random constant
	Ionospheric delay	Random walk

3.1.1. Observational Noises of Pseudo-Range and Carrier-Phase Observations

Figure 3 shows the estimated variance factors of the pseudo-range observations for all stations, which are transformed to the unit of meter. The factors of the three systems vary around 0.1 m for most stations with visible temporal variations, which shows that the observational conditions are overall stable with slight fluctuations through time.

Furtherly, Figure 4 shows the estimated variance factors of the carrier-phase observations for all stations. The curves of most stations vary around 1 mm for all three systems and the temporal variations are also preserved.

Overall, the ASM can estimate the variance factors of both pseudo-range and carrier-phase observations and the temporal variations are preserved. It should be noted that the estimated factors are influenced by not only the observational noise but also systematic errors such as the precision of orbit/clock products. The factors reflect the overall observational conditions, which are influenced by multiple error sources; therefore, they can enhance the performance of positioning.

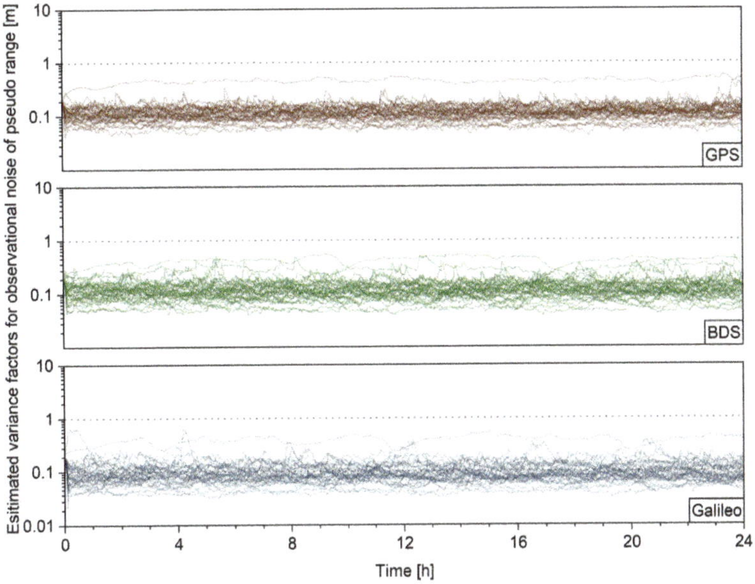

Figure 3. Estimated variance factors of the pseudo-range observations for all stations.

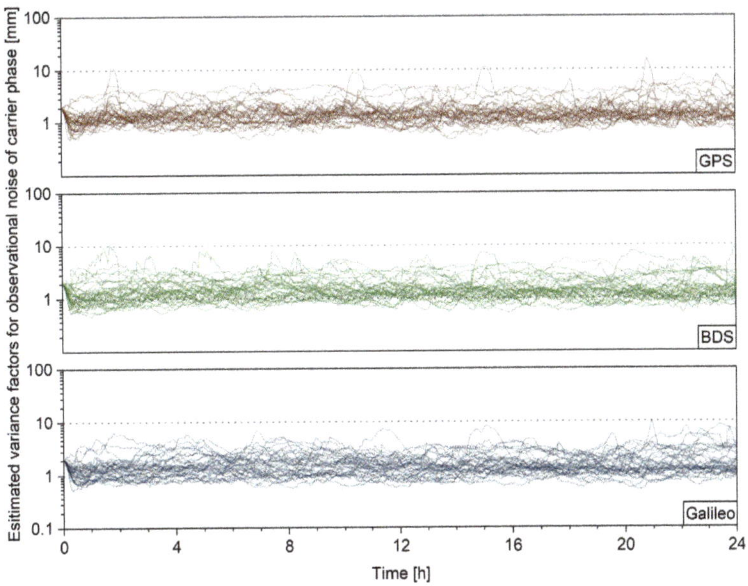

Figure 4. Estimated variance factors of the carrier-phase observations for all stations.

3.1.2. Process Noises of Atmospheric Parameters

Besides refining the observational stochastic model, the stochastic model of process noise is also improvable with an ASM. Figure 5 shows the estimated variance factors of ionospheric delay for all stations, where the x-axis is local time. The curves are correlated with local time and reach peak value at 12–18 every day, which matches the fact that the ionosphere is more active in daytime. It shows that the ASM is very helpful in adjusting the variance factor of ionospheric delay and could enhance cycle-slip detection, which requires refined ionospheric modeling for effective detection [43,44]. The enhancement of the ASM of cycle-slip detection with DIA will be examined in Section 3.4.

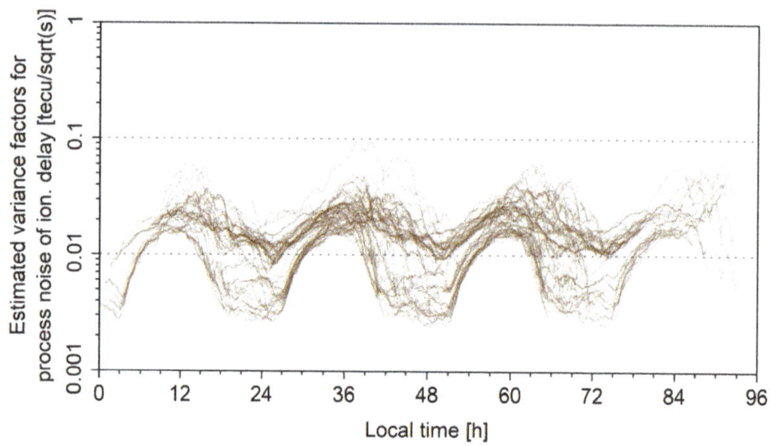

Figure 5. Estimated variance factors of ionospheric delay for all stations.

Furtherly, Figure 6 shows the estimated variance factors of zenith tropospheric delay. The curves are almost straight lines, which shows that the factor of tropospheric delay is not estimable with the ASM and keeps its initial value. This is due to the fact that the variations of tropospheric delay between epochs are so small that they are overwhelmed by observational noises which are at least 10 times larger than them. However, the estimated factors do not diverge either, which shows that the algorithm is stable.

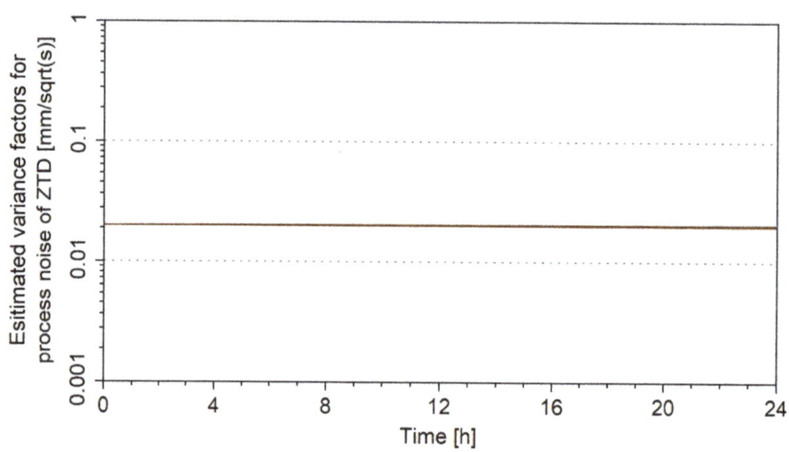

Figure 6. Estimated variance factors of zenith tropospheric delay for all stations.

3.1.3. Process Noises of ISBs

Finally, Figure 7 shows the estimated variance factors of ISB. Like the factors of tropospheric delay, the curves are almost straight lines, which shows that the factors of ISB are not estimable with the ASM. This is due to the fact that the ISBs are linear combinations of receiver code biases, which are mostly stable.

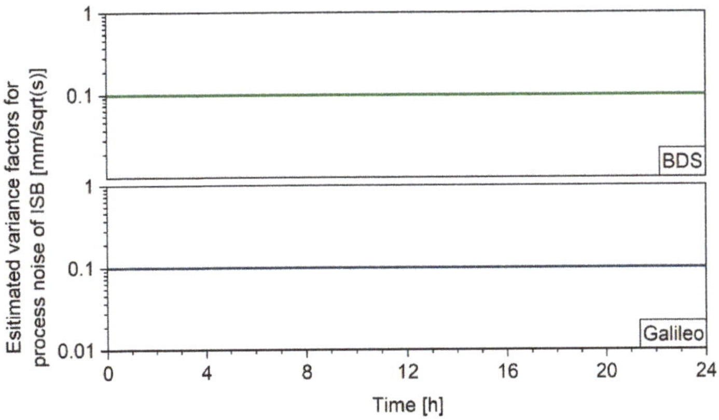

Figure 7. Estimated variance factors of ISB for all stations.

3.2. Standard PPP Tests

In this section, PPP tests are carried out with an ASM (Adaptive) and without an ASM (Original) to show whether the positioning performance is improved with the proposed model.

3.2.1. PPP Results for Global Stations

Figure 8 shows the PPP positioning errors for IGS stations CHPG, MET3, and YAR3. The plots on the left and the right panels are very similar, which shows that the ASM has little impact on the performance.

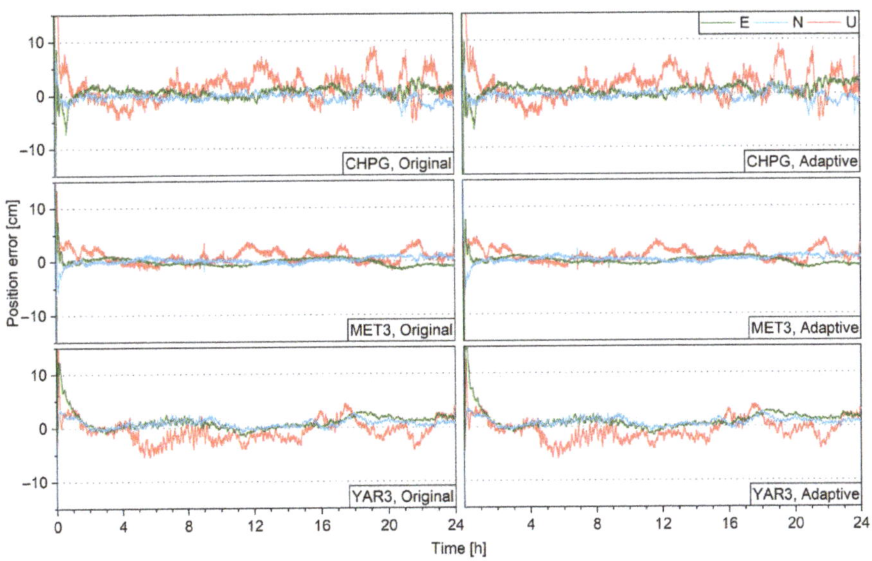

Figure 8. PPP positioning errors without (**left**) and with (**right**) ASM.

Furtherly, with the solutions of global stations, the percentage of converged stations for different thresholds are shown in Figure 9 and the hists for 3D positioning errors are given in Figure 10. It shows that the ASM has little effect on both convergence time and positioning precision. This is due to that the estimated pseudo-range and carrier-phase variance factors in Figures 3 and 4 are all close to their initial values and show little inter-station differences. Therefore, the ASM is not beneficial to PPP float solution when the stochastic model is suitable.

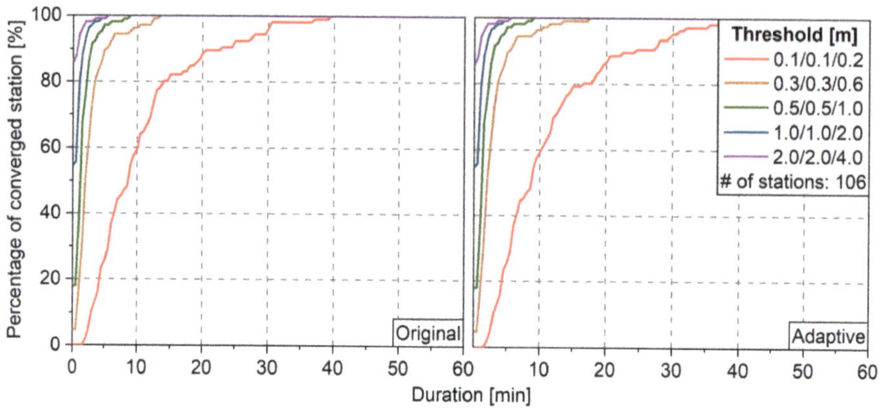

Figure 9. Percentage of converged stations without (**left**) and with (**right**) ASM.

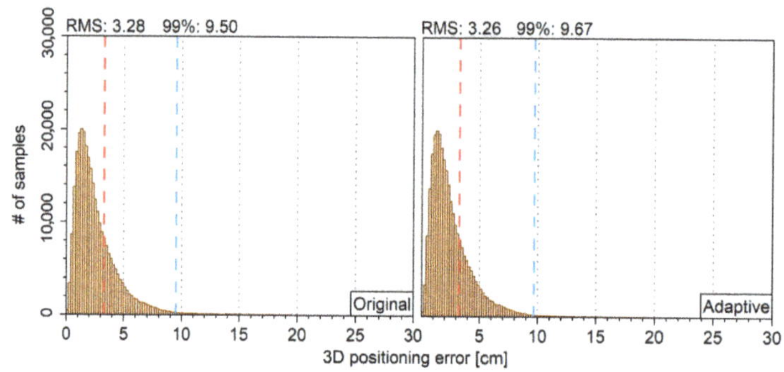

Figure 10. Hists for 3D positioning error without (**left**) and with (**right**) ASM.

3.2.2. PPP Results for Unsuitable Stochastic Model

The results for the IGS station TONG are specifically analyzed, which is the only station whose observational errors are obviously larger than other stations. Figure 11 shows the estimated observational variance factors. The factors for pseudo-range and carrier-phase observations are about 0.6 m and 6 mm, respectively, which are larger than other stations. Though the exact cause for the large factors remains unknown, enhancing the stochastic model could still be beneficial to the performance. Figure 12 shows the positioning errors with and without the ASM and Table 2 shows the RMSs of positioning errors on east, north, and up directions after convergence. Note that the stochastic model for results without the ASM is the same as those in Section 3.2.1. The positioning precision is improved with the help of the ASM for 57%, 51%, and 25% for east, north, and up directions. It shows that when the predetermined stochastic model is unsuitable, the ASM can be adjusted according to the actual statistical properties of the observational noise and improve the positioning performance. Unfortunately, the data quality for most IGS stations is good, and it is difficult to find another example to prove the benefits of ASM to PPP float solutions. Non-IGS data under bad observational conditions is needed for future study.

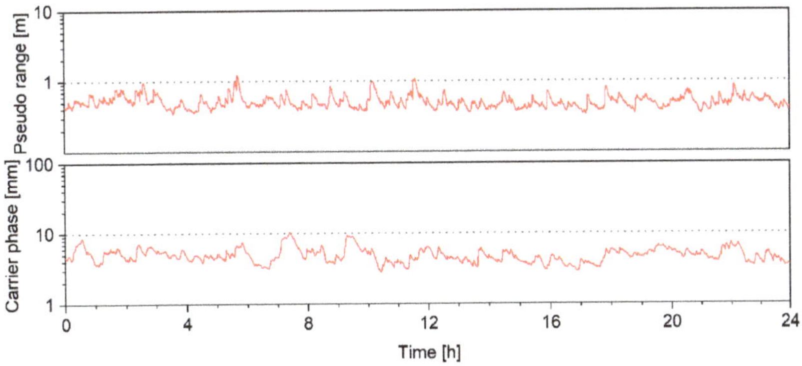

Figure 11. Estimated pseudo-range and carrier-phase variance factors for TONG station.

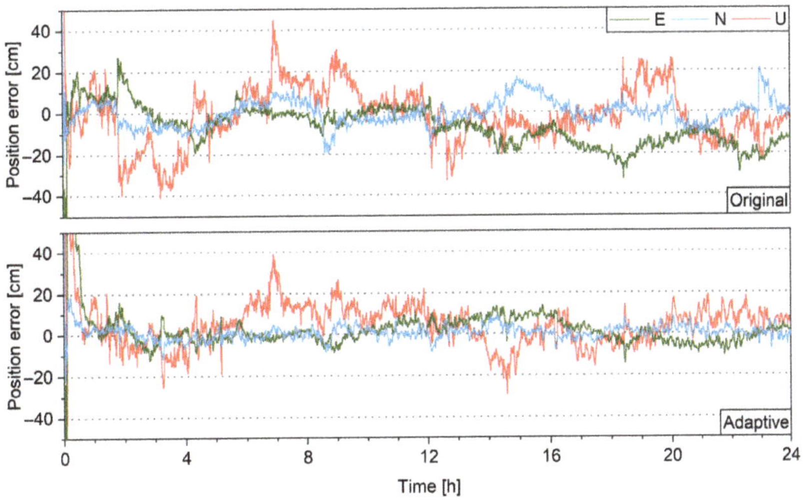

Figure 12. Positioning error for TONG station without (**top**) and with (**bottom**) ASM.

Table 2. RMS of the position errors for TONG station.

Processing Scheme	RMS on Three Directions (m)		
	East	North	Up
Original	0.116	0.061	0.136
Adaptive	0.050	0.030	0.102

To further show the benefits of the ASM, PPP without an ASM is carried out again for the TONG station with standard deviations of pseudo-range and carrier-phase observations as 0.6 m and 6 mm, whose positioning error is shown in Figure 13. The result is different from the top panel of Figure 12 while similar to the bottom panel, which shows that the positioning performance is improved when the stochastic model is adjusted. This proves that the benefits of the ASM to positioning come from the ability of adjusting the stochastic model on-the-fly.

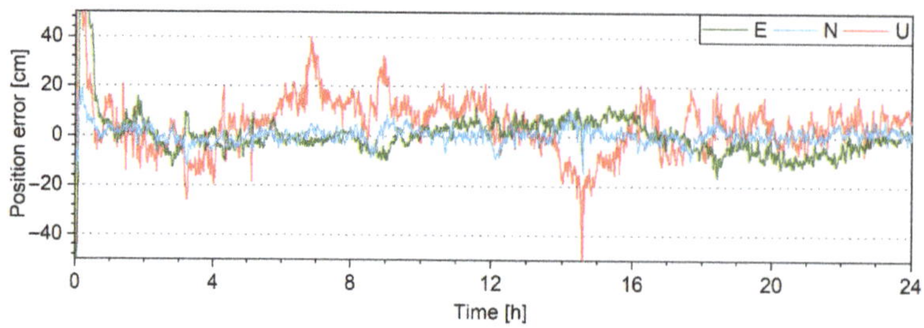

Figure 13. Positioning error for TONG station with standard deviations of pseudo-range and carrier-phase observations as 0.6 m and 6 mm.

3.3. PPP-AR Test

Besides PPP float solution, ambiguity resolution is also influenced by the stochastic model, which will be analyzed in this section. The ambiguity resolution with the Integer Bootstrapping (IB) strategy will be introduced first; then, the performance of PPP-AR with an ASM will be analyzed.

3.3.1. Ambiguity Resolution

The IB strategy is a widely used ambiguity resolution strategy which fixes one ambiguity at a time and corrects the remaining float ambiguities with the fixed one [45]. For integer ambiguity parameter $\boldsymbol{x}_A$ and its float estimation $\hat{\boldsymbol{x}}_A$ with variance matrix $\boldsymbol{Q}_{\hat{x}_A}$, the ith ambiguity is fixed with the IB strategy as

$$\check{x}_{A,i} = \left[\hat{x}_{A,i|i-1}\right] \tag{29}$$

where $\hat{x}_{A,i|i-1}$ is $\hat{x}_{A,i}$ corrected with the fixed solution of its previous $i-1$ ambiguities, i.e., $\hat{x}_{A,i}$ under the conditions of $\check{x}_{A,1}, \cdots, \check{x}_{A,i-1}$. The recursive evaluation equation for $\hat{x}_{A,i|i-1}$ reads

$$\hat{x}_{A,i|k} = \hat{x}_{A,i|k-1} - \frac{Q_{\hat{x}_A,i,k|k-1}\left(\hat{x}_{A,k|k-1} - \check{x}_{A,k}\right)}{Q_{\hat{x}_A,k,k|k-1}} \tag{30}$$

where $Q_{\hat{x}_A,i,k|k-1}$ and $Q_{\hat{x}_A,k,k|k-1}$ are $Q_{\hat{x}_A,i,k}$ and $Q_{\hat{x}_A,k,k}$ corrected with the fixed solution of the previous $i-1$ ambiguities, $Q_{\hat{x}_A,i,k}$ is the entry on the ith row and kth column of $\boldsymbol{Q}_{\hat{x}_A}$ and $Q_{\hat{x}_A,k,k}$ is the kth diagonal entry of $\boldsymbol{Q}_{\hat{x}_A}$. For arbitrary i, j, and k that $i \leq n, j \leq n$, $k < i$, and $k < j$ hold, where n is the number of ambiguities, one has

$$Q_{\hat{x}_A,i,j|k} = Q_{\hat{x}_A,i,j|k-1} - \frac{Q_{\hat{x}_A,i,k|k-1}Q_{\hat{x}_A,k,j|k-1}}{Q_{\hat{x}_A,k,k|k-1}} \tag{31}$$

where for $k = 0$, one has $\hat{x}_{A,i|0} = \hat{x}_{A,i}$ and $Q_{\hat{x}_A,i,j|0} = Q_{\hat{x}_A,i,j}$.

3.3.2. PPP-AR Results for Global Stations

The ambiguity parameters of PPP are not integer due to the existence of phase biases of the satellites and receiver. To obtain ambiguities with integer properties, the satellite phase biases need to be corrected with IGS product and single difference should be applied between satellites to eliminate receiver phase biases. The BIA product from CODE is used to correct the satellite phase biases [41]. Due to the lack of a BDS phase bias product, only the ambiguities of GPS and Galileo are fixable. The AR process starts with float solutions of ambiguities and the corresponding variance matrix. Z-transformation is applied to obtain decorrelated ambiguities. Then, the widely used partial AR (PAR) procedure with the IB strategy is conducted, where a subset of decorrelated ambiguities is selected with a success rate threshold of 99.99% [29,46–48]. Finally, the position parameters are corrected with the fixed decorrelated ambiguities. Figure 14 shows the GPS+Galileo PPP-AR positioning errors for IGS stations AJAC, CHPG, and FAIR. The plots in the left and the right panels are very similar, which shows the ASM has little impact on the performance of PPP-AR.

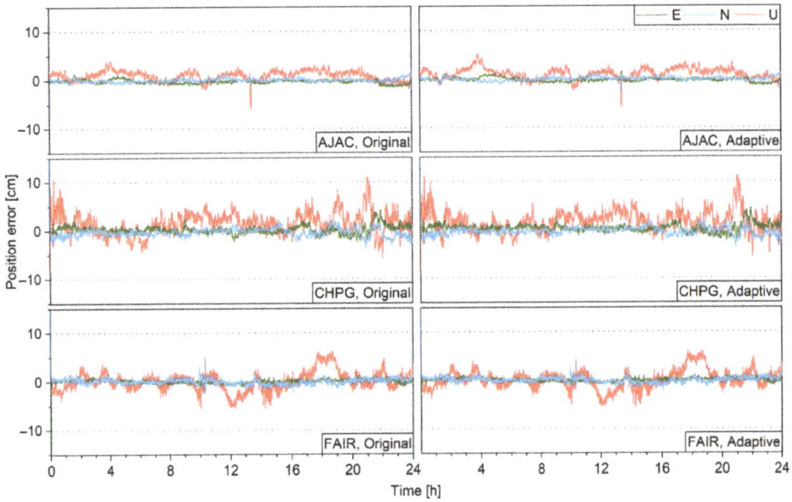

Figure 14. PPP-AR positioning errors without (**left**) and with (**right**) ASM.

With the solutions of all stations, Figure 15 shows the hists for 3D positioning errors. The plots for results with the ASM are similar to those without the ASM, which also happens to the float solution. This shows that, if the stochastic model is suitable, the ASM is not beneficial to the performance of PPP-AR.

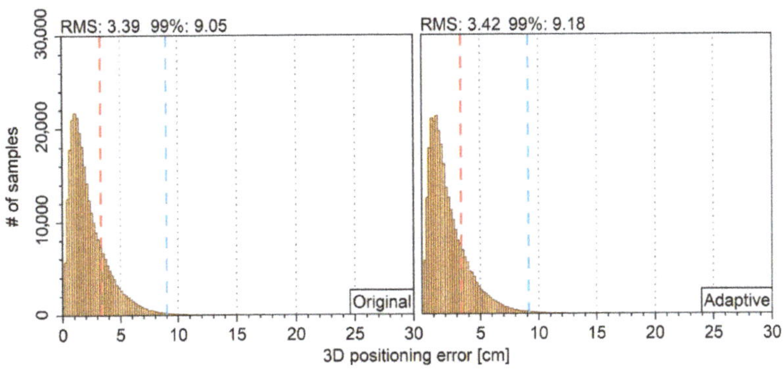

Figure 15. Hist for PPP-AR 3D positioning error without (**left**) and with (**right**) ASM.

3.3.3. PPP-AR Results for Unsuitable Stochastic Model

The results for IGS station TONG are specifically analyzed, whose observational errors are larger than other stations, as is shown in Section 3.2.2. Figure 16 shows the PPP-AR positioning errors for the TONG station and the RMSs on three directions are given in Table 3. With the ASM, the positioning errors on east, north, and up directions are reduced for 36%, 28%, and 36%, respectively. This shows when the predetermined stochastic model is unsuitable, an ASM can enhance the performance of PPP-AR greatly.

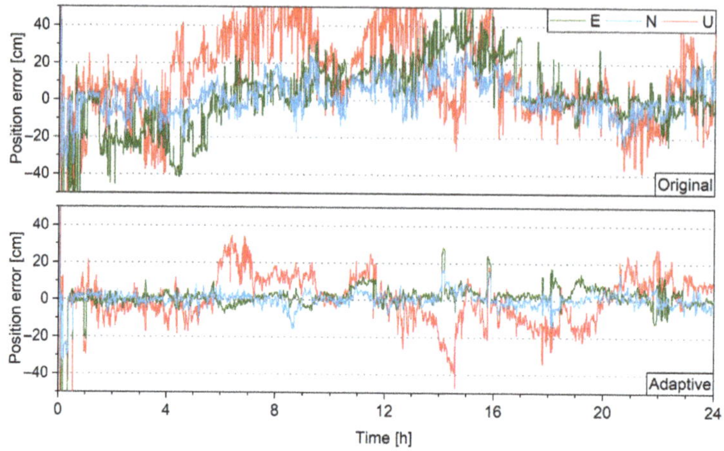

Figure 16. PPP-AR positioning error for TONG station without (**top**) and with (**bottom**) ASM.

Table 3. RMS of the PPP-AR position errors for TONG station.

Processing Scheme	RMS on Three Directions (m)		
	East	**North**	**Up**
Original	0.211	0.109	0.380
Adaptive	0.134	0.078	0.244

To further show the benefits of the ASM, PPP-AR without an ASM is carried out again for the TONG station with standard deviations of pseudo-range and carrier-phase observations as 0.6 m and 6 mm, whose positioning error is shown in Figure 17. The result is better than the top panel of Figure 16 while slightly worse than the bottom panel, which shows the benefits of the ASM to positioning come from the ability of adjusting the stochastic model on-the-fly.

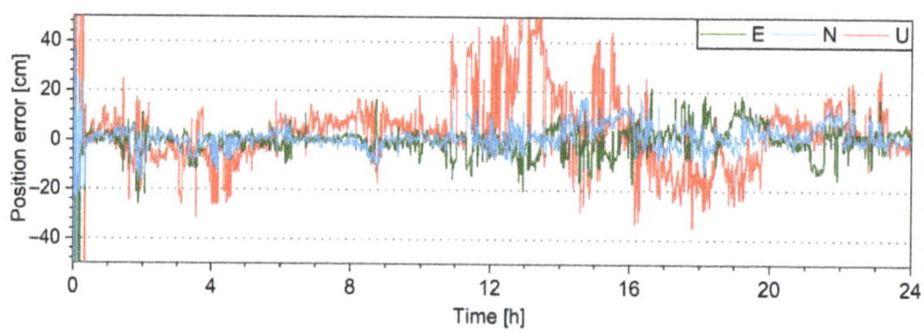

Figure 17. PPP-AR positioning error for TONG station with standard deviations of pseudo-range and carrier-phase observations as 0.6 m and 6 mm.

3.4. Tests on Cycle-Slip Detection

The detection of unmodeled bias with DIA is another algorithm that could benefit from an ASM. DIA could use all the redundant information and construct an optimal procedure for anomaly detection [32]. Since DIA uses the stochastic model, it can show the ability of an ASM to enhance the detection effectiveness when the predefined stochastic model is improper. Among the anomalies of PPP, the cycle slip is the most important one due to its impact on positioning; therefore, cycle-slip detection with DIA will be tested to examine the benefit of the ASM. The DIA algorithm will be introduced; then, the PPP cycle-slip detection with DIA will be designed; finally, the performance will be analyzed.

3.4.1. DIA Algorithm

For an arbitrary observational model

$$y = Ax + \varepsilon_m \tag{32}$$

where $D(\varepsilon_m) = Q_m$, one has the least square estimation (LSQ)

$$\hat{x} = N^{-1}W, \quad Q_{\hat{x}} = N^{-1} \tag{33}$$

where $N = A^T Q_m^{-1} A$ and $W = A^T Q_m^{-1} y$ form the normal equation $W = N\hat{x}$. The estimated value of the LSQ residual reads

$$\hat{\varepsilon}_m = P_A^{\perp} y, \quad Q_{\hat{\varepsilon}_m} = Q_m - AQ_{\hat{x}}A^T \tag{34}$$

where $P_A^{\perp} = E - P_A$, $P_A = AN^{-1}A^T Q_m^{-1}$. Equation (32) is also known as the null hypothesis. When there is unmodeled bias in the observational model, one has the alternative hypothesis

$$y = Ax + Bb + \varepsilon_m \tag{35}$$

where b is the unmodeled bias parameter with corresponding design matrix B. Since Equation (35) is solved without considering b, the solution of Equation (35) reads

$$\hat{x} = A^{+} y + A^{+} Bb \tag{36}$$

$$\hat{\varepsilon}_m = P_A^{\perp} y + P_A^{\perp} Bb \tag{37}$$

where $A^{+} = N^{-1} A^T Q_m^{-1}$. This shows that the unmodeled bias is absorbed by both $\hat{x}$ and $\hat{\varepsilon}_m$, causing both increased positioning error and a larger residual. Note that there could be different types of unmodeled biases. The DIA algorithm is carried out in three steps: it determines whether there is unmodeled bias in the detection step, distinguishes what type of bias it is in the identification step, and finally modifies the observational model to include the unmodeled bias for better positioning performance in the adaptation step. For the detection, an overall test for determining whether there is unmodeled bias is carried out as

$$\hat{\varepsilon}_m^T Q_m^{-1} \hat{\varepsilon}_m > \chi_\alpha^2(r, 0) \tag{38}$$

where $\chi_\alpha^2(r, 0)$ is the upper α-quantile of a central chi-squared distribution with r degrees of freedom, α is the false alarm probability of the overall test, and r is the rank of the residual estimation. If Equation (38) holds, there is unmodeled bias during the observation, which needs to be identified. For the identification with n different alternative hypothesis, T_i is calculated for the ith alternative hypothesis $y = Ax + B_i b_i + \varepsilon_m$ as

$$T_i = \hat{\varepsilon}_m^T P_{\overline{B}_i}^T Q_m^{-1} P_{\overline{B}_i} \hat{\varepsilon}_m \tag{39}$$

where $P_{\overline{B}_i} = \overline{B}_i (\overline{B}_i^T Q_m^{-1} \overline{B}_i)^{-1} \overline{B}_i^T Q_m^{-1}$, $\overline{B}_i = P_A^{\perp} B_i$. When the dimensions of each b_i are the same, the alternative hypothesis with the biggest T_i is selected as the most likely one. Finally for the adaptation, the solution $\hat{x}$ and its variance matrix $Q_{\hat{x}}$ are updated as

$$\hat{x} = \hat{x} - A^{+} B_i \hat{b}_i, Q_{\hat{x}} = Q_{\hat{x}} + A^{+} B_i Q_{\hat{b}_i} B_i^T A^{+,T} \tag{40}$$

where $\hat{b}_i = (\overline{B}_i^T Q_m^{-1} \overline{B}_i)^{-1} \overline{B}_i^T Q_m^{-1} y$, $Q_{\hat{b}_i} = (\overline{B}_i^T Q_m^{-1} \overline{B}_i)^{-1}$.

3.4.2. PPP Cycle-Slip Detection with DIA

To detect cycle slip with DIA, the model of the Kalman filter should be expressed in the form of Equation (32) and the alternative hypothesis for cycle slip should be designed. With the estimated system state at epoch $k - 1$ in Equation (7), dynamic model in Equation (5), and observational model in Equation (6), the single-epoch model of the Kalman filter can be summarized as

$$\underbrace{\begin{bmatrix} \hat{x}_{k-1} \\ 0 \\ y_k \end{bmatrix}}_{y} = \underbrace{\begin{bmatrix} E & \\ F_k & -E \\ & A_k \end{bmatrix}}_{A} \underbrace{\begin{bmatrix} x_{k-1} \\ x_k \end{bmatrix}}_{x} + \underbrace{\begin{bmatrix} \varepsilon_{\hat{x}_{k-1}} \\ \varepsilon_{p,k} \\ \varepsilon_{m,k} \end{bmatrix}}_{\varepsilon_m} \tag{41}$$

The alternative hypothesis for the dynamic model with unmodeled bias reads

$$x_k = F_k x_{k-1} + c_i b + \varepsilon_{p,k} \tag{42}$$

where c_i is a unit vector with the ith entry as 1, and b is the bias that introduced for the ith parameter of x_k. To detect the cycle slip, an alternative hypothesis in the form of Equation (42) is set up for each ambiguity parameter of PPP, where b is actually the cycle-slip parameter. Then, DIA algorithm is applied to detect the cycle slip.

3.4.3. Results for the Cycle-Slip Detection

The test is carried out for observational data with simulated cycle slips. For the observation of each station, two carrier-phase observations are randomly selected every 10 min and simulated with cycle slips of one cycle. Note that the observations are modified for not only the epoch that the cycle slips happen in but also subsequent epochs. GPS + BDS + Galileo PPP is performed for global stations with DIA detecting the simulated cycle slips, where either a predefined stochastic model or an ASM is used. Figure 17 shows the positioning error for the IGS station AJAC. Without the ASM, the positioning error degrades due to some undetected cycle slips; with the help of the ASM, the positioning error looks normal and free of undetected cycle slips. The main reason could be that the enhanced stochastic model for ionospheric delay makes the detection with dual-frequency data easier.

Figure 18 shows the hists for 3D positioning error. Though not comparable with those without simulated cycle slips, the positioning error of the ASM is dramatically reduced compared with that of the predefined stochastic model. Table 4 shows the numbers of simulated and detected cycle slips of all stations. The detection rate is improved from 81% to 92% with the help of the ASM, which indeed shows that the enhanced stochastic model is helpful to cycle-slip detection.

Table 4. Numbers of simulated and detected cycle slips.

	Original	**Adaptive**
Simulated	30,528	30,528
Detected	24,711	28,047
Detection rate	81%	92%

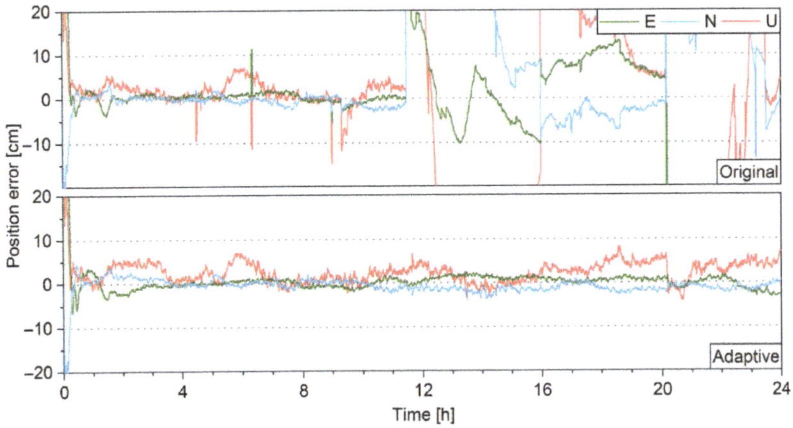

Figure 18. PPP positioning error with simulated cycle slips which are detected with DIA.

3.5. Tests on Reconvergence and Interruption Repair

The reconvergence and interruption repair of PPP after data interruption are also influenced by the stochastic model. Here the interruption means all satellites are not visible for a few minutes and then reappear, which could happen in a city environment. Direct reconvergence is the basic method for handling interrupted data, which is to keep the parameters of the satellites and continue updating them once the satellites reappear. However, the ambiguities have to be reset due to the interrupted tracking of carrier-phase signals, whose reconvergences take dozens of minutes. Furthermore, the difference between the ambiguities before and after the interruption is actually a cycle-slip parameter with integer properties, which could be fixed with AR algorithm to accelerate the reconvergence [49–53]. The latter method is called interruption repair. The direct reconvergence requires a reliable time update of the filter to preserve the information, which would require a suitable stochastic model; the interruption repair is based on the AR algorithm, which is also influenced by the stochastic model. The variance factor of ionospheric delay is the only time-variant factor among all factors. The enhanced stochastic model of ionospheric delay should be beneficial to convergence. Unfortunately, the estimated factor also takes time to converge, which makes the ASM not beneficial to the convergence. However, the factor is already converged when interruptions occur; therefore, the stochastic model of ionospheric delay refined with an ASM could be beneficial to reconvergence and interruption repair, which will be tested in this section.

3.5.1. Reconvergences After 3 Min Interruptions

For the observation data of each station, 3 min of data interruptions are simulated every 2 h by simply removing all observations during the interruptions. GPS PPP tests are carried out, where the ionospheric delay parameters are kept during the interruption while ambiguity parameters are reset. Figure 19 shows the positioning errors for the IGS station BUCU. For data segments such as 2~4, 4~6, and 6~8, the ASM improves the reconvergence. This is mainly because of the enhanced stochastic model for ionospheric delay. Figure 20 shows the percentage of reconverged data segments for all stations. The time for 60% of the segments to reconverge to the threshold of 0.1/0.1/0.2 m is reduced from 19 min to 15.5 min, showing the benefits of ASM.

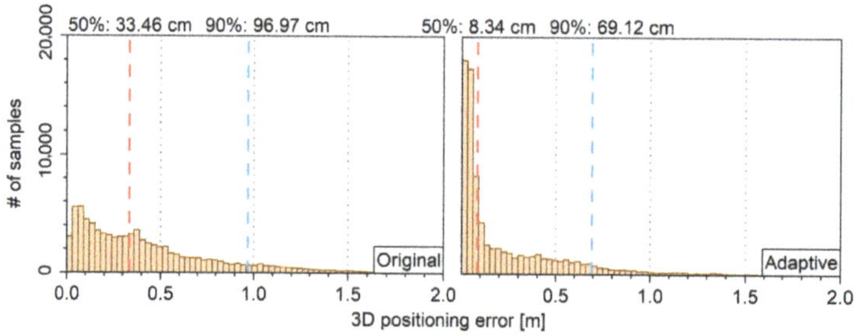

Figure 19. Hists for 3D positioning error with simulated cycle slips which are detected with DIA.

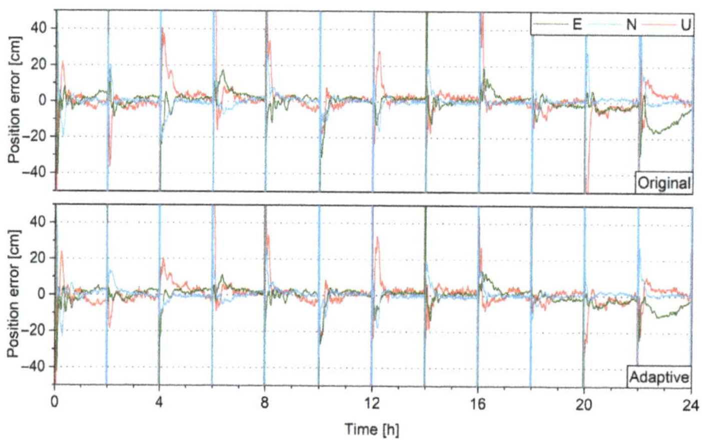

Figure 20. PPP positioning errors with 3 min data interruptions every 2 h, without (**top**) and with (**bottom**) ASM.

To show the impact of refined ionospheric stochastic model to reconvergence, Figure 21 shows the PPP positioning errors for the IGS station BUCU where ionospheric delay is modeled as white noise process. It shows that, as the model of ionospheric delay cannot be enhanced by the ASM, the reconvergence is not improved at all.

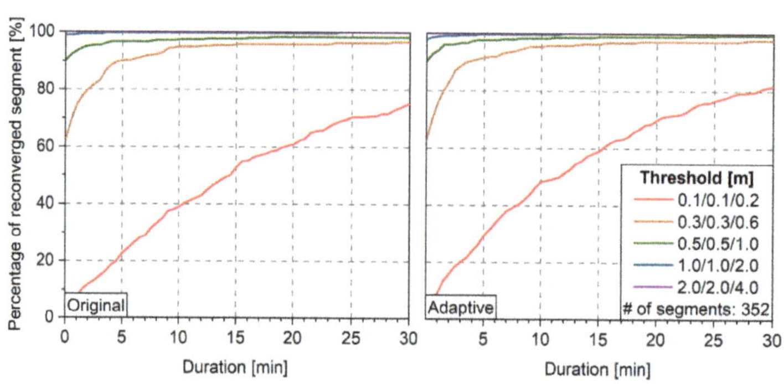

Figure 21. Percentage of reconverged data segments without (**left**) and with (**right**) ASM.

3.5.2. Interruption Repair for 3 Min Interruptions

The algorithm for repairing interruption will be introduced first. The basic idea is to fix the cycle slip between the ambiguity before the interruption and the one after the interruption. The ionospheric delay and ambiguity parameters of all satellites are kept during the interruption. Once the satellites reappear, new ambiguity parameters are introduced and updated during the filter update. Then cycle slips are derived from the ambiguities before the interruption and the newly introduced one. The ambiguity resolution method in Section 3.3 is applied to the cycle-slip parameters and the system state is updated with the fixed cycle slips. This attempt of repairing repeats for each subsequent epoch. After 30 min, the ambiguities before the interruption are removed and the remaining "float" cycle slips will not be fixed anymore. Since the cycle slips are obtained via subtraction between old and new ambiguities, positive correlation between ambiguities would yield more precise and easy-to-fix cycle slips. With the ionospheric information preserved, the new ambiguity is linked with the old one through ionospheric parameters; therefore, the ASM could help the repair through precise modeling of the ionospheric delay.

PPP tests are carried out for global stations with interruption repair. Figure 22 shows the positioning errors with interruption repair for the IGS station CPVG. With the predefined stochastic model, at least six data segments are not repaired; while with the ASM, only two data segments are not repaired and the remaining segments instantly reconverge once the satellites reappear. This shows that with the ionospheric model refined, the repair of interruption becomes more effective. Finally, Figure 23 shows the percentage of reconverged data segments for all stations. The time for 80% of the segments to converge to the threshold of 0.1/0.1/0.2 m is greatly reduced from 30 min to 13.5 min, which proves the enhanced ionospheric stochastic model is indeed beneficial to the interruption repair.

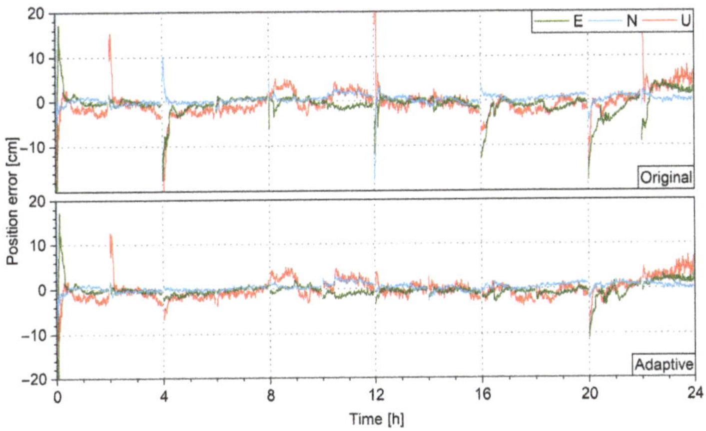

Figure 22. PPP positioning errors with interruption repair, without (**top**) and with (**bottom**) ASM.

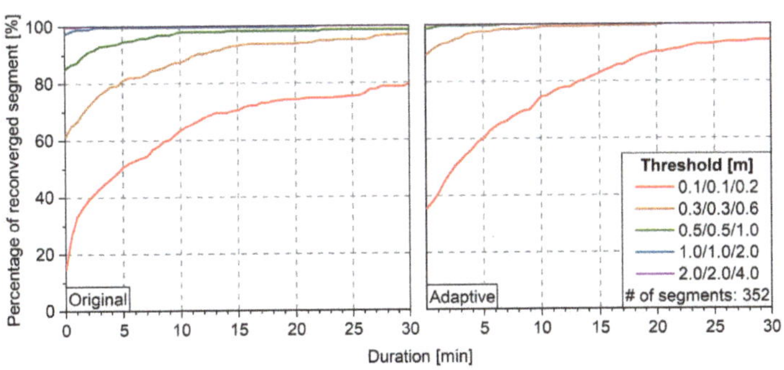

Figure 23. Percentage of reconverged data segments with interruption repair, without (**left**) and with (**right**) ASM.

4. Discussion

An adaptive stochastic model (ASM) significantly enhances various aspects of Precise Point Positioning (PPP), necessitating a comprehensive assessment of its benefits. This paper implements an ASM by integrating the least-squares variance component estimation (LS-VCE) algorithm with the Kalman filter model and evaluates its contributions to PPP performance from multiple perspectives, including observational condition reflection, PPP float and fixed solutions, cycle-slip detection using DIA, reconvergence, and interruption repair. The main findings demonstrate that the ASM effectively improves the observational variance matrix when the predefined model is inadequate, benefiting both PPP float and fixed solutions. Additionally, the ASM enhances the ionospheric stochastic model, captures its temporal variations, and contributes to higher DIA success rates, faster reconvergence after data interruptions, and improved interruption repair. The detailed conclusions are summarized as follows:

1. Variance Factor Estimation: Variance factors for the observational noise of pseudo-range and carrier-phase measurements are estimable and capture time-varying observational conditions. The variance factor for the ionospheric delay process noise is estimable and reflects the temporal variation of the ionosphere. The variance factor for tropospheric delay is not estimable due to its minimal variation between epochs, which is overwhelmed by the noise.
2. PPP Float and Fixed Solutions: The ASM offers no significant advantage when the predefined stochastic model is appropriate. However, when observational noise is large, rendering the predefined model unsuitable, the ASM adjusts the variance factors to deliver superior PPP float and fixed solutions.
3. Cycle-Slip Detection with DIA: With the predefined stochastic model, 19% of simulated cycle slips remain undetected, resulting in degraded positioning precision at the meter level. With the ASM, the undetected cycle slips are reduced to 8%, and positioning precision improves. Notably, the 50th percentile error is reduced by 75%.
4. Reconvergence and Interruption Repair: Simulated 3 min data interruptions every 2 h reveal that the ASM reduces the reconvergence time for 60% of data segments to a threshold of 0.1/0.1/0.2 m from 19 min (predefined stochastic model) to 15.5 min. This improvement is attributed to the enhanced variance factor for ionospheric delay process noise. Interruption repair tests show that reconvergence time for 80% of segments to the same threshold decreases from 30 min (predefined stochastic model) to 13.5 min (ASM), underscoring the benefits of the improved ionospheric stochastic model.

In summary, the adaptive stochastic model not only refines variance factor estimation to reflect observational conditions but also significantly enhances PPP float and fixed solutions under challenging conditions, improves cycle-slip detection, and accelerates reconvergence and interruption repair. These findings highlight the broad applicability and effectiveness of an ASM in improving PPP performance. However, the IGS stations used for our tests are mostly under good observational conditions. The ASM will be tested for challenging applications such as maritime positioning in the future to show its ability to improve positioning in poor observational conditions.

Author Contributions: Methodology, Y.Z. (Yanning Zheng) and Y.S.; software, Y.Z. (Yanning Zheng); validation, Y.Z. (Yanning Zheng), Y.Z. (Yubin Zhou), and S.W.; resources, Y.S. and Y.Z. (Yubin Zhou); writing—original draft preparation, Y.Z. (Yanning Zheng) and Y.L.; writing—review and editing, Y.S. and S.W.; funding acquisition, Y.Z. (Yubin Zhou). All authors have read and agreed to the published version of the manuscript.

Funding: This research was supported by the Key Project of Marine Service Industry in Shandong Province (37000025050809990005X).

Data Availability Statement: The data presented in this study are available on the IGS data server of ESA at ftp://gssc.esa.int (accessed on 12 August 2024).

Conflicts of Interest: The authors declare no conflicts of interest.

References

1. Zumberge, J.F.; Heflin, M.B.; Jefferson, D.C.; Watkins, M.M.; Webb, F.H. Precise Point Positioning for the Efficient and Robust Analysis of GPS Data from Large Networks. *J. Geophys. Res.* **1997**, *102*, 5005–5017. [CrossRef]
2. Jin, S.; Wang, J.; Park, P.-H. An Improvement of GPS Height Estimations: Stochastic Modeling. *Earth Planets Space* **2005**, *57*, 253–259. [CrossRef]
3. Li, Y.; Li, B.; Gao, Y. Improved PPP Ambiguity Resolution Considering the Stochastic Characteristics of Atmospheric Corrections from Regional Networks. *Sensors* **2015**, *15*, 29893–29909. [CrossRef] [PubMed]
4. Li, B.; Lou, L.; Shen, Y. GNSS Elevation-Dependent Stochastic Modeling and Its Impacts on the Statistic Testing. *J. Surv. Eng.* **2016**, *142*, 04015012. [CrossRef]
5. Li, B.; Zhang, L.; Verhagen, S. Impacts of BeiDou Stochastic Model on Reliability: Overall Test, w-Test and Minimal Detectable Bias. *GPS Solut.* **2017**, *21*, 1095–1112. [CrossRef]
6. Yang, L.; Li, B.; Li, H.; Rizos, C.; Shen, Y. The Influence of Improper Stochastic Modeling of Beidou Pseudoranges on System Reliability. *Adv. Space Res.* **2017**, *60*, 2680–2690. [CrossRef]
7. Zaminpardaz, S.; Teunissen, P.J.G. DIA-Datasnooping and Identifiability. *J. Geod.* **2019**, *93*, 85–101. [CrossRef]
8. Kim, D.; Langley, R.B. GPS Ambiguity Resolution and Validation: Methodologies, Trends and Issues. In Proceedings of the 7th GNSS Workshop–International Symposium on GPS/GNSS, Seoul, Korea, 30 November–2 December 2000.
9. Amiri-Simkooei, A.R.; Jazaeri, S.; Zangeneh-Nejad, F.; Asgari, J. Role of Stochastic Model on GPS Integer Ambiguity Resolution Success Rate. *GPS Solut.* **2016**, *20*, 51–61. [CrossRef]
10. Han, S. Quality-Control Issues Relating to Instantaneous Ambiguity Resolution for Real-Time GPS Kinematic Positioning. *J. Geod.* **1997**, *71*, 351–361. [CrossRef]
11. Mehra, R. Approaches to Adaptive Filtering. *IEEE Trans. Automat. Contr.* **1972**, *17*, 693–698. [CrossRef]
12. Chin, L. Advances in Adaptive Filtering. In *Control and Dynamic Systems, Advanced in Theory and Applications*; Leondes, C.T., Ed.; Elsevier: Amsterdam, The Netherlands, 1979; Volume 15, pp. 277–356.
13. Zhu, Z.; Liu, S. An Improved Sage-Husa Adaptive Filtering Algorithm. In Proceedings of the 31st Chinese Control Conference, Hefei, China, 25–27 July 2012; pp. 5113–5117.
14. Xu, T.; Xu, X.; Xu, D.; Zou, Z.; Zhao, H. A New Robust Filtering Method of GNSS/MINS Integrated System for Land Vehicle Navigation. *IEEE Trans. Veh. Technol.* **2022**, *71*, 11443–11453. [CrossRef]
15. Elzoghby, M.; Arif, U.; Li, F.; Zhi Yu, X. Investigation of Adaptive Robust Kalman Filtering Algorithms for GPS/DR Navigation System Filters. *IOP Conf. Ser. Mater. Sci. Eng.* **2017**, *187*, 012019. [CrossRef]
16. Xu, S.; Zhou, H.; Wang, J.; He, Z.; Wang, D. SINS/CNS/GNSS Integrated Navigation Based on an Improved Federated Sage–Husa Adaptive Filter. *Sensors* **2019**, *19*, 3812. [CrossRef]
17. Zhou, Z.; Shen, Y.; Li, B. Stochastic Model of GPS Doppler: Evaluation, Modeling and Performance. In *China Satellite Navigation Conference (CSNC) 2012 Proceedings*; Sun, J., Liu, J., Yang, Y., Fan, S., Eds.; Lecture Notes in Electrical Engineering; Springer: Berlin/Heidelberg, Germany, 2012; Volume 159, pp. 395–406. ISBN 978-3-642-29186-9.
18. Li, B.; Hao, Q. Towards Reliable and Precise BeiDou Positioning with Stochastic Modelling. In *International Symposium on Earth and Environmental Sciences for Future Generations*; Freymueller, J.T., Sánchez, L., Eds.; International Association of Geodesy Symposia; Springer International Publishing: Cham, Switzerland, 2016; Volume 147, pp. 229–235. ISBN 978-3-319-69169-5.
19. Li, B. Stochastic Modeling of Triple-Frequency BeiDou Signals: Estimation, Assessment and Impact Analysis. *J. Geod.* **2016**, *90*, 593–610. [CrossRef]
20. Hu, H.; Zhou, F.; Jin, S. Improved Stochastic Modeling of Multi-GNSS Single Point Positioning with Additional BDS-3 Observations. *Meas. Sci. Technol.* **2021**, *32*, 045105. [CrossRef]
21. Wang, J.-G.; Gopaul, N.; Scherzinger, B. Simplified Algorithms of Variance Component Estimation for Static and Kinematic GPS Single Point Positioning. *J. Glob. Position. Syst.* **2009**, *8*, 43–52. [CrossRef]
22. Zhang, X.; Lu, X. Online Algorithm for Variance Components Estimation. *Commun. Nonlinear Sci. Numer. Simul.* **2021**, *97*, 105722. [CrossRef]
23. Hou, P.; Zha, J.; Liu, T.; Zhang, B. LS-VCE Applied to Stochastic Modeling of GNSS Observation Noise and Process Noise. *Remote Sens.* **2022**, *14*, 258. [CrossRef]
24. Zhang, Q.; Zhao, L.; Zhou, J. A Novel Weighting Approach for Variance Component Estimation in GPS/BDS PPP. *IEEE Sens. J.* **2019**, *19*, 3763–3771. [CrossRef]

25. Zhang, Q.; Zhao, L. Adaptive Stochastic Model Based on LS-VCE for Observations in Kinematic Precise Point Positioning. In Proceedings of the 2022 41st Chinese Control Conference (CCC), Hefei, China, 25–27 July 2022; IEEE: Piscataway, NJ, USA, 2022; pp. 3410–3414.
26. Yang, X.; Chang, G.; Wang, Q.; Zhang, S.; Mao, Y.; Chen, X. An Adaptive Kalman Filter Based on Variance Component Estimation for a Real-Time ZTD Solution. *Acta Geod. Geophys.* **2019**, *54*, 89–121. [CrossRef]
27. Chang, G.; Xu, T.; Yao, Y.; Wang, Q. Adaptive Kalman Filter Based on Variance Component Estimation for the Prediction of Ionospheric Delay in Aiding the Cycle Slip Repair of GNSS Triple-Frequency Signals. *J. Geod.* **2018**, *92*, 1241–1253. [CrossRef]
28. Teunissen, P.J.G. Least-Squares Estimation of the Integer GPS Ambiguities. In *Invited Lecture. Section IV Theory and Methodology, IAG General Meeting, Beijing, China*; Delft Geodetic Computing Centre: Delft, The Netherlands, 1993.
29. Psychas, D.; Verhagen, S.; Teunissen, P.J.G. Precision Analysis of Partial Ambiguity Resolution-Enabled PPP Using Multi-GNSS and Multi-Frequency Signals. *Adv. Space Res.* **2020**, *66*, 2075–2093. [CrossRef]
30. De Jonge, P.; Tiberius, C.C.J.M. *The LAMBDA Method for Integer Ambiguity Estimation: Implementation Aspects*; Publications of the Delft Computing Centre, LGR-Series; Delft Computing Centre: Delft, The Netherlands, 1996; Volume 12, pp. 1–47.
31. Verhagen, S.; Li, B.; Teunissen, P.J.G. Ps-LAMBDA: Ambiguity Success Rate Evaluation Software for Interferometric Applications. *Comput. Geosci.* **2013**, *54*, 361–376. [CrossRef]
32. Baarda, W. *A Testing Procedure for Use in Geodetic Networks*; Netherlands Geodetic Commission: Amersfoort, The Netherlands, 1968; Volume 2.
33. Chen, D. A Method for the Repair of Cycle Slip Using Double-Differenced Velocity Estimation for GNSS RTK Positioning. *Adv. Space Res.* **2019**, *63*, 2809–2821. [CrossRef]
34. Momoh, J.A. Robust GNSS Point Positioning in the Presence of Cycle Slips and Observation Gaps. Ph.D. Thesis, University College London, London, UK, 2013.
35. Li, B.; Qin, Y.; Liu, T. Geometry-Based Cycle Slip and Data Gap Repair for Multi-GNSS and Multi-Frequency Observations. *J. Geod.* **2019**, *93*, 399–417. [CrossRef]
36. Rao, C.R. Estimation of Variance and Covariance Components-MINQUE Theory. *J. Multivar. Anal.* **1971**, *1*, 257–275. [CrossRef]
37. Koch, K. Schätzung von Varianzkomponenten. *Allg. Vermess. Nachrich.* **1978**, *85*, 264–269.
38. Teunissen, P.J.G.; Amiri-Simkooei, A.R. Least-Squares Variance Component Estimation. *J. Geod.* **2008**, *82*, 65–82. [CrossRef]
39. Zhang, B.; Hou, P.; Liu, T.; Yuan, Y. A Single-Receiver Geometry-Free Approach to Stochastic Modeling of Multi-Frequency GNSS Observables. *J. Geod.* **2020**, *94*, 37. [CrossRef]
40. Teunissen, P.J.G. Zero Order Design: Generalized Inverses, Adjustment, the Datum Problem and S-Transformations. In *Optimization and Design of Geodetic Networks*; Grafarend, E.W., Sansò, F., Eds.; Springer: Berlin/Heidelberg, Germany, 1985; pp. 11–55, ISBN 978-3-642-70661-5.
41. Schaer, S.; Villiger, A.; Arnold, D.; Dach, R.; Prange, L.; Jäggi, A. The CODE Ambiguity-Fixed Clock and Phase Bias Analysis Products: Generation, Properties, and Performance. *J. Geod.* **2021**, *95*, 81. [CrossRef]
42. Su, K.; Jiao, G. Estimation of BDS Pseudorange Biases with High Temporal Resolution: Feasibility, Affecting Factors, and Necessity. *Satell. Navig.* **2023**, *4*, 17. [CrossRef]
43. de Lacy, M.C.; Reguzzoni, M.; Sansò, F.; Venuti, G. The Bayesian Detection of Discontinuities in a Polynomial Regression and Its Application to the Cycle-Slip Problem. *J. Geod.* **2008**, *82*, 527–542. [CrossRef]
44. Cai, C.; Liu, Z.; Xia, P.; Dai, W. Cycle Slip Detection and Repair for Undifferenced GPS Observations under High Ionospheric Activity. *GPS Solut.* **2013**, *17*, 247–260. [CrossRef]
45. Teunissen, P.J.G. GNSS Ambiguity Bootstrapping: Theory and Application. In Proceedings of the International Symposium on Kinematic Systems in Geodesy, Geomatics and Navigation, Banff, AB, Canada, 5–8 June 2001; pp. 246–254.
46. Teunissen, P.J.G. An Optimality Property of the Integer Least-Squares Estimator. *J. Geod.* **1999**, *73*, 587–593. [CrossRef]
47. Odijk, D.; Arora, B.S.; Teunissen, P.J.G. Predicting the Success Rate of Long-Baseline GPS+Galileo (Partial) Ambiguity Resolution. *J. Navig.* **2014**, *67*, 385–401. [CrossRef]
48. Nardo, A.; Li, B.; Teunissen, P.J.G. Partial Ambiguity Resolution for Ground and Space-Based Applications in a GPS+Galileo Scenario: A Simulation Study. *Adv. Space Res.* **2016**, *57*, 30–45. [CrossRef]
49. Dai, Z.; Knedlik, S.; Loffeld, O. Instantaneous Triple-Frequency GPS Cycle-Slip Detection and Repair. *Int. J. Navig. Obs.* **2009**, *2009*, 407231. [CrossRef]
50. Zhang, W.; Wang, J. A Real-Time Cycle Slip Repair Method Using the Multi-Epoch Geometry-Based Model. *GPS Solut.* **2021**, *25*, 60. [CrossRef]
51. Zhang, Z.; Zeng, J.; Li, B.; He, X. Principles, Methods and Applications of Cycle Slip Detection and Repair under Complex Observation Conditions. *J. Geod.* **2023**, *97*, 50. [CrossRef]

52. Zhang, X.; Li, X. Instantaneous Re-Initialization in Real-Time Kinematic PPP with Cycle Slip Fixing. *GPS Solut.* **2012**, *16*, 315–327. [CrossRef]
53. Yao, Y.-F.; Gao, J.-X.; Wang, J.; Hu, H.; Li, Z.-K. Real-Time Cycle-Slip Detection and Repair for BeiDou Triple-Frequency Undifferenced Observations. *Surv. Rev.* **2016**, *48*, 367–375. [CrossRef]

remote sensing

MDPI

Article

Cascaded Ambiguity Resolution for Pseudolite System-Augmented GNSS PPP

Caoming Fan [1], Zheng Yao [1,*], Jinling Wang [2] and Mingquan Lu [1]

1 Department of Electronic Engineering, Tsinghua University, Beijing 100084, China; fancm@tsinghua.edu.cn (C.F.); lumq@tsinghua.edu.cn (M.L.)

2 School of Civil and Environmental Engineering, University of New South Wales, Sydney 2006, Australia; jinling.wang@unsw.edu.au

* Correspondence: yaozheng@tsinghua.edu.cn

Abstract: Global navigation satellite System (GNSS) precise point positioning (PPP) enables high-precision global positioning using a single receiver, yet its widespread application is hindered by long convergence times. In contrast, pseudolite system (PLS) transmitters are located relatively close to receivers, and the movement of receivers induces rapid spatial geometry changes, which greatly facilitate fast parameter convergence. Therefore, leveraging the fast-converging PLS to augment GNSS PPP presents a promising solution. This study proposes a tightly coupled PLS and GNSS observation-level integration model. A key factor influencing the augmentation effectiveness is the strategy of ambiguity resolution. In this work, we design a novel strategy of ambiguity resolution, in which the fast convergence property of PLS is taken into account, and the PLS ambiguities are picked out to be fixed independently. This strategy can resolve the PLS ambiguities, GNSS wide-lane (WL) ambiguities, and GNSS L1 ambiguities cascadingly. Further, the fixed ambiguities can be treated as constraints in the filtering process. The experimental results demonstrate that the proposed strategy substantially improves the ambiguity fixing rates, especially in short-duration augmentation.

Keywords: global navigation satellite systems; precise point positioning; ambiguity resolution; pseudolite system (PLS); augmentation

1. Introduction

Mass-market applications such as autonomous vehicles and unmanned aerial vehicles are increasingly demanding fast, high-accuracy, and low-bandwidth location services [1]. Precise point positioning (PPP) is one of the mainstream global navigation satellite system (GNSS) positioning technologies that can achieve centimeter-level positioning accuracy globally with a single receiver [2]. However, slow geometry variations, atmospheric delays, and uncalibrated phase delays result in PPP usually taking tens of minutes to converge, which hinders its wide application.

Integer ambiguity resolution (AR) can effectively shorten the convergence time of GNSS PPP. This technique utilizes precise satellite orbit, clock, and code/phase bias provided by the server side to recover the integer nature of ambiguity, thus achieving integer ambiguity resolution [3–5]. Once integer ambiguities are fixed correctly, the positional parameters will converge instantaneously, which greatly reduces the convergence time of the PPP [6,7]. In addition, precise atmospheric delay corrections obtained from a regional network can also be applied to constrain the relevant parameters to further accelerate the ambiguity resolution and convergence [8–10].

Some scholars utilize other positioning systems such as the pseudolite system (PLS) to augment GNSS to accelerate its convergence. PLS transmits GNSS-like signals near the ground [11–14]. It can not only work independently but can also be used to augment GNSS. Due to PLS transmitters being close to receivers, moving receivers can produce quick spatial geometry variations, which is conducive to the rapid convergence of parameters [15,16]. Jiang et al. [17] proposed a Locata (a ground-based pseudolite system) augmented GPS PPP, which was referred to as Locata/GPS-PPP. The convergence time of the integrated Locata/GPS-PPP was only about 10 s. However, neither Locata nor GPS ambiguities were fixed to integers. If the ambiguity can be fixed to an integer, the convergence time is expected to be further reduced. In order to fix the integer ambiguity of PLS, our previous work proposed an on-the-fly method to estimate the PLS transmitter phase bias (TPB) and recover the integer nature of PLS ambiguities [18]. Furthermore, we presented a concept of PLS short-duration augmented GNSS PPP and realized the integer ambiguity resolution of PLS and GNSS [19].

However, in prior research, PLS has been typically treated as a subsystem of GNSS, with its ambiguities resolved jointly. This approach overlooks a critical characteristic: PLS ambiguities converge significantly faster than those of GNSS. Simultaneous ambiguity resolution may inadvertently degrade the PLS ambiguity fixing rate. Therefore, it is worthwhile to further investigate how to perform ambiguity resolution and how to utilize the fixed ambiguities under the PLS augmented GNSS PPP model. In this work, a strategy of ambiguity resolution has been proposed based on the characteristics of PLS and GNSS, and its performance has been analyzed.

The structure of this paper is organized as follows: Firstly, the observation models for GNSS and PLS are presented; a tightly coupled model for PLS-augmented GNSS PPP is then established. Based on this tightly coupled model, a strategy of ambiguity resolution has been proposed. Subsequently, real-time PPP experiments are conducted to evaluate the performance under different strategies of ambiguity resolution. Finally, conclusions are presented.

2. Observation Model

Here, the GNSS and PLS observation models are presented, respectively. The GNSS pseudorange observation ($\rho_{r,j}^{G,s}$) and phase observation ($L_{r,j}^{G,s}$) are described as follows:

$$\begin{cases} \rho_{r,j}^{G,s} = R_r^{G,s} + c\left(\delta t_r^G - \delta t^{G,s}\right) + \gamma_j^G I_{r,1}^{G,s} + T_r^{G,s} + b_{r,j}^G - b_j^{G,s} + e_{r,j}^{G,s} \\ L_{r,j}^{G,s} = R_r^{G,s} + c\left(\delta t_r^G - \delta t^{G,s}\right) - \gamma_j^G I_{r,1}^{G,s} + T_r^{G,s} + \lambda_j^G\left(N_{r,j}^{G,s} + B_{r,j}^G - B_j^{G,s}\right) + \varepsilon_{r,j}^{G,s} \end{cases} \quad (1)$$

where indices G, s, r, and j refer to the GNSS, satellite, receiver, and frequency band, respectively. $R_r^{G,s}$ is the geometric distance between the phase center of the receiver antenna and the satellite antenna. c is the speed of light in vacuum. δt_r^G and $\delta t^{G,s}$ denote the clock offsets of the GNSS receiver and satellite, respectively. $I_{r,1}^{G,s}$ represents the first-order slant ionospheric delay on the first frequency, and the slant ionospheric delay on the j-th frequency can be obtained by a multiplier factor $\gamma_j^G = \left[\lambda_j^G/\lambda_1^G\right]^2$ that depends on the wavelength λ_j^G. $T_r^{G,s}$ is the slant tropospheric delay. $b_{r,j}^G$ and $b_j^{G,s}$ are the code biases of the GNSS receiver and satellite, respectively. $B_{r,j}^G$ and $B_j^{G,s}$ are the phase biases of the GNSS receiver and satellite, respectively. $N_{r,j}^{G,s}$ denotes the GNSS integer ambiguity. $e_{r,j}^{G,s}$ and $\varepsilon_{r,j}^{G,s}$ are the sum of measurement noise and multipath error of code and phase observations, respectively.

In GNSS PPP, the errors caused by the Sagnac effect, relativistic effect, polar tide, ocean tide, solid earth tide, phase center offset, phase center variations, and phase wind-up are corrected by models [20]. The tropospheric delays consist of dry and wet components. The dry component can be accurately modeled, while the wet component is difficult to eliminate by models, and thus the zenith wet delay is usually estimated with mapping functions. The satellite orbit, clock, and code/phase bias are corrected by precise products generated by analysis centers.

Compared with GNSS observations, the PLS observations have relatively fewer errors. The tropospheric delay can be ignored or corrected by a simplified model [21,22] if the signal propagation distance is short. Furthermore, due to PLS signals propagating near the surface of the Earth, PLS observations are not affected by ionospheric delays. The coordinates of transmitters are accurately measured in advance. For synchronized PLS, the clock offsets of transmitters are equivalent and can be absorbed by the clock offset of the receiver. The transmitter code/phase biases can be estimated by the service end and broadcast to the user [18]. Therefore, the PLS observations equation can be simplified from Equation (1) as follows:

$$\begin{cases} \rho_{r,j}^{\mathrm{P},s} = R_r^{\mathrm{P},s} + c\delta t_r^{\mathrm{P}} + b_{r,j}^{\mathrm{P}} + e_{r,j}^{\mathrm{P},s} \\ L_{r,j}^{\mathrm{P},s} = R_r^{\mathrm{P},s} + c\delta t_r^{\mathrm{P}} + \lambda_j^{\mathrm{P}}\left(N_{r,j}^{\mathrm{P},s} + B_{r,j}^{\mathrm{P}}\right) + \varepsilon_{r,j}^{\mathrm{P},s} \end{cases} \tag{2}$$

where the superscript P denotes PLS. $\rho_{r,j}^{\mathrm{P},s}$ and $L_{r,j}^{\mathrm{P},s}$ are the PLS pseudorange and phase observations, respectively. δt_r^{P} is the clock offset of the PLS receiver. It should be noted that δt_r^{P} absorbs the clock offsets of the transmitters so that it drifts differently from that of the GNSS receiver. This should be taken care of in the setting of the clock offset parameter within the integration model.

3. PLS-Augmented GNSS PPP Model

In this work, we implement PLS-augmented GNSS PPP based on a tightly coupled (TC) model [19]. The diagram of the GNSS/PLS TC model is shown in Figure 1. The input data include GNSS observations, State Space representation (SSR) of GNSS corrections, PLS observations, and PLS ephemeris. SSR corrections include satellite orbit, clock, and code/phase biases, which are used for GNSS real-time PPP. PLS ephemeris contains the position, attitude, and code/phase biases of the pseudolite [18]. The attitude of the pseudolite antenna is primarily used to calculate the wind-up correction, and the phase bias is mainly employed to recover the integer nature of the PLS ambiguity. In this study, dual-frequency GNSS observations and single-frequency PLS observations are integrated with an Extended Kalman Filter (EKF). The undifferenced and uncombined PPP model is very convenient for system and frequency expansion [23]. Therefore, it is adopted in the GNSS/PLS TC model.

In the EKF, the state equation and measurement equation at the k-th epoch are as follows:

$$\begin{cases} \mathbf{x}_k = \mathbf{x}_{k-1} + \mathbf{w}_{k-1} \\ \mathbf{z}_k = \mathbf{h}(\mathbf{x}_k) + \mathbf{v}_k \end{cases} \tag{3}$$

where the unknown state vector $\mathbf{x}_k = \left[\mathbf{r}_r{}^T, c\delta t_r^{\mathrm{G}}, c\delta t_r^{\mathrm{P}}, \left(\mathbf{I}_{r,1}^{\mathrm{G}}\right)^T, T_{r,w}^{\mathrm{G}}, \left(\mathbf{N}_{r,1}^{\mathrm{G}}\right)^T, \left(\mathbf{N}_{r,2}^{\mathrm{G}}\right)^T, \left(\mathbf{N}_{r,1}^{\mathrm{P}}\right)^T\right]_k^T$. $\mathbf{r}_r$ represents the receiver position; $\mathbf{I}_{r,1} = \left[I_{r,1}^{\mathrm{G},1}, I_{r,1}^{\mathrm{G},2}, \ldots I_{r,1}^{\mathrm{G},m}\right]^T$ is the slant ionosphere delay, $\mathbf{N}_{r,1}^{\mathrm{G}} = \left[N_{r,1}^{\mathrm{G},1}, N_{r,1}^{\mathrm{G},2}, \ldots, N_{r,1}^{\mathrm{G},m}\right]^T$, $\mathbf{N}_{r,2}^{\mathrm{G}} = \left[N_{r,2}^{\mathrm{G},1}, N_{r,2}^{\mathrm{G},2}, \ldots, N_{r,2}^{\mathrm{G},m}\right]^T$, and $\mathbf{N}_{r,1}^{\mathrm{P}} = \left[N_{r,1}^{\mathrm{P},1}, N_{r,1}^{\mathrm{P},2}, \ldots, N_{r,1}^{\mathrm{P},n}\right]^T$ are the GNSS first frequency band, GNSS second frequency band,

and PLS first frequency band ambiguity vectors, respectively. The measurement model vector reads $\mathbf{h}(\mathbf{x}_k) = \left[\left(\mathbf{h}^{\mathrm{G}}_{\rho,1}\right)^T, \left(\mathbf{h}^{\mathrm{G}}_{L,1}\right)^T, \left(\mathbf{h}^{\mathrm{G}}_{\rho,2}\right)^T, \left(\mathbf{h}^{\mathrm{G}}_{L,2}\right)^T, \left(\mathbf{h}^{\mathrm{P}}_{\rho,1}\right)^T, \left(\mathbf{h}^{\mathrm{P}}_{L,1}\right)^T\right]^T$, with

$$\mathbf{h}^{\mathrm{G}}_{\rho,j} = \begin{bmatrix} R_r^{\mathrm{G},1} + c\left(\delta t_r^{\mathrm{G}} - \delta t^{\mathrm{G},1}\right) + \gamma_j^{\mathrm{G}} I_{r,1}^{\mathrm{G},1} + T_r^{\mathrm{G},1} + b_{r,j}^{\mathrm{G}} - b_j^{\mathrm{G},1} \\ R_r^{\mathrm{G},2} + c\left(\delta t_r^{\mathrm{G}} - \delta t^{\mathrm{G},2}\right) + \gamma_j^{\mathrm{G}} I_{r,1}^{\mathrm{G},2} + T_r^{\mathrm{G},2} + b_{r,j}^{\mathrm{G}} - b_j^{\mathrm{G},2} \\ \vdots \\ R_r^{\mathrm{G},m} + c\left(\delta t_r^{\mathrm{G}} - \delta t^{\mathrm{G},m}\right) + \gamma_j^{\mathrm{G}} I_{r,1}^{\mathrm{G},m} + T_r^{\mathrm{G},m} + b_{r,j}^{\mathrm{G}} - b_j^{\mathrm{G},m} \end{bmatrix}, \tag{4}$$

$$\mathbf{h}^{\mathrm{G}}_{L,j} = \begin{bmatrix} R_r^{\mathrm{G},1} + c\left(\delta t_r^{\mathrm{G}} - \delta t^{\mathrm{G},1}\right) - \gamma_j^{\mathrm{G}} I_{r,1}^{\mathrm{G},1} + T_r^{\mathrm{G},1} + \lambda_j^{\mathrm{G}}\left(B_{r,j}^{\mathrm{G}} - B_j^{\mathrm{G},1} + N_{r,j}^{\mathrm{G},1}\right) \\ R_r^{\mathrm{G},2} + c\left(\delta t_r^{\mathrm{G}} - \delta t^{\mathrm{G},2}\right) - \gamma_j^{\mathrm{G}} I_{r,1}^{\mathrm{G},2} + T_r^{\mathrm{G},2} + \lambda_j^{\mathrm{G}}\left(B_{r,j}^{\mathrm{G}} - B_j^{\mathrm{G},2} + N_{r,j}^{\mathrm{G},2}\right) \\ \vdots \\ R_r^{\mathrm{G},m} + c\left(\delta t_r^{\mathrm{G}} - \delta t^{\mathrm{G},m}\right) - \gamma_j^{\mathrm{G}} I_{r,1}^{\mathrm{G},m} + T_r^{\mathrm{G},m} + \lambda_j^{\mathrm{G}}\left(B_{r,j}^{\mathrm{G}} - B_j^{\mathrm{G},m} + N_{r,j}^{\mathrm{G},m}\right) \end{bmatrix}, \tag{5}$$

and

$$\mathbf{h}^{\mathrm{P}}_{\rho,1} = \begin{bmatrix} R_r^{\mathrm{P},1} + c\delta t_r^{\mathrm{P}} + b_{r,1}^{\mathrm{P}} \\ R_r^{\mathrm{P},2} + c\delta t_r^{\mathrm{P}} + b_{r,1}^{\mathrm{P}} \\ \vdots \\ R_r^{\mathrm{P},n} + c\delta t_r^{\mathrm{P}} + b_{r,1}^{\mathrm{P}} \end{bmatrix}, \mathbf{h}^{\mathrm{P}}_{L,1} = \begin{bmatrix} R_r^{\mathrm{P},1} + c\delta t_r^{\mathrm{P}} + \lambda_1^{\mathrm{P}}\left(N_{r,1}^{\mathrm{P},1} + B_{r,1}^{\mathrm{P}}\right) \\ R_r^{\mathrm{P},2} + c\delta t_r^{\mathrm{P}} + \lambda_1^{\mathrm{P}}\left(N_{r,1}^{\mathrm{P},1} + B_{r,1}^{\mathrm{P}}\right) \\ \vdots \\ R_r^{\mathrm{P},n} + c\delta t_r^{\mathrm{P}} + \lambda_1^{\mathrm{P}}\left(N_{r,1}^{\mathrm{P},1} + B_{r,1}^{\mathrm{P}}\right) \end{bmatrix}. \tag{6}$$

The partial derivatives matrix of $\mathbf{h}(\mathbf{x}_k)$ reads

$$\begin{aligned} \mathbf{H}(\mathbf{x}_k) &= \frac{\partial \mathbf{h}(\mathbf{x}_k)}{\partial \mathbf{x}_k}\Big|_{\mathbf{x}_k=\hat{\mathbf{x}}_k} \\ &= \begin{bmatrix} -\mathbf{g}_r^{\mathrm{G}} & \mathbf{E}_m & \mathbf{0} & \mathbf{I}_m & \mathbf{M}_T & \mathbf{0} & \mathbf{0} & \mathbf{0} \\ -\mathbf{g}_r^{\mathrm{G}} & \mathbf{E}_m & \mathbf{0} & -\mathbf{I}_m & \mathbf{M}_T & \lambda_1^{\mathrm{G}}\mathbf{I}_m & \mathbf{0} & \mathbf{0} \\ -\mathbf{g}_r^{\mathrm{G}} & \mathbf{E}_m & \mathbf{0} & \gamma_2^{\mathrm{G}}\mathbf{I}_m & \mathbf{M}_T & \mathbf{0} & \mathbf{0} & \mathbf{0} \\ -\mathbf{g}_r^{\mathrm{G}} & \mathbf{E}_m & \mathbf{0} & -\gamma_2^{\mathrm{G}}\mathbf{I}_m & \mathbf{M}_T & \mathbf{0} & \lambda_1^{\mathrm{G}}\mathbf{I}_m & \mathbf{0} \\ -\mathbf{g}_r^{\mathrm{P}} & \mathbf{0} & \mathbf{E}_n & \mathbf{0} & \mathbf{0} & \mathbf{0} & \mathbf{0} & \mathbf{0} \\ -\mathbf{g}_r^{\mathrm{P}} & \mathbf{0} & \mathbf{E}_n & \mathbf{0} & \mathbf{0} & \mathbf{0} & \mathbf{0} & \lambda_1^{\mathrm{P}}\mathbf{I}_n \end{bmatrix} \end{aligned} \tag{7}$$

where $\mathbf{g}_r^{\mathrm{G}}$ and $\mathbf{g}_r^{\mathrm{P}}$ are the line-of-sight unit vector from the receiver to GNSS satellite and pseudolite, respectively; $\mathbf{E}_m = \begin{bmatrix} 1 & 1 & \cdots & 1 \end{bmatrix}_{1\times m}^T$; $\mathbf{I}_m$ is a m-dimensional identity matrix; $\mathbf{M}_T = \begin{bmatrix} m_r^{\mathrm{G},1} & m_r^{\mathrm{G},2} & \cdots & m_r^{\mathrm{G},m} \end{bmatrix}^T$ is the mapping matrix of zenith wet delay. The covariance matrixes of process noise $\mathbf{w}_k$ read

$$\mathbf{Q} = \begin{bmatrix} \infty_r & & & & & & & \\ & \infty_{\delta t}^{\mathrm{G}} & & & & & & \\ & & \infty_{\delta t}^{\mathrm{P}} & & & & & \\ & & & \mathbf{Q}_I^{\mathrm{G}} & & & & \\ & & & & \mathbf{Q}_T^{\mathrm{G}} & & & \\ & & & & & \mathbf{0}_{m\times m} & & \\ & & & & & & \mathbf{0}_{m\times m} & \\ & & & & & & & \mathbf{0}_{n\times n} \end{bmatrix} \tag{8}$$

Adequately large process noises ($\infty = 10^6$) of receiver position and receiver clock offset are added to the variance at every epoch. $\mathbf{Q}_I^{\mathrm{G}}$ and $\mathbf{Q}_T^{\mathrm{G}}$ is the process noise covariance slant ionosphere and zenith wet troposphere terms, respectively [24]. The covariance matrixes of measurement error $\mathbf{v}_k$ read

$$\mathbf{R}_k = \begin{bmatrix} \mathbf{R}^{G}_{\rho,k} & & & \\ & \mathbf{R}^{G}_{L,k} & & \\ & & \mathbf{R}^{P}_{\rho,k} & \\ & & & \mathbf{R}^{P}_{L,k} \end{bmatrix} \quad (9)$$

where $\mathbf{R}^{G}_{\rho,k} = \mathrm{diag}\left((\sigma^{G,1}_{\rho,k})^2, (\sigma^{G,2}_{\rho,k})^2, \cdots (\sigma^{G,m}_{\rho,k})^2\right)$, $\mathbf{R}^{G}_{L,k} = \mathrm{diag}\left((\sigma^{G,1}_{L,k})^2, (\sigma^{G,2}_{L,k})^2, \cdots (\sigma^{G,m}_{L,k})^2\right)$, $\mathbf{R}^{P}_{\rho,k} = \mathrm{diag}\left((\sigma^{P,1}_{\rho,k})^2, (\sigma^{P,2}_{\rho,k})^2, \cdots (\sigma^{P,n}_{\rho,k})^2\right)$, and $\mathbf{R}^{P}_{L,k} = \mathrm{diag}\left((\sigma^{P,1}_{L,k})^2, (\sigma^{P,2}_{L,k})^2, \cdots (\sigma^{P,n}_{L,k})^2\right)$. $\sigma^{G,s}_{\rho,k}$ and $\sigma^{G,s}_{L,k}$ are the standard deviations of GNSS pseudorange and phase measurement errors, respectively. $\sigma^{P,s}_{\rho,k}$ and $\sigma^{P,s}_{L,k}$ are the standard deviations of PLS pseudorange and phase measurement errors, respectively. The setting of the standard deviations of GNSS measurement errors employed an elevation-dependent model [24]. However, for PLS measurements, their standard deviations are primarily related to environmental factors. Based on empirical values, the standard deviations of PLS pseudorange and phase measurement errors are set to constants of 2 m and 0.01 m, respectively.

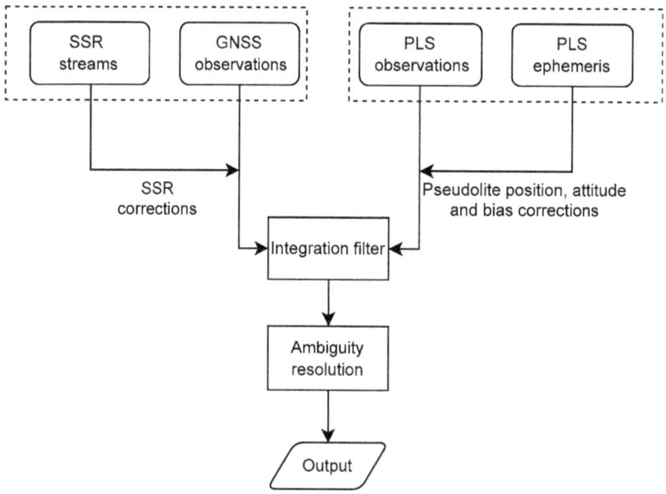

Figure 1. The diagram of GNSS/PLS tightly coupled model.

4. Strategies of Ambiguity Resolution

In the undifferenced and uncombined PPP model, ambiguities are affected by both initial phase biases and clock offsets, thus lacking the integer nature. Specifically, satellite-dependent initial phase biases and precise clock offsets are typically provided by the International GNSS Service (IGS) analysis center, which can be removed from the observations. Nevertheless, the raw zero-difference ambiguity parameters estimated within the filter are still affected by receiver-dependent initial phase biases and clock offsets. Before fixing the ambiguities using the Least-squares Ambiguity Decorrelation Adjustment (LAMBDA) algorithm [25], the receiver clock offsets and initial phase biases can be eliminated through the between-satellites single difference. Here, the first satellite is assumed as the reference satellite, thus defining the single-difference transformation matrix as follows:

$$\mathbf{D} = \begin{bmatrix} -1 & 1 & & & \\ -1 & & 1 & & \\ \vdots & & & \ddots & \\ -1 & & & & 1 \end{bmatrix} \tag{10}$$

Single-difference ambiguities are obtained through the single-difference transformation matrix as follows:

$$\Delta\hat{\mathbf{N}} = \mathbf{D}\hat{\mathbf{N}} = \begin{bmatrix} -1 & 1 & & & \\ -1 & & 1 & & \\ \vdots & & & \ddots & \\ -1 & & & & 1 \end{bmatrix} \begin{bmatrix} \hat{N}^1 \\ \hat{N}^2 \\ \vdots \\ \hat{N}^s \end{bmatrix} \tag{11}$$

The float single-difference ambiguity vector $\Delta\hat{\mathbf{N}}$ is processed by the LAMBDA algorithm for integer ambiguity resolution. Upon successful resolution, the fixed single-difference ambiguity vector $\Delta\breve{\mathbf{N}}$ is obtained. The $\Delta\breve{\mathbf{N}}$ can be used as a pseudo-observation to constrain other parameters to obtain fixed solutions.

This constraint can be distinguished between temporary constraint and tight constraint, which correspond to the modes termed "Continuous" (Figure 2a) and "Fix and Hold" (Figure 2b) in the open-source software RTKLIB [24]. In "Continuous" mode, upon successful ambiguity resolution, the fixed integer ambiguity vector constrains a temporarily copied filter, yielding a fixed solution. Crucially, the integer ambiguity vector is not fed back to the original filter, whereas, in "Fix and Hold" mode, the fixed integer ambiguity vector is fed back to the original filter, constraining its ambiguity parameters to integers to obtain the fixed solution.

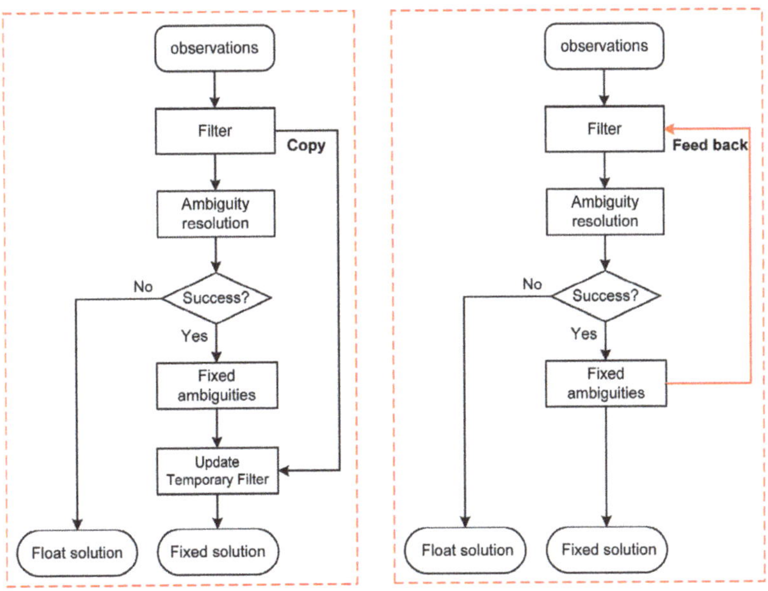

(a) Continuous **(b)** Fix and Hold

Figure 2. Two modes after ambiguities are fixed successfully.

The fundamental distinction between the two modes is as follows: in the "Continuous" mode, the fixed ambiguities influence only the current epoch, while in the "Fix and Hold" mode, they persist to constrain subsequent epochs. Notably, "Fix and Hold" mode carries higher operational risk. Should an incorrect ambiguity vector be fed back into the original filter, it propagates systematic errors, causing deviations in positioning solutions for the following epochs.

In the GNSS/PLS TC model, PLS can be regarded as a subsystem of GNSS, in which PLS observations and GNSS observations are processed together. However, the ambiguity convergence rate of PLS is significantly faster than that of GNSS. Ignoring this characteristic and jointly fixing their ambiguities simultaneously may compromise the optimal fixation rate, thereby compromising augmentation effectiveness. Therefore, it can be considered to prioritize fixing the PLS ambiguities, and the fixed PLS ambiguity vector can subsequently be employed as a virtual observation to constrain the remaining parameters. This strategy reduces the correlation between GNSS ambiguities and other parameters (e.g., atmospheric delays), facilitating the subsequent GNSS ambiguity resolution.

Furthermore, GNSS dual-frequency data enables the formation of wide-lane (WL) ambiguity. Owing to the fact that the WL ambiguities have substantially longer wavelength (~86 cm for GPS WL) compared to fundamental-frequency ambiguities (~19 cm for GPS L1), WL ambiguities can be fixed rapidly and reliably. The WL ambiguity can be constructed by the following transformation:

$$\hat{N}_{\mathrm{WL}}^{\mathrm{G}} = \begin{bmatrix} 1 & -1 \end{bmatrix} \begin{bmatrix} \hat{N}_{\mathrm{L1}}^{\mathrm{G}} \\ \hat{N}_{\mathrm{L2}}^{\mathrm{G}} \end{bmatrix} \tag{12}$$

where $\hat{N}_{\mathrm{L1}}^{\mathrm{G}}$ and $\hat{N}_{\mathrm{L2}}^{\mathrm{G}}$ represent the estimated GNSS ambiguities of the first and second frequency bands, respectively. Once WL ambiguities are fixed, they can be used as a pseudo-observation to constrain the fundamental-frequency (L1/L2) ambiguities, reducing parameter correlations and accelerating their resolution.

Combining the aforementioned ambiguity resolution sequence and the constrained mode after the ambiguity is fixed, this paper proposes a new strategy of ambiguity resolution, as illustrated in Figure 3. In the traditional strategy, the PLS is treated as a subsystem of GNSS, and its ambiguities are fixed together. However, in the proposed strategy, the PLS ambiguities are fixed first, and the fixed PLS integer ambiguity vector is used to update the temporary filter; subsequently, the GNSS WL ambiguities are fixed, and the temporary filter is updated with the fixed WL ambiguity vector; finally, the GNSS L1 (first frequency band) ambiguities are fixed. Updating the temporary filter only affects the current epoch, meaning it solely constrains the remaining parameters at that epoch. To propagate the influence of fixed ambiguities across all subsequent epochs, these fixed ambiguities should be fed back into the original filter. This feedback mechanism is implemented through the "Fix and Hold" mode, which operates as an optional strategy in the processing workflow.

It should be noted that in PPP mode, attempting to fix the full set of ambiguities is highly challenging due to high-dimensional correlations and residual atmospheric errors. Consequently, partial ambiguity resolution (PAR), which only fixes a suitable subset of the ambiguities, is typically employed [26–29]. Several kinds of strategies have been proposed to select ambiguity subsets, such as the elevation order strategy, the Signal-to-Noise Ratio (SNR) order strategy. Pseudolites typically have low elevation angles; using the elevation order strategy is likely to exclude them. As for the SNR order strategy, pseudolites generally exhibit significantly higher SNR than GNSS satellites. Once a new pseudolite appears, if it cannot be fixed, it will block the fixing of subsequent GNSS satellites. Furthermore, considering the fast geometric variations between pseudolites and moving users, the correlation among the PLS ambiguity parameters decreases fast, and theoretically, PLS ambiguities will be fixed faster than GNSS ambiguities. Therefore, selecting a subset of ambiguities based on the ordering of ambiguity correlation ensures that the PLS ambiguities can be effectively retained during the subset selection process, thereby improving the fixing rate.

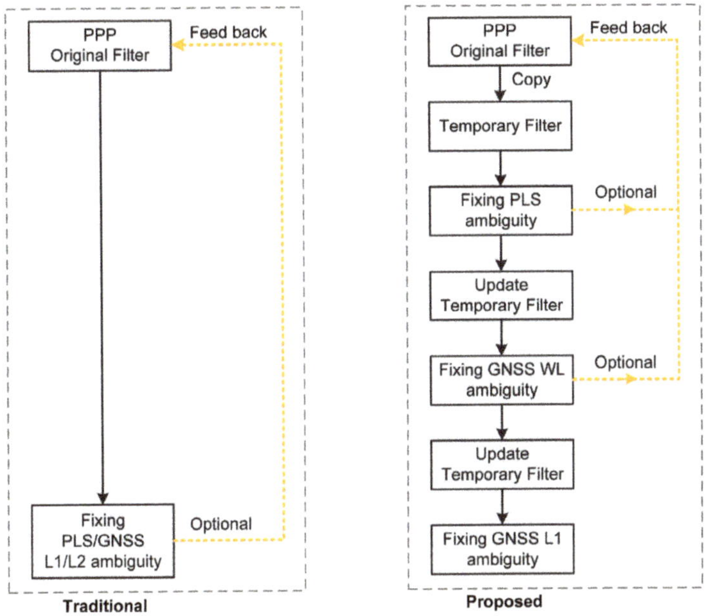

Figure 3. The traditional and proposed strategies of ambiguity resolution.

Figure 4 shows the flowchart of the PAR procedure adopted in this study. The core of this method involves sorting the decorrelated ambiguity variances in ascending order and sequentially removing ambiguities with the largest variances until the remaining subset satisfies the minimum predefined success rate. First, the PAR process starts with the decorrelation of the ambiguities [25]. Then, the full set of ambiguity resolution success rates is computed as P_s [30]. If $P_s > 99\%$, a normal full AR process is performed; otherwise, PAR is performed. It should be noted that the diagonal elements of $\mathbf{D}_z$ have been sorted in ascending order during the decorrelation process. Therefore, we sequentially remove the ambiguities with the largest covariances from the first to the last until the success rate exceeds 99%. If the number of selected ambiguities is more than 4, the AR search is performed. In this PAR, a fixed solution is obtained only if both the success rate and the ratio-test [31] are satisfied.

In summary, the proposed strategy employs a PAR algorithm based on decorrelated ambiguity variances and leverages the rapid convergence advantage of PLS. The PLS ambiguities are first sequentially resolved, followed by the GNSS WL ambiguities, and finally the GNSS L1 ambiguities. Furthermore, during the ambiguity fixing process, the already fixed ambiguities are utilized to constrain the remaining ambiguities, thereby enhancing the success rate of fixing the remaining ambiguities.

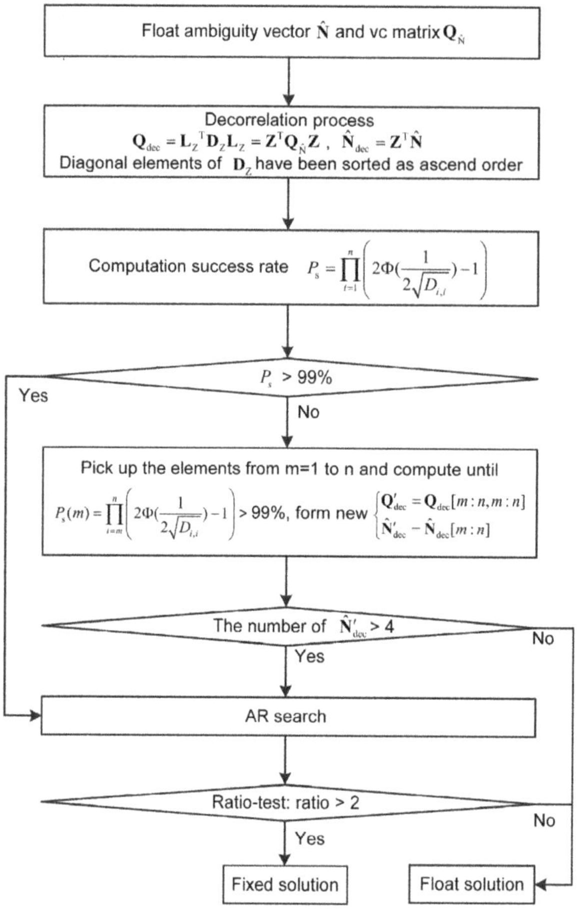

Figure 4. The flowchart of partial ambiguity resolution.

5. Experiment and Analysis

5.1. Overview of the Experiment

To evaluate the performance of different ambiguity resolution strategies on real-time kinematic PPP augmented by PLS, we conducted a two-hour kinematic experiment on a building rooftop. Due to frequency regulation standards, the PLS transmits signals in the 2.4 GHz band, which is license-free for industrial, scientific, and medical (ISM). Therefore, GNSS and PLS work in different frequency bands, meaning that separate antennas need to be used to receive GNSS and PLS signals, respectively. The experimental environment and platform are illustrated in Figure 5. The rover was equipped with one PLS receiver and two GNSS receivers. The PLS receiver and its antenna are custom developed, while the model of the GNSS receivers is Septentrio Mosaic X5, with two low-cost quadrifilar helix GNSS antennas. The PLS receiving antenna was located in the middle of the two GNSS antennas. This dual-GNSS-antenna configuration enabled lever-arm correction, thereby correcting measurements from the main GNSS antenna to the PLS antenna phase center [19]. Additionally, the average coordinates of two GNSS antennas, derived from moving-RTK solutions, served as the reference trajectory for assessing positioning results during the experiment.

Figure 5. Experimental environment and platform.

The coordinates of the GNSS RTK base station were determined through long-term static GNSS PPP processing, while the coordinates of the 6 PLs were obtained from long-term static GNSS RTK and total station measurements. Figure 6 presents the planimetric view of the layout of the 6 PLs and the rover trajectory during the experiment. Within this figure, the PLs are denoted by yellow circles. The two-hour walking trajectory (about 1 min to walk one circuit) is indicated by the blue line.

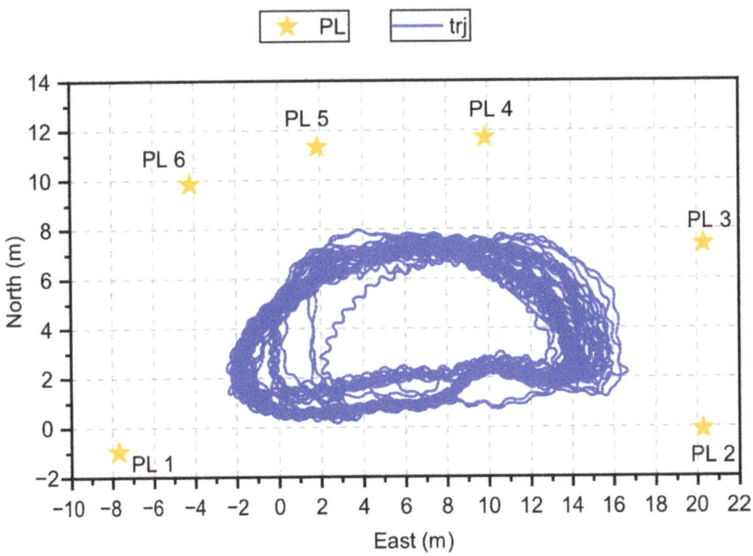

Figure 6. Planimetric views of the moving trajectory and PLs. The blue line indicates the 2 h trajectory, and the yellow circles indicate PLs.

Figure 7 illustrates the variation in the number of tracked GNSS satellites. The black, red, blue, and green lines represent the number of satellites of GPS, Galileo, GLONASS, and BDS, respectively, while the purple line denotes the total number of all satellites. The number of satellites changes frequently, due to the fact that some of the satellites' signals are blocked when the user approaches the wall. At certain moments, the total number of GNSS satellites drops to approximately 20, primarily caused by the loss of BDS observations output from the receiver. Overall, the total number of GNSS satellites is about 35 throughout the experiment.

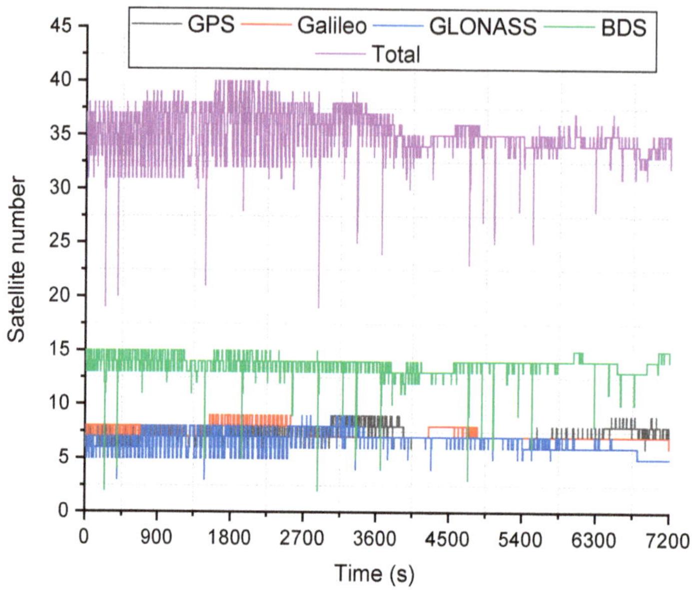

Figure 7. The variation in the number of tracked GNSS satellites.

5.2. Analysis of the Experimental Results

In this experiment, we analyzed the GNSS/PLS ambiguity fixing rate and positioning accuracy. The ambiguity fixing rate was defined as the ratio of the number of epochs with successful ambiguity resolution to the total number of epochs. The experiment duration spanned 2 h, subdivided into 8 sequential segments, each lasting 15 min (900 s), with a reset procedure applied at the start of each segment. To enhance visual clarity in illustrating the positioning errors and the number of fixed ambiguities over time, only the data from the first 15 min segment (900 s) is presented graphically. However, for the statistical evaluation of the ambiguity fixing rate and root mean square error (RMSE) of positioning, data from all eight segments (the full 2 h) was utilized.

In the following experiments, we analyze both the traditional and the proposed strategies of ambiguity resolution. Based on the constraint mode after ambiguity is fixed, the traditional strategy is categorized into "Traditional (C)" and "Traditional (H)", while the proposed strategy is similarly categorized into "Proposed (C)" and "Proposed (H)". In "Traditional (C)" and "Proposed (C)", the fixed ambiguities are not fed back to the original filter; that is, "Continuous" mode is adopted. Conversely, in "Traditional (H)" and "Proposed (H)", the fixed ambiguities are fed back to the original filter, corresponding to the "Fix and Hold" mode. Additionally, based on the duration of PLS-augmented GNSS, we divided the experiments into long-duration augmentation and short-duration augmentation.

5.2.1. Long-Duration Augmentation

Figure 8 presents the number of fixed ambiguities with different strategies of ambiguity resolution, including the number of PLS ambiguities and GNSS L1 (first frequency band) ambiguities. First, we will focus on the PLS ambiguity resolution performance. It can be seen that PLS ambiguities are fixed in about 4 s for all strategies. Note that in the traditional strategy, the PLS and GNSS ambiguities are fixed together without deliberate separation, but it achieves fast resolution comparable to the proposed strategy. The traditional strategy achieved fast ambiguity resolution, primarily due to the PAR algorithm, which ranks ambiguity parameters based on their variance after decorrelation. The rapid geometric variations result in PLS ambiguity variances decreasing faster than that of GNSS ambiguities. Consequently, the PAR algorithm first removes GNSS ambiguities with large variances and preserves the PLS ambiguities with smaller variances to be fixed. This process inherently performs a selection mechanism functionally similar to the explicit PLS ambiguity selection employed in the proposed strategy. However, after approximately 160 s, the PLS ambiguities in "Traditional (C)" and "Traditional (H)" fail to be fixed. After careful analysis, it was found that after 160 s, the variances of GNSS ambiguities also decrease. Partial GNSS ambiguities satisfied the success rate threshold yet failed the ratio-test threshold, resulting in the failure of ambiguity resolution. Since the PLS and GNSS ambiguities are fixed together in this strategy, the overall failure consequently prevents the resolution of the PLS ambiguities. As the GNSS ambiguities continued to converge and passed the "ratio-test" threshold, PLS ambiguities were therefore fixed again, as shown in "Traditional (H)". There is no obvious difference between the "Proposed (C)" and "Proposed (H)" in terms of PLS ambiguity resolution. In a word, the slow convergence of GNSS ambiguities may compromise the fixing rate of PLS ambiguities when they are resolved together. Therefore, it is advisable to resolve them separately.

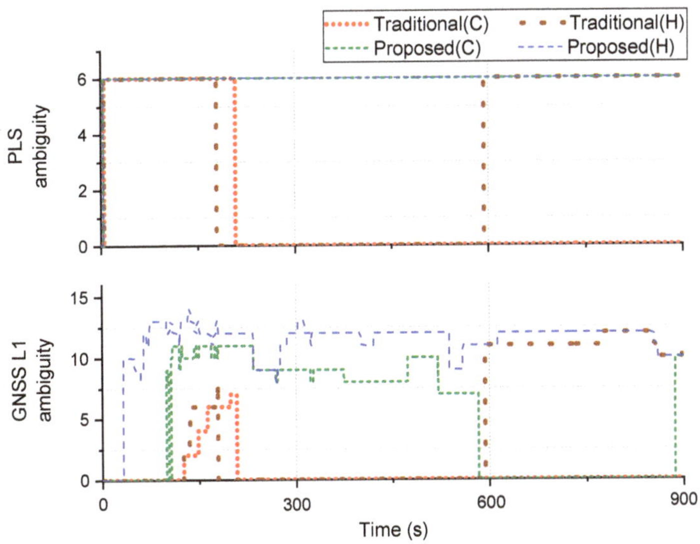

Figure 8. The number of fixed ambiguities of PLS and GNSS L1 under different strategies of ambiguity resolution.

Regarding GNSS L1 ambiguity resolution, the proposed strategy achieves a higher fixing rate compared to the traditional strategy in both "Continuous" and "Fix and Hold" modes. Notably, the fixing rate of the "Proposed (H)" strategy significantly exceeds that of other strategies, demonstrating the effectiveness of feeding fixed ambiguities back into the original filter for enhancing the ambiguity fixing rate.

Figure 9 presents the statistical results of the ambiguity fixing rates under different strategies of ambiguity resolution. The performance is consistent with the previous analysis.

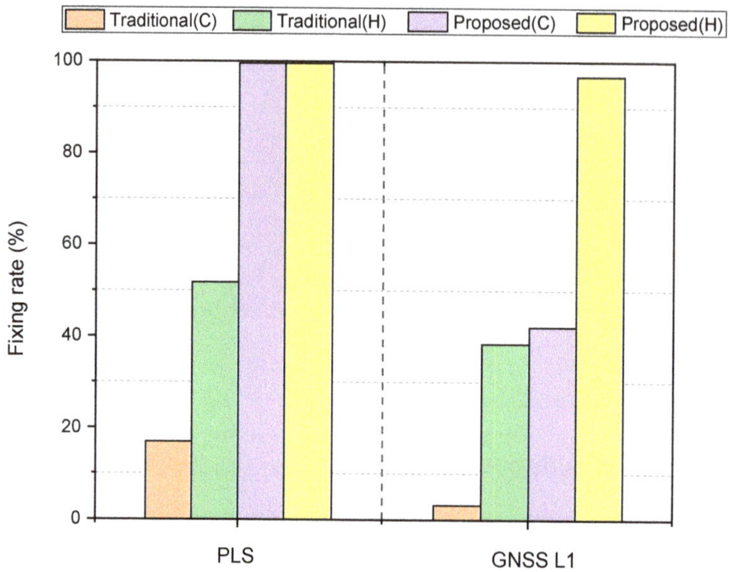

Figure 9. Ambiguity fixing rates of PLS and GNSS L1 under five strategies of ambiguity resolution.

Figure 10 depicts the positioning errors under these strategies. The statistical results of the positioning errors under strategies Traditional (C), Traditional (H), Proposed (C), and Proposed (H) are 0.021, 0.020, 0.019, and 0.020 cm, respectively. It can be seen that the positioning errors obtained by these strategies show no significant differences, with positioning errors consistently around 0.02 m. This consistency arises primarily because the PLS converges rapidly and stays converged throughout, which plays a major role in positioning.

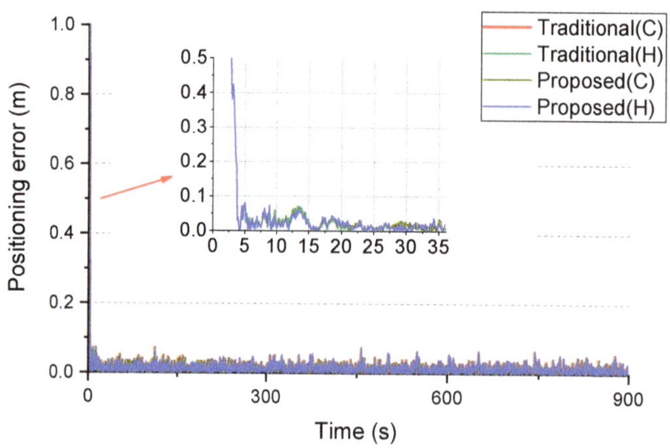

Figure 10. The positioning errors under different strategies of ambiguity resolution.

5.2.2. Short-Duration Augmentation

On the one hand, it is difficult to see the effect of these strategies of ambiguity resolution on the positioning accuracy because the PLS augments GNSS throughout the whole period. On the other hand, due to the limited coverage area of PLS, sometimes GNSS can only be augmented for a short period of time. Therefore, we specifically investigated the performance of the short-duration (15 s) PLS augmentation to GNSS.

Figure 11 depicts the positioning errors under these strategies of ambiguity resolution with short-duration PLS augmentation. It can be seen that the positioning error in the "Proposed (H)" strategy remains consistently convergent, while other strategies exhibit a little divergence tendency. The statistical results of the positioning errors under strategies Traditional (C), Traditional (H), Proposed (C), and Proposed (H) are 0.111, 0.059, 0.097, and 0.048 cm, respectively. It can be seen that the positioning accuracy in the "Proposed (C)" strategy is higher than that achieved by the "Traditional (C)" strategy. When applying the "Fix and Hold" mode, the positioning accuracy is further improved, and the "Proposed (H)" mode still achieves higher positioning accuracy than the "Traditional (H)". This demonstrates that regardless of whether the "Continuous" or "Fix and Hold" mode is employed, the proposed strategy of ambiguity resolution yields consistently higher positioning accuracy. However, the positioning accuracy of "Traditional (H)" is better than that of Proposed (C)". This indicates that feeding back the fixed PLS integer ambiguities to the original filter is crucial in short-duration augmentation scenarios.

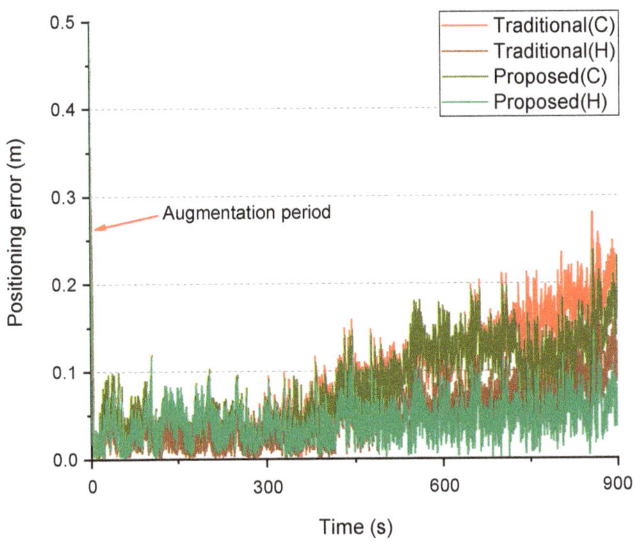

Figure 11. The positioning errors under different strategies of ambiguity resolution with short-duration PLS augmentation.

From the previous experiment, it can be seen that applying "Fix and Hold" mode to fixed ambiguities does not yield significant improvements in terms of positioning accuracy under long-duration augmentation scenarios. However, in short-duration augmentation scenarios, "Fix and Hold" mode provides considerable benefit. This is because in the case of long-duration augmentation, the PLS ambiguity parameters typically converge near integer values. As a result, whether they are fixed or not yields no significant difference in performance. However, in the case of short-duration augmentation, the PLS ambiguity parameters have not yet converged near integer values. Therefore, imposing additional integer constraints ("Fix and Hold" mode) leads to a significant augmentation improvement.

The effectiveness of ambiguity resolution under PLS short-duration augmentation to GNSS is further evaluated in Figures 12 and 13. In the short-duration augmentation,

PLS is available only for the initial 15 s and becomes unavailable thereafter. Therefore, the number of fixed PLS ambiguities dropped after 15 s. Here, the PLS ambiguity fixing rate is defined as the ratio of fixed epochs to the total number of available epochs. As can be seen, the PLS ambiguity fixing rate exhibits very little difference across these strategies. As we analyzed earlier, the PAR algorithm can automatically select PLS ambiguities for priority fixing. The previous experiments confirmed that GNSS ambiguities primarily impair the PLS ambiguity fixing rate after 100 s. However, in this experiment, PLS was unavailable after 100 s, thus its fixing rate was hardly affected by GNSS. Consequently, the PLS ambiguity fixing rates show comparable levels across all strategies under short-duration augmentation.

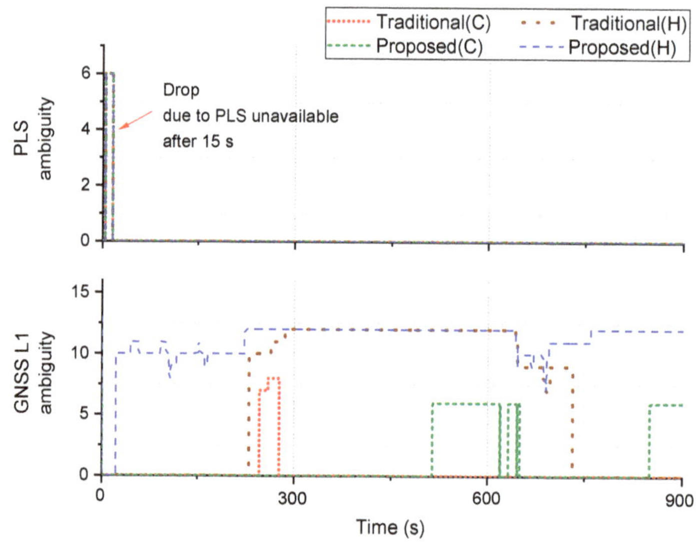

Figure 12. The number of fixed ambiguities of PLS and GNSS L1 under different strategies of ambiguity resolution with short-duration augmentation.

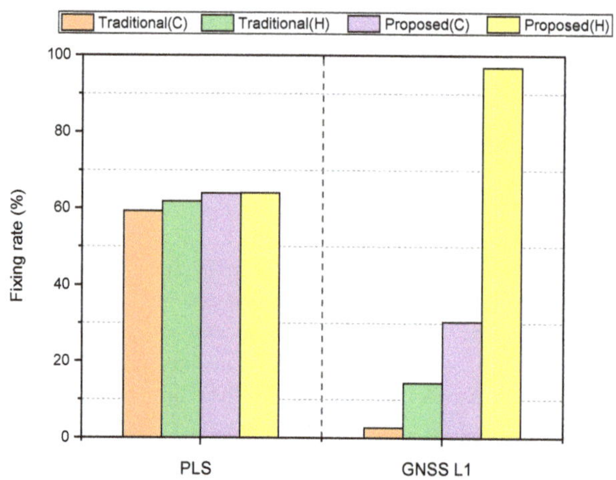

Figure 13. Ambiguity fixing rates of PLS and GNSS L1 under different strategies of ambiguity resolution with short-duration augmentation.

In contrast, significant differences exist in the GNSS L1 ambiguity fixing rates. In the "Proposed (C)" strategy, the L1 ambiguity fixing rate is approximately 30%, while in the "Traditional (C)" strategy, it is only approximately 3%. Furthermore, under the "Fix and Hold" mode, the "Proposed (H)" strategy achieves a remarkable fixing rate of approximately 97%, vastly exceeding the 15% attained by "Traditional (H)". These results demonstrate that the cascaded ambiguity resolution approach and the "Fix and Hold" mode both greatly improve the GNSS ambiguity fixing rate.

Fixed PLS integer ambiguities adopt the "Fix and Hold" mode to provide a significant improvement in short-duration augmentation scenarios. However, the wavelength of PLS ambiguity (approximately 12 cm) is really short. Consequently, incorrectly fixed PLS ambiguities may introduce substantial deviations into positioning solutions. Figure 14 shows a positioning experiment, in which the duration of PLS-augmented GNSS is still 15 s. At approximately the 4th s, the PLS ambiguities were initially erroneously fixed in the "Proposed (H)" strategy and fed back into the original filter, contaminating the state parameters and resulting in significant positioning deviations. In contrast, the "Proposed (C)" strategy—which avoids feedback of wrong PLS ambiguities to the original filter—maintains relatively better positioning accuracy. As previously analyzed, in the long-duration augmentation, the "Fix and Hold" mode offers limited benefits yet carries substantial risks. Therefore, it is not recommended to apply the "Fix and Hold" mode to PLS. In the case of short-duration augmentation, feedback of PLS ambiguities to the original filter yields significant benefits but requires cautious validation mechanisms.

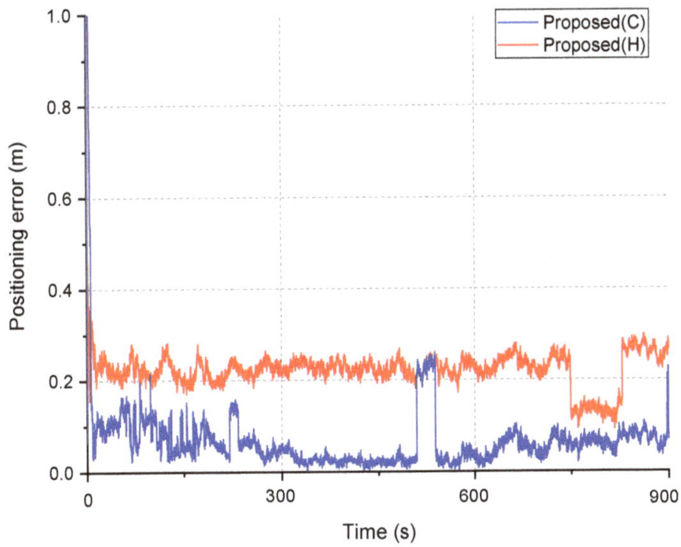

Figure 14. Positioning error when the PLS ambiguities are erroneously fixed under "Proposed (C)" and "Proposed (H)" strategies.

6. Conclusions

Based on the PLS-augmented GNSS PPP model, this study proposes a new strategy of ambiguity resolution. Within this proposed strategy, by leveraging the rapid convergence characteristics of PLS ambiguities, the PLS ambiguities and GNSS ambiguities are fixed cascadingly. Experiments were conducted to analyze the ambiguity fixing rate and positioning accuracy under different strategies of ambiguity resolution in both long-duration and short-duration PLS-augmented GNSS scenarios.

In the long-duration augmentation, if the rapid convergence characteristic of PLS ambiguities is not considered and they are fixed together with GNSS ambiguities, the PLS ambiguity fixing rate is only 16.9%. However, in the proposed strategy, the PLS ambiguity fixing rate can reach 99.5%. Furthermore, employing the proposed strategy also substantially enhanced the GNSS ambiguity fixing rate.

For short-duration PLS augmentation, the proposed strategy can also improve the GNSS ambiguity fixing rate and positioning accuracy. Additionally, during the short-duration augmentation, where PLS ambiguities have not yet converged sufficiently close to integer values, feeding back the fixed PLS integer ambiguities to the original filter effectively introduces additional integer constraints. This process significantly boosts positioning accuracy and the GNSS ambiguity fixing rate. However, the feedback of fixed PLS ambiguities to the original filter requires extreme caution, since the ambiguity with short wavelength is more prone to being incorrectly fixed, resulting in systematic deviations in the positioning solutions.

Given the rapid convergence characteristics of Low Earth Orbit (LEO) satellites, this study thus provides valuable reference for ambiguity resolution in LEO-augmented GNSS.

Author Contributions: Conceptualization, C.F. and Z.Y.; methodology, C.F., Z.Y. and J.W.; software, C.F.; validation, C.F.; resources, Z.Y. and M.L.; data curation, C.F.; writing—original draft preparation, C.F.; writing—review and editing, C.F., Z.Y., M.L. and J.W.; visualization, C.F.; supervision, M.L.; project administration, Z.Y.; funding acquisition, Z.Y. All authors have read and agreed to the published version of the manuscript.

Funding: This work is supported by the National Natural Science Foundation of China under Grant 42274018 and the Tsinghua University Initiative Scientific Research Program.

Data Availability Statement: The original contributions presented in this study are included in the article. Further inquiries can be directed to the corresponding author.

Acknowledgments: We are very grateful to the reviewers for their helpful remarks for improving this manuscript. We would also like to express our gratitude to CNES/Nav for providing real-time multi-GNSS products.

Conflicts of Interest: The authors declare no conflicts of interest.

References

1. European GNSS Agency. *PPP-RTK Market and Technology Report*; European GNSS Agency: Prague, Czechia, 2019.
2. Zumberge, J.; Heflin, M.; Jefferson, D.; Watkins, M.; Webb, F. Precise point positioning for the efficient and robust analysis of GPS data from large networks. *J. Geophys. Res. Solid. Earth* **1997**, *102*, 5005–5017. [CrossRef]
3. Ge, M.; Gendt, G.; Rothacher, M.; Shi, C.; Liu, J. Resolution of GPS carrier-phase ambiguities in Precise Point Positioning (PPP) with daily observations. *J. Geod.* **2008**, *82*, 389–399. [CrossRef]
4. Laurichesse, D.; Mercier, F.; Berthias, J.-P.; Broca, P.; Cerri, L. Integer Ambiguity Resolution on Undifferenced GPS Phase Measurements and Its Application to PPP and Satellite Precise Orbit Determination. *Navigation* **2009**, *56*, 135–149. [CrossRef]
5. Collins, P. Isolating and Estimating Undifferenced GPS Integer Ambiguities. In Proceedings of the 2008 National Technical Meeting of The Institute of Navigation, San Diego, CA, USA, 28–30 January 2008.
6. Liu, S.; Yuan, Y.; Guo, X.; Wang, K.; Xiao, G. Multi-GNSS Precise Point Positioning with Ambiguity Resolution Based on the Decoupled Clock Model. *Remote Sens.* **2024**, *16*, 2999. [CrossRef]
7. Liu, Y.; Hugentobler, U.; Duan, B. Improving Performance of Uncombined PPP-AR Model with Ambiguity Constraints. *Remote Sens.* **2024**, *16*, 4537. [CrossRef]
8. Li, X.; Wang, B.; Li, X.; Huang, J.; Lyu, H.; Han, X. Principle and performance of multi-frequency and multi-GNSS PPP-RTK. *Satell. Navig.* **2022**, *3*, 7. [CrossRef]
9. Hong, J.; Tu, R.; Liu, Y.; Ge, Y.; Li, F. LEO-Enhanced BDS-3 PPP Performance Based on B2b Signal. *Remote Sens.* **2025**, *17*, 2183. [CrossRef]
10. Kiliszek, D. GPS and Galileo Precise Point Positioning Performance with Tropospheric Estimation Using Different Products: BRDM, RTS, HAS, and MGEX. *Remote Sens.* **2025**, *17*, 2080. [CrossRef]

11. Tang, W.; Chen, J.; Yu, C.; Ding, J.; Wang, R. A New Ground-Based Pseudolite System Deployment Algorithm Based on MOPSO. *Sensors* **2021**, *21*, 5364. [CrossRef]
12. Cobb, H.S. GPS Pseudolites: Theory, Design, and Applications. Ph.D. Thesis, Stanford University, Stanford, CA, USA, 1997.
13. Wang, T.; Yao, Z.; Lu, M. Combined difference square observation-based ambiguity determination for ground-based positioning system. *J. Geod.* **2019**, *93*, 1867–1880. [CrossRef]
14. Wang, T.; Yao, Z.; Lu, M. Autonomous coordinate establishment of local reference frames for ground-based positioning systems without known points. *J. Geod.* **2023**, *97*, 4. [CrossRef]
15. Wang, T.; Yao, Z.; Lu, M. On-the-fly ambiguity resolution involving only carrier phase measurements for stand-alone ground-based positioning systems. *GPS Solut.* **2019**, *23*, 36. [CrossRef]
16. Liu, X.; Yao, Z.; Lu, M. A rapid convergent positioning algorithm based on projected cancellation technique for pseudolite positioning systems. *GPS Solut.* **2021**, *26*, 15. [CrossRef]
17. Jiang, W.; Li, Y.; Rizos, C. Locata-based precise point positioning for kinematic maritime applications. *GPS Solut.* **2014**, *19*, 117–128. [CrossRef]
18. Fan, C.; Yao, Z.; Yun, S.; Xing, J. Ground-based PPP-RTK for pseudolite systems. *J. Geod.* **2021**, *95*, 133. [CrossRef]
19. Fan, C.; Yao, Z.; Wang, S.; Xing, J. Pseudolite system-augmented GNSS real-time kinematic PPP. *J. Geod.* **2022**, *96*, 77. [CrossRef]
20. Kouba, J.; Heroux, P. Precise point positioning using IGS orbit and clock products. *GPS Solut.* **2001**, *5*, 12–28. [CrossRef]
21. Choudhury, M.; Harvey, B.; Rizos, C. Tropospheric correction for Locata when known point ambiguity resolution technique is used in static survey-Is it required? In Proceedings of the International Global Navigation Satellite Systems Society IGNSS Symposium 2009, Gold Coast, Australia, 1–3 December 2009.
22. Tang, W.; Chen, J.; Zhang, Y.; Ding, J.; Song, Z. Refined troposphere delay models by NWM ray-tracing for pseudolite positioning system and their performance assessment. *Adv. Space Res.* **2024**, *73*, 5017–5030. [CrossRef]
23. Zhang, B.; Teunissen, P.; Odijk, D. A Novel Un-differenced PPP-RTK Concept. *J. Navig.* **2011**, *64*, 180–191. [CrossRef]
24. Takasu, T.; Yasuda, A. Development of the low-cost RTK-GPS receiver with an open source program package RTKLIB. In Proceedings of the International Symposium on GPS/GNSS, (International Convention Center, Jeju, Republic of Korea, 4–6 November 2009.
25. Teunissen, P.J.G. The least-square ambiguity decorrelation adjustment: A method for fast GPS ambiguity estimation. *J. Geod.* **1995**, *70*, 65–82. [CrossRef]
26. Verhagen, S.; Teunissen, P.J.G.; Marel, H.v.d.; Li, B. GNSS ambiguity resolution: Which subset to fix? In Proceedings of the International Symposium on GPS/GNSS, University of New South Wales, Sydney, Australia, 15–17 November 2011.
27. Massarweh, L.; Verhagen, S.; Teunissen, P.J.G. New LAMBDA toolbox for mixed-integer models: Estimation and evaluation. *GPS Solut.* **2024**, *29*, 14. [CrossRef]
28. Wang, J.; Feng, Y. Reliability of partial ambiguity fixing with multiple GNSS constellations. *J. Geod.* **2013**, *87*, 1–14. [CrossRef]
29. Teunissen, P.J.G.; Joosten, P.; Tiberius, C.C.J.M. Geometry-free Ambiguity Success Rates in Case of Partial Fixing. In Proceedings of the National Technical Meeting of The Institute of Navigation, San Diego, CA, USA, 25–27 January 1999.
30. Teunissen, P.J.G. Success probability of integer GPS ambiguity rounding and bootstrapping. *J. Geod.* **1998**, *72*, 606–612. [CrossRef]
31. Teunissen, P.J.G.; Verhagen, S. The GNSS ambiguity ratio-test revisited: A better way of using it. *Surv. Rev.* **2009**, *41*, 138–151. [CrossRef]

remote sensing

MDPI

Review

Review of Research on Satellite Clock Bias Prediction Models in GNSS

Yinhong Lv, Zhijun Meng *, Guangming Wang, Mingkai Liu and Enqi Yan

College of Intelligence Science and Technology, National University of Defense Technology, Changsha 410000, China; lvyinhong23@nudt.edu.cn (Y.L.); wgming@nudt.edu.cn (G.W.); liumingkai22@nudt.edu.cn (M.L.); yanenqi@nudt.edu.cn (E.Y.)

* Correspondence: mengzhijun@nudt.edu.cn

Highlights:

1. This review systematically summarizes and compares both classical and AI-based satellite clock bias prediction models, providing a comprehensive reference for model selection in real-time high-precision GNSS applications.
2. We propose generalized modeling frameworks for classical and AI-driven approaches and analyze error sources from systematic, data, and applicability perspectives to enhance prediction robustness.
3. This review identifies that AI models (e.g., LSTM, Transformer) outperform in complex nonlinear scenarios, whereas classical models (e.g., polynomial, Kalman) excel under short-term or stable conditions.
4. We outline promising research directions including multi-source data fusion, integration of short- and long-term prediction, and enhanced model generalizability to advance next-generation GNSS timing services.

Abstract: As foundational infrastructure for spatiotemporal information, the Global Navigation Satellite System (GNSS) delivers high-precision positioning, navigation, and timing (PNT) services worldwide. However, satellite atomic clock drift causes satellite clock bias, degrading PNT service quality. Compared to post-processed clock bias products and real-time estimation, satellite clock bias prediction offers a key advantage: it provides high-precision real-time clock bias even in scenarios with limited real-time data or poor communication. Through analysis and summarization of error sources in prediction models, this paper proposed generalized modeling frameworks for both classical and AI-based approaches. We reviewed current research on classical mathematical models—including polynomial, grey, Kalman filter, and time series models—and AI-based models such as machine learning (ML), multilayer perceptron (MLP), recurrent neural networks (RNN), and Transformer architectures. Technical characteristics, applicability, and limitations of each model were discussed. While AI-based models demonstrate superior flexibility and adaptability in complex scenarios compared to classical approaches, they require extensive datasets and computational resources. In conclusion, we summarized the advantages, disadvantages, and future research directions, offering insights for developing next-generation real-time high-precision GNSS PNT services.

Keywords: high-precision PNT; real-time satellite clock bias; clock bias prediction model; artificial intelligence

1. Introduction

As fundamental infrastructure for spatiotemporal information, the Global Navigation Satellite System (GNSS) provides high-precision global positioning, navigation, and timing (PNT) services. It is extensively utilized in transportation logistics, agriculture, environmental monitoring, scientific research, and military applications, delivering critical foundational support for developing information-based intelligent societies [1–4]. With technological advancements, emerging high-precision applications—including autonomous driving, unmanned aerial vehicle (UAV) collaborative operations, smart grids, and communication networks—increasingly demand real-time GNSS positioning accuracy at the centimeter level and timing precision at the nanosecond level. These requirements impose heightened demands on next-generation GNSS real-time high-precision PNT services. Consequently, real-time precise satellite clock bias prediction has become a prerequisite for achieving high-precision GNSS time synchronization and PNT services, establishing it as a key research focus in the GNSS field.

Autonomous GNSS navigation and sustained high-precision PNT services require precise satellite clock bias products. GNSS ground control centers establish and maintain the navigation system's time reference using high-performance atomic clock ensembles. These centers continuously monitor and periodically correct clock deviations—termed satellite clock bias—between GNSS satellite atomic clocks and ground-system time references. For example, the US GPS system conducts satellite-to-ground time comparisons via a global network of ground stations performing continuous 24-h GPS satellite observations [5]. Similarly, China's BeiDou System (BDS) employs inter-satellite and satellite-to-ground links for real-time monitoring of on-orbit satellite clock bias [6,7]. However, disruptions to ground stations or communication links may permit natural onboard atomic clock drift to induce clock bias, potentially degrading PNT service quality. Notably, a satellite clock bias of merely 1 nanosecond (10^{-9} s) can introduce pseudo-range errors up to 30 cm [8]. Achieving centimeter-level positioning thus demands clock bias products with sub-nanosecond accuracy. User Range Error (URE), a key determinant of positioning accuracy, quantifies errors in measured satellite-to-user ranges caused by orbital and clock bias inaccuracies. Current GNSS open service performance indicates average UREs (95% reliability) of 1.0 m for BDS, 7.8 m for GPS, 7.0 m for Galileo, and 18.0 m for GLONASS [9].

Currently, precise satellite clock bias is primarily obtained from four sources: broadcast ephemeris clock bias, post-processed precise products from data analysis centers, real-time estimation, and prediction [10]. GNSS broadcast clock bias satisfies real-time requirements but exhibits limited accuracy, typically around 5 ns. Post-processed precise products achieved tens of picoseconds accuracy, yet suffer from significant latency—often released hours to days after observation. The International GNSS Service (IGS) final products currently provided 75 ps accuracy but require 12–18 days for availability. To address this, IGS offered ultra-rapid products comprising a 48-h arc: the first 24-h observed segment achieves <0.15 ns accuracy with 3–9 h latency, while the subsequent 24-h predicted segment attains only ~3 ns accuracy. This marginally exceeded broadcast ephemeris performance but remains substantially below the IGS-recommended 0.3 ns threshold for real-time high-precision positioning [11]. Domestically, Wuhan University's WUM products demonstrate consistency with IGS final products. When referenced against IGS final GPS solutions, WUM achieves daily average RMS and STD values of 154.9 ps and 55.1 ps, respectively, though requiring 2–3 days for release [12].

Real-time satellite clock bias prediction employs a global network of GNSS observation stations, utilizing efficient data communication and rapid processing to generate clock bias in real time. This approach provides near-real-time precise solutions with significantly enhanced accuracy compared to broadcast products, forming the foundation for contemporary high-precision positioning services. In 2013, the International GNSS Service (IGS) launched its comprehensive Real-Time Service (RTS) through collaboration with Natural Resources Canada (NRCan), the German Federal Agency for Cartography and Geodesy (BKG), and ESA's European Space Operations Centre (ESOC). Supported by 160+ globally distributed stations, multiple data centers, and 10 analysis centers, RTS delivered products with 0.300 ns accuracy at ~25 s latency [13]. Nevertheless, practical implementations may encounter service interruptions, clock bias anomalies (errors/jumps), and network failures [14].

Satellite clock bias prediction involves predicting future clock bias variations over a certain period by establishing mathematical models or employing deep learning methods. Compared to post-processed precise clock bias products and real-time estimation, clock bias prediction offers the advantage of providing real-time high-precision clock bias services even in scenarios lacking real-time data or under poor communication conditions. This capability is particularly critical in specific applications, such as environments with limited ground stations, poor data transmission conditions, or strict mission planning requirements. Due to these advantages, extensive research on satellite clock bias prediction has been conducted both domestically and internationally. However, the field currently lacks a systematic and comparative review that synthesizes the diverse modeling approaches—from classical mathematical methods to modern artificial intelligence techniques—and evaluates their performance, applicability, and limitations under varying operational conditions. This gap hinders the optimization and selection of appropriate prediction strategies for emerging high-stakes applications. Therefore, the primary motivation of this review is to comprehensively summarize and analyze existing satellite clock bias prediction models, clarify their theoretical foundations and practical challenges, and provide a clear reference for future research and operational deployments. This paper summarizes the factors influencing the errors of prediction models and presents generalized modeling flowcharts for classical prediction models and artificial intelligence-based prediction models. It reviews the research progress, technical characteristics, and limitations of satellite clock bias prediction models from both classical mathematical and artificial intelligence perspectives, while also outlining future research directions for reference by researchers in related fields worldwide.

2. Model Error Analysis

Errors in clock bias prediction models mainly arise from model construction methods, data quality, and applicability constraints. As illustrated in Figure 1, we analyze these errors through three dimensions: systematic model errors, input data error propagation, and scenario-specific applicability limitations.

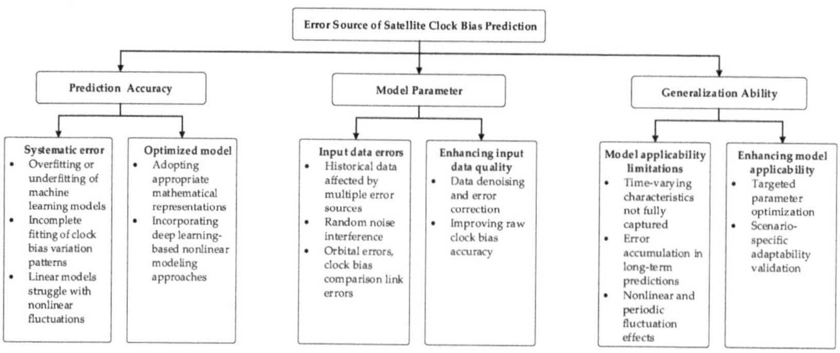

Figure 1. Error analysis of satellite clock bias.

Systematic errors in clock bias prediction models originate from inherent design limitations, primarily determined by whether the mathematical formulation accurately captures variation patterns. Current mainstream models include time series methods (e.g., Autoregressive Integrated Moving Average, Kalman filtering) and machine learning approaches (e.g., Support Vector Machines, Long Short-Term Memory networks). Time series models typically assume autocorrelation and trend characteristics in clock bias data. However, empirical clock signals often exhibit nonlinear fluctuations that models cannot fully characterize, introducing systematic errors. For example, in a study predicting the clock bias of BDS satellites, the ARIMA model exhibited an average prediction error of approximately 3.5 ns over 6 h due to unmodeled periodic variations and thermal-induced fluctuations [15]. Although machine learning models excel at handling complex data, their dependence on historical datasets creates vulnerabilities: insufficient training data or suboptimal feature selection frequently causes overfitting or underfitting, degrading prediction accuracy [16]. When using LSTM for Galileo satellite clock error prediction, model performance degraded significantly during solar radiation periods because the training set lacked sufficient examples of space weather-induced anomalies [17].

Input data error propagation constitutes another major error source in clock bias prediction models. Historical clock bias data—essential for model training—contain inherent inaccuracies from acquisition and processing stages [18]. Errors originating from onboard atomic clocks and post-processed products directly propagate into prediction models, compromising prediction precision. Satellite orbit inaccuracies and signal propagation delays, for example, introduce additional noise that degrades data integrity. A notable case is the error amplification observed in BeiDou-3 satellites: orbital determination errors of 20 cm resulted in additional clock bias prediction inaccuracies of up to 1.8 ns over a 12-h prediction horizon [19]. Since models train on historical data, both systematic and random errors affect parameter estimation, propagating into future predictions. Furthermore, satellite clock bias exhibits random noise characteristics, particularly power-law noise components (e.g., frequency random walk, phase flicker noise) [20], which destabilize data and reduce predictability, further diminishing prediction accuracy. Experimental support comes from a study on clocks of Compass constellations where ignoring noise type mismatches (e.g., treating flicker noise as white noise) increased prediction RMS error by 40% compared to noise-adaptive filtering methods [21].

Model applicability limitations constitute a third significant error source in clock bias prediction. The generalization ability of the prediction model depends on multiple factors, including the time-varying nature of clock variations and complex environmental influences. During operation, satellite clock bias is affected not only by onboard atomic clock performance but also by periodic orbital perturbations, attitude variations, and space environmental conditions. These factors exhibit nonlinear periodic fluctuations that challenge model adaptability [22]. Failure to accurately capture such dynamics induces error accumulation, particularly in long-term predictions. For instance, during geomagnetic storms, the prediction error of a quadratic polynomial model for GPS clocks increased by 230% due to unmodeled ionospheric delay variations and increased atomic frequency instability [23]. Consequently, adaptive validation across satellites and timeframes, coupled with dynamic parameter adjustments, becomes essential for enhancing model robustness and accuracy in diverse scenarios.

In summary, clock bias prediction model errors originate from three primary sources: inherent systematic errors in model design, propagated input data inaccuracies, and scenario-specific applicability limitations. Mitigating these errors requires not only judicious model parameter selection and optimization during the modeling phase but also rigorous input data preprocessing through noise reduction and error correction techniques to enhance prediction accuracy and reliability.

3. Classical Prediction Models

Classical prediction models rely on precise mathematical formulations for modeling, achieving forecasts through data feature extraction and fitting. Figure 2 illustrates the general modeling process of these models, demonstrating sequential relationships between stages and incorporating a closed-loop feedback mechanism to support continuous optimization and improvement.

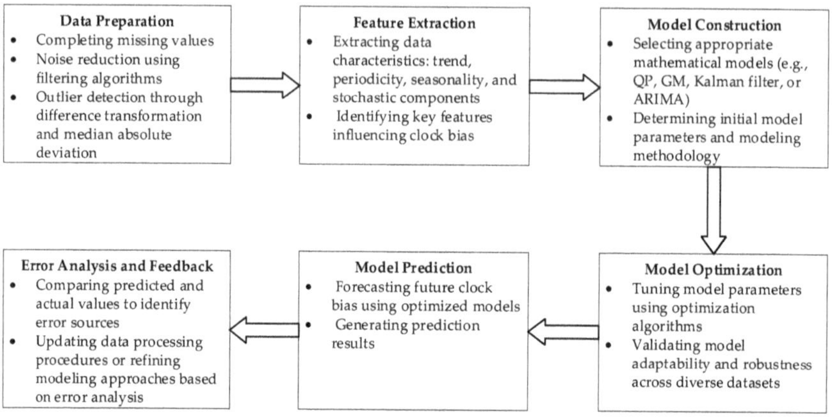

Figure 2. General modeling flow of a classical prediction model.

This section will introduce the following four categories of classical prediction models: Polynomial Model (PM), Grey Model (GM), Kalman Filter Model (KFM), and Autoregressive Integrated Moving Average model (ARIMA).

3.1. Polynomial Model

The polynomial model for satellite clock bias prediction originates from studies on the operational behavior of onboard atomic clocks. Its primary objective is to forecast clock bias variations using mathematical modeling techniques, thereby improving the accuracy of navigation systems. The earliest implementation of such a model was proposed by developers of the U.S. GPS system, where clock bias variations were represented using a quadratic polynomial. This method characterizes clock bias changes through three parameters: clock bias, frequency bias, and frequency drift [24,25]. The mathematical formulation is presented in Equation (1).

$$x(t_i) = a_0 + a_1(t_i - t_0) + a_2(t_i - t_0)^2 + \psi(t_i) \tag{1}$$

where $x(t_i)$ represents the clock bias at epoch t_i, t_0 is the reference time, $\psi(t_i)$ denotes the residual error of the clock bias at epoch t_i, a_0, a_1 and a_2 are the parameters representing clock bias, frequency bias, and frequency drift, respectively.

To more accurately capture the characteristics of onboard atomic clocks, the periodic variations of satellite clocks should also be taken into account. Spectral analysis techniques can be applied to identify significant periodic components within the data sequence. By incorporating these components into the quadratic polynomial model, a spectral analysis–based model can be derived [26], whose mathematical formulation is presented in Equation (2).

$$x(t_i) = a_0 + a_1(t_i - t_0) + a_2(t_i - t_0)^2 + \sum_{k=1}^{p} A_k \sin[2\pi f_k(t_i - t_0) + \varphi_k] + \psi(t_i) \tag{2}$$

where p represents the number of main periodic functions, A_k, f_k and φ_k denote the amplitude, frequency, and phase of corresponding periodic terms, respectively. Other parameters maintain the same meaning as in Equation (1).

Recent research has introduced numerous enhancements to polynomial models for satellite clock bias prediction. Han et al. [24] combined IGS precise orbit data with clock bias observations from IGS tracking stations in 2001 to predict post-processed precise clock bias at 30-s intervals, achieving an accuracy of 0.1 ns. Bernier et al. [25] investigated the characteristics of new passive hydrogen masers in 2005 using standardized quadratic difference prediction algorithms, obtaining results second only to those of energy spectrum noise models. In 2008, Panfilo et al. [26] conducted targeted research on noise components in onboard atomic clock bias, extracting random walk noise and developing a prediction model that produced objective results. Allan et al. [27], in their 2018 study, analyzed optimal estimation methods for onboard atomic clocks under five common noise conditions, deriving extrapolation expressions for these noise models and providing corresponding prediction confidence levels and trends. Lei et al. [28] employed polynomial fitting based on the physical characteristics of onboard atomic clocks to extract clock bias trend terms, followed by phase space reconstruction and Gaussian process prediction for the fitting residuals. For processing random terms in polynomial models, Wang et al. [29] adopted robust least-squares collocation methods, whereas Liao et al. [30] proposed iterative approaches for random error correction and modeling, achieving a 1-day RMS accuracy of 0.57 ns and a maximum deviation of 1 ns.

Polynomial models exhibit strong data-fitting capabilities and are effective for short-term clock bias prediction; however, they face challenges in capturing complex nonlinear variations and generally perform poorly in extrapolation. While these models can provide basic fitting and prediction of clock bias, their accuracy deteriorates markedly over time. Therefore, regularly updating the polynomial model parameters in navigation messages is essential for maintaining prediction precision.

3.2. Grey Model

The grey model was first applied to GPS satellite clock bias prediction in the 1990s. A grey prediction system is a prediction framework that operates with partially known and partially unknown information, representing an incompletely certain system. Based on grey models, it generates new data sequences by accumulating or differencing the original series, and then models these sequences to more objectively capture useful information. Atomic clocks onboard navigation satellites are highly sensitive to both external and internal factors, making it challenging to characterize their complex and detailed variation patterns. These characteristics closely correspond to the principles of grey system theory.

Among grey system models, the GM(1,1) model is the most commonly used grey model, with its expression shown in Equation (3).

$$l^{(0)}(k) = (1 - e^{a})[l^{(0)}(1) - \frac{u}{a}]e^{-a(k-1)} \tag{3}$$

where $l^{(0)}(k)$ represents the original clock bias sequence, a denotes the grey coefficient, u indicates the grey action quantity, e is a natural number.

A comprehensive review of literature from the past decade reveals substantial research efforts aimed at improving the prediction accuracy of grey models from multiple perspectives. In clock bias data processing, Liang et al. [31] employed difference sequences, whereas Mei et al. [32] proposed ratio sequences for clock bias modeling. Jiang Yu et al. [33] applied a power function transformation to clock bias data, and Yu et al. [34] implemented Vondrák filtering for preprocessing. Regarding coefficient determination and optimization,

Zheng et al. [35] analyzed the relationship between prediction accuracy, GM exponential coefficients, and different sampling intervals, and compared the results with those from polynomial models. Yu et al. [36] improved the grey prediction model by employing least absolute deviation to enhance performance under abrupt-change conditions. Yu Ye et al. [37] developed a particle swarm optimization–weighted grey regression hybrid model, achieving average 24-h prediction residuals below 5.71 ns. Li et al. [38] proposed an adaptive twin-subpopulation improved particle swarm optimization algorithm (TS-IPSO), whereas Yang et al. [39] optimized grey model parameters using the Levenberg–Marquardt algorithm. Yuan et al. [40] introduced an enhanced firefly algorithm (SAFA-FD), achieving 3–6 h prediction errors below 1 ns and 9–12 h errors under 2 ns. Additionally, Zhao et al. [41] developed a dynamic grey model that continuously updates by discarding outdated data and incorporating new observations, thereby improving the whiteness of prediction intervals. Tan et al. [42] optimized the initial conditions of grey models by deriving them from components of the original clock bias sequence.

The grey model demonstrates effective prediction capability with limited samples, low data requirements, and fast modeling speed, yet exhibits instability in long-term predictions, making it suitable for monotonic data but less effective for highly fluctuating datasets.

3.3. Kalman Filter Model

The Kalman filter, introduced by Rudolf E. Kalman in 1960, is a recursive optimal estimation algorithm originally developed for navigation systems. The National Institute of Standards and Technology (NIST) employed the Kalman filter algorithm to establish Atomic Time. However, due to the incomplete observability of the system during the filtering process, the error covariance matrix diverged continuously, leading to unbounded growth in the clock bias estimates. Additionally, the Kalman filter's estimation of clock offset relied on the long-term stability of atomic clocks while overlooking their short-term stability. To address these limitations, several improved Kalman filter algorithms were proposed. For instance, Davis et al. [43] introduced two distinct methods to reduce the estimation error covariance matrix and demonstrated the optimality of the reduced Kalman scale. Greenhall et al. [44] developed an approach that combined the Kalman filter with a weighted average algorithm, along with a method for modeling flicker noise, to constrain the divergence in the estimated clock error covariance matrix.

Kalman filter enables real-time computation of satellite clock bias, frequency bias, and frequency drift, thereby functioning as both a clock bias prediction model and an estimation model. In this framework, the current state prediction depends on the previous state estimate and the latest observations. The state equation of the Kalman filter for onboard atomic clocks is presented in Equation (4) [45].

$$\begin{bmatrix} x(t+\tau) \\ y(t+\tau) \\ z(t+\tau) \end{bmatrix} = \begin{bmatrix} 1 & \tau & \tau^2/2 \\ 0 & 1 & \tau \\ 0 & 0 & 1 \end{bmatrix} \begin{bmatrix} x(t) \\ y(t) \\ z(t) \end{bmatrix} + \begin{bmatrix} \varepsilon_x \\ \varepsilon_y \\ \varepsilon_z \end{bmatrix} \tag{4}$$

where: τ represents the sampling interval; $x(t)$, $y(t)$ and $z(t)$ denote the clock bias, frequency bias, and frequency drift parameters of the onboard clock, respectively. ε_x, ε_y and ε_z are random errors with zero mean.

The key to constructing a Kalman filter model lies in determining the process noise covariance matrix and the measurement noise covariance matrix. Current approaches include the Hadamard total variance method proposed by Guo and Yang [46], the Allan variance introduced by Howe et al. [47], and the variance recursion method developed by Hutsell et al. [48]. Further research indicates that Davis J. et al. [49] implemented a Kalman filter–based prediction algorithm capable of achieving near-optimal performance, maintaining prediction errors at the 1-nanosecond level for next-generation Block IIR and IIF satellites over a one-day prediction horizon. Pihlajasalo et al. [50] used grid search to identify optimal noise parameters, selecting those that minimized prediction errors in experimental tests. Pratt et al. [51] incorporated both satellite-to-ground and satellite-to-satellite measurement data into a Kalman filter framework to estimate each Iridium satellite's clock bias, with typical results showing a current filter accuracy within 200 ns, consistently meeting the original system specification of 0.5 μs. Various improvements to the Kalman filter model have been proposed: Zhang et al. [52] integrated prediction penalties and amplitude variation constraints into clock bias estimation, while Wang et al. [53] applied wavelet denoising to clock bias observations and proposed a linearly weighted combination Kalman filter model.

The Kalman filter model can update predictions in real-time and is suitable for clock bias prediction in dynamic environments. However, its performance may degrade for nonlinear systems or when noise distributions deviate from Gaussian assumptions, as it strictly requires accurate noise statistical characteristics—prediction quality decreases with inaccurate noise models, and demands prior knowledge of both system models and noise statistics.

3.4. Autoregressive Integrated Moving Average Model

The autoregressive integrated moving average (ARIMA) model, proposed by Box and Jenkins in 1970, is a widely used time series analysis method applied in fields such as economics, meteorology, and engineering for prediction purposes. ARIMA modeling and prediction rely on analyzing the characteristics of the time series data to determine an appropriate model type and select suitable orders. The ARIMA (p, d, q) model, which incorporates differencing, is denoted as ARIMA (p, d, q), where p and q represent the autoregressive and moving average orders, respectively, and denotes the number of differencing operations. When $d = 0$, the ARIMA model reduces to the autoregressive moving average (ARMA) model, whose formulation is given in Equation (5) [54].

$$x_t = \sum_{i=1}^{p} a_i x_{t-i} + \varepsilon_t + \sum_{j=1}^{q} b_j \varepsilon_{t-j} \tag{5}$$

where: a_i and b_j represent the autoregressive and moving average parameters to be estimated; $\{\varepsilon_t\} \sim WN(0, \sigma^2)$ denotes the white noise sequence, and σ^2 is the white noise variance; when p and q equal 0, respectively, the model becomes an MA model and an AR model correspondingly.

Mutaz et al. [54] applied time series models to predict and correct time-delay errors in precise point positioning (PPP) single-point solutions, achieving a 28% improvement in prediction accuracy. Zhou et al. [55] optimized the ARIMA model by introducing a novel fitting approach and refined criteria for determining the p and q orders, applying it to short-term satellite clock bias prediction, with most GPS rubidium clocks exhibiting prediction errors below 0.6 ns for 6-h forecasts and below 0.25 ns for 1-h forecasts. Jiang et al. [56] evaluated the performance of ARIMA models under varying data-fitting lengths and prediction horizons. Several enhancements to ARIMA models have been reported: Yan et al. [57] applied sparsity theory to autoregressive model parameters to mitigate the influence of redundant parameters; Gonzalez et al. [58] incorporated seasonal components into prediction residuals, achieving nanosecond-level root-mean-square errors (RMSE) throughout the day; Kim et al. [59] integrated neural networks with ARIMA models for predicting IGS real-time service corrections, where the ARIMA coefficients were determined via neural networks. In addition, time series models are widely employed for detecting and correcting AO-type outliers in clock bias data. Han et al. [60] developed an algorithm for detecting AO-type outliers in autoregressive (AR) models, effectively addressing masking and swamping issues in clustered outlier detection. Ma et al. [61] proposed a variance-inflation model based on ARIMA for anomaly detection in BDS satellite clock bias, employing likelihood ratio methods to identify outliers in BDS clock bias time series. Zhang et al. [62] introduced a method that combines ARMA time series modeling with Bayesian statistics for both precise satellite clock bias prediction and historical anomaly detection.

The ARIMA model effectively captures temporal dependencies and autocorrelations in clock bias data, making it suitable for long-term prediction with solid theoretical foundations and broad applicability. However, parameter selection and order determination can be complex, often requiring substantial prior knowledge, and while the model works well for linear variations, its performance degrades with nonlinear data patterns.

3.5. Summary of Classical Models

Classical mathematical modeling approaches rely on explicitly defined mathematical formulas and parameters, where the predictive accuracy is strongly influenced by the trend, periodicity, stochastic characteristics, and noise properties of the clock bias data. However, such models encounter challenges when addressing complex nonlinear relationships and typically impose stringent requirements on parameter selection and prior knowledge. As the prediction horizon extends, errors tend to accumulate progressively, thereby constraining the models' capability for long-term prediction. Moreover, the inherent linear assumptions and fixed parameter configurations in classical methods reduce their adaptability to the dynamic variations and intricate noise patterns frequently observed in satellite clock behavior—particularly over extended horizons, where error propagation becomes increasingly significant. These limitations have driven researchers to investigate more advanced modeling techniques, such as machine learning approaches, to more effectively capture nonlinear dynamics and enhance the long-term prediction stability of satellite clock bias.

4. Artificial Intelligence Prediction Models

The general modeling workflow for artificial intelligence (AI)-based prediction models is illustrated in Figure 3. Compared with classical mathematical modeling approaches, the AI prediction framework places greater emphasis on model complexity and the importance of iterative optimization. At present, AI models applied to satellite clock bias time series prediction can be broadly classified into four categories: machine learning (ML)-based models, multilayer perceptron (MLP)-based models, recurrent neural network (RNN)-based models, and Transformer-based models that employ self-attention mechanisms.

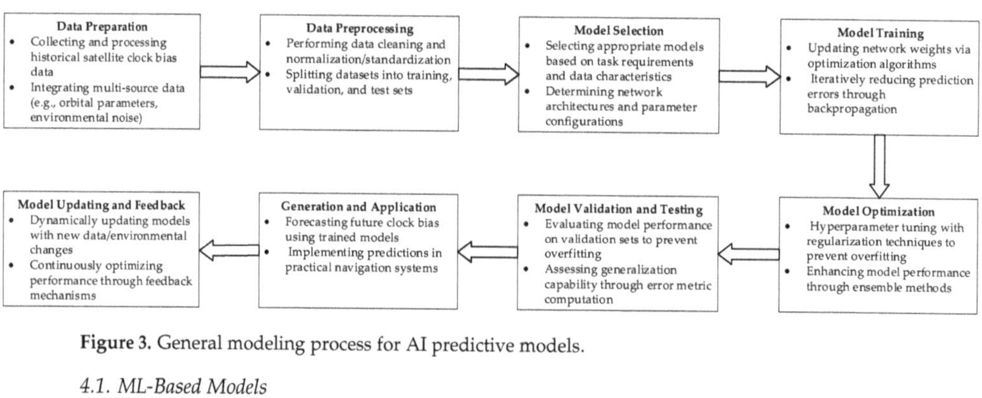

Figure 3. General modeling process for AI predictive models.

4.1. ML-Based Models

Classical machine learning algorithms have well-established theoretical foundations, providing strong mathematical interpretability and reliable performance guarantees. These conventional models depend on manual feature extraction and generally adopt relatively simple architectures with fewer layers, which enable faster training and make them particularly suitable for small- to medium-sized datasets.

In Support Vector Machine (SVM) prediction models, the radial basis function (RBF) is commonly employed as the kernel function, enabling nonlinear mapping of input variables into a high-dimensional feature space where linear optimization functions are constructed for prediction. Xiao et al. [63] developed a refined quadratic polynomial prediction algorithm based on Particle Swarm Optimization–SVM (PSO-SVM), which performs rolling predictions by modeling residuals during the fitting stage. He et al. [64] proposed an enhanced approach that integrates spectral analysis models (SAM) with least squares support vector machines (LS-SVM). As illustrated in Figure 4, their satellite clock bias data processing framework demonstrated that the improved model can maintain prediction deviations within ±1.0 ns for most Medium Earth Orbit (MEO) satellites over 3-h prediction periods. Zhu et al. [65] introduced a least squares support vector machine (LSSVM) prediction algorithm for hydrogen atomic clocks, achieving improvements in prediction accuracy of 50% and 29% compared with linear prediction algorithms and standard SVM prediction algorithms, respectively.

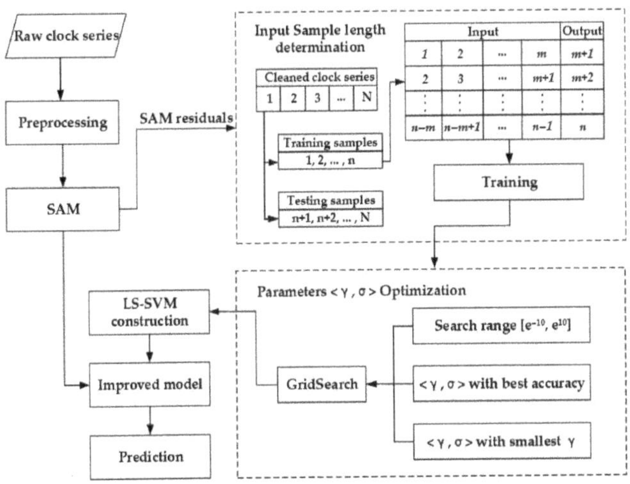

Figure 4. Clock bias data processing flowchart using an improved SVM model [56].

The Extreme Learning Machine (ELM), proposed by Huang and colleagues in 2004, is a single-hidden-layer feedforward neural network distinguished from classical neural networks by its random initialization of input weights and hidden layer biases, which remain fixed during training, while the output weights are computed directly via least squares methods. This design significantly accelerates the training process. He et al. [66] developed an enhanced clock prediction method by integrating spectral analysis models with ELM, demonstrating that prediction accuracy remains consistently high, with errors below 1.0 ns as the prediction horizon extends from 0 to 18 h. Lei et al. [67] proposed an ELM network structure design based on Adaptive Resonance Theory (ART) networks, leveraging ART's superior self-organizing classification capabilities. Ya et al. [68] implemented a Sparrow Search Algorithm (SSA)-optimized ELM prediction model that utilizes 12-h historical satellite clock bias data to forecast the subsequent 6-h bias. The SSA-ELM model achieved improvements of 53.16% and 52.09% over QP models, and 40.66% and 46.38% over GM models, respectively, as illustrated in Figure 5.

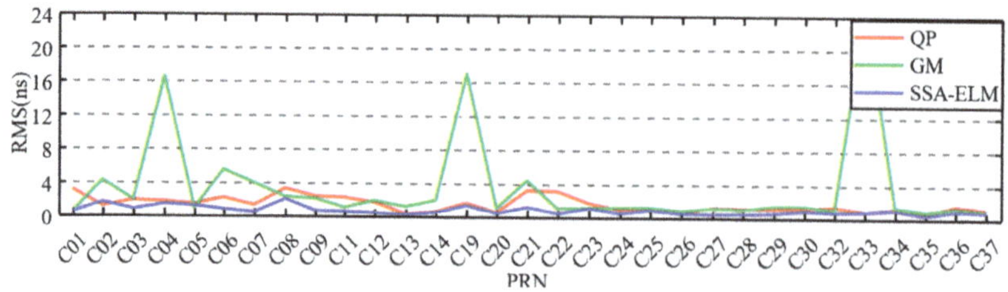

Figure 5. Predicting BDS Satellite Clock Bias Accuracy for the Next 6 Hours Based on Historical 12-Hour Satellite Clock Bias Data [60].

SVMs are well-suited for small sample sizes and have strong theoretical foundations; however, they face challenges in parameter selection and exhibit high computational complexity. ELMs offer advantages such as rapid training speeds and fewer tunable parameters, facilitating relatively straightforward implementation. Nonetheless, ELMs suffer from limited generalization capability and comparatively poor robustness to noise and anomalous data. Moreover, classical machine learning methods may encounter computational and storage limitations when processing large-scale datasets, restricting their applicability in scenarios that require real-time processing of massive GNSS observation data.

4.2. MLP-Based Models

The Multilayer Perceptron (MLP) is a feedforward neural network consisting of multiple fully connected layers. Input data is processed through successive layers of linear transformations followed by nonlinear activation functions, resulting in hierarchical feature extraction. In time series prediction tasks, historical values of the time series are typically employed as inputs, allowing the MLP to learn underlying patterns and trends for prediction purposes [69]. The MLP demonstrates strong adaptability, robustness, and associative memory capabilities, rendering it especially effective in capturing complex nonlinear relationships within satellite clock bias data.

Du et al. [70] employed empirical mode decomposition combined with backpropagation neural networks to predict GNSS receiver clock biases, demonstrating excellent performance in predicting receiver clock bias time series. However, MLP models are susceptible to becoming trapped in local optima during backpropagation training and exhibit slow convergence rates, which ultimately affect final convergence accuracy. To optimize MLP neural network weights and thresholds while enhancing prediction accuracy and convergence speed, numerous scholars have proposed various optimization algorithms, including genetic algorithms [71], mind evolutionary algorithms [72,73], sparrow search algorithms [69], whale optimization algorithms [74], beetle antennae search algorithms [75], and improved particle swarm optimization algorithms [76]. These optimized models demonstrate varying degrees of improvement in prediction accuracy compared to classical MLP models and conventional prediction methods. Specifically, Bai et al. [73] achieved a 12-h prediction accuracy of 1.56 ns using mind evolutionary algorithm optimization, while Lyu et al. [76] attained average prediction accuracies better than 0.15 ns for 20-min and 60-min forecasts with their improved particle swarm optimization model, representing approximately 85% improvement over conventional models. Building upon the MLP neural network framework, researchers have further enhanced hidden layer neural units by replacing standard activation functions with wavelet basis functions or radial basis functions, resulting in wavelet neural networks [77–79] and radial basis function neural networks [80,81], both of which provide notable improvements in clock bias prediction accuracy.

MLP neural networks exhibit strong adaptability in modeling complex nonlinear relationships; however, their training processes require multiple iterations and are time-consuming, with a propensity to converge to local optima. Wavelet neural networks leverage the multiscale analysis capabilities of wavelet transforms to more effectively capture time-frequency characteristics in clock bias data, as wavelet transforms can compress data efficiently and enhance computational performance. Nevertheless, these networks involve complex architectural designs and wavelet basis function selections, necessitating substantial training data and computational resources. RBF neural networks offer rapid training speeds suitable for real-time prediction and demonstrate strong generalization abilities; however, they are highly dependent on sample quality and present challenges in parameter selection. Each of these neural network approaches presents distinct advantages and limitations in satellite clock bias prediction, with their relative efficacy contingent upon specific operational requirements, available computational resources, and desired prediction horizons within GNSS.

4.3. RNN-Based Models

Recent research on clock bias prediction has increasingly focused on Recurrent Neural Networks (RNNs), which possess inherent "memory" capabilities that differentiate them from conventional Multilayer Perceptron (MLP) networks. Designed specifically for sequential data processing, RNNs are characterized by their recurrent connections, which enable the retention of historical information. At each time step, RNNs incorporate the hidden state from the previous step, allowing them to capture temporal dependencies within sequential data. Long Short-Term Memory (LSTM) networks and Gated Recurrent Unit (GRU) networks represent enhanced variants of RNNs, addressing the vanishing gradient problem via more sophisticated and more streamlined gating mechanisms, respectively, thereby improving their capacity to model long-term dependencies. The architectures of these three models are depicted in Figure 6 [82], illustrating their progressive complexity and efficiency in handling time series data. The Tanh function maps any real-valued number to a value between −1 and 1. The Sigma function is a mathematical function that maps any real-valued number to a value between 0 and 1. In a standard RNN cell, the Tanh function is typically used to activate the combined input and hidden state to produce the new hidden state. The Sigma function is not used in the simplest RNN cells but becomes crucial in gated architectures like LSTMs to manage information flow over long sequences.

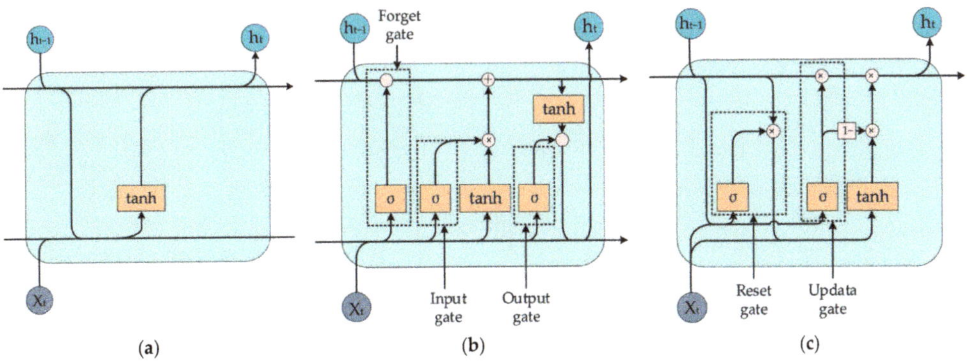

Figure 6. Three Typical Recurrent Neural Network Models. (**a**) The Basic RNN Model Architecture. (**b**) The LSTM Model Architecture. (**c**) The GRU Model Architecture.

Piotr et al. [82] employed LSTM models for real-time clock bias prediction of GPS satellites, achieving results superior to current state-of-the-art methods. Huang et al. [83] proposed a supervised learning LSTM network (SL-LSTM) that separately models the periodic and random components of clock bias data, demonstrating improved prediction accuracy and stability in long-term prediction. He et al. [84] utilized LSTM models to predict BDS-3 satellite clock biases, attaining short-term (0–6 h) prediction accuracies of 0.5 ns and 0.3 ns for IGSO and MEO satellites, respectively. For long-term clock bias prediction, the LSTM model exhibited improvements of 72.0% and 64.0% over ARIMA and QP models, respectively. Cai et al. [85] developed an integrated LSTM-attention neural network incorporating self-attention mechanisms, as illustrated in Figure 7, which effectively balances global attention with local feature extraction, achieving 62.51% improvement over standard LSTM models in 12-h prediction tasks and 71.16% enhancement in 24-h prediction.

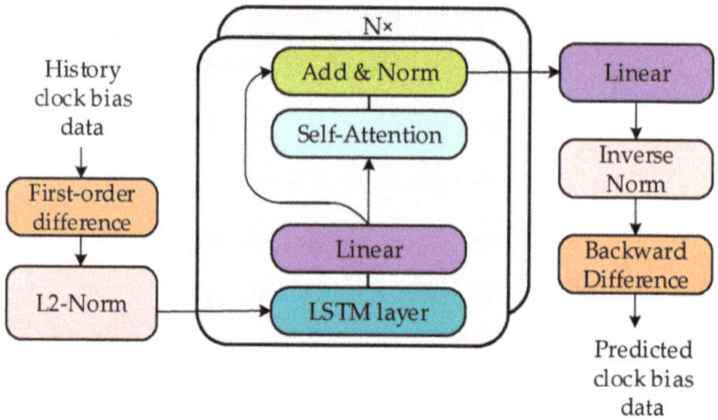

Figure 7. The LSTM-Attention Neural Network Model [77].

Yang et al. [86] employed a sparsely sampled LSTM network algorithm to enhance the long-term prediction accuracy of clock biases for space-based time scales, achieving onboard frequency stabilities of 1.35×10^{-15}, 3.37×10^{-16}, and 2.81×10^{-16} at sampling intervals of 300 s, 8.64×10^4 s, and 1×10^6 s, respectively. Liu et al. [87] developed a BDS-3 satellite clock bias prediction model using GRU neural networks, implementing an improved Sparrow Search Algorithm (SSA)-GRU hybrid model (ITSSA-GRU) that demonstrates superior generalization capability and prediction performance across three types of satellite clocks. Additionally, Liang et al. proposed the Nonlinear Autoregressive Exogenous (NARX) neural network in 2021 [88], an enhanced recurrent neural network variant capable of incorporating feedback from output neurons, and introduced the Elman neural network in 2022 [89], which incorporates a "context layer" to memorize previous states, rendering it particularly effective for time series processing and dynamic system modeling.

RNN-based models excel at capturing long-term dependencies and are well-suited for time series prediction. By employing gating mechanisms to regulate information flow, these models demonstrate strong memory capabilities for extended sequential data. However, they require lengthy training periods, involve complex training procedures, and demand careful tuning of network architectures and parameters. Despite these challenges in practical implementation, RNN-based models outperform conventional approaches in capturing the intricate temporal patterns of satellite clock behavior.

4.4. Transformer-Based Models

The Transformer neural network model, first proposed by Vaswani et al. [90] in 2017 was illustrated in Figure 8. It was originally designed to address sequence modeling challenges in natural language processing (NLP) and to overcome the limitations of classical sequence models such as RNNs and LSTMs in capturing long-range dependencies. The core innovation of the Transformer is the self-attention mechanism, which fundamentally differs from the recurrent structures of RNNs and LSTMs by processing input sequences entirely through attention mechanisms. This design enables effective modeling of long-distance dependencies and allows parallel data processing, resulting in higher computational efficiency. The Transformer architecture comprises two main components: an encoder and a decoder, each consisting of multiple identical layers.

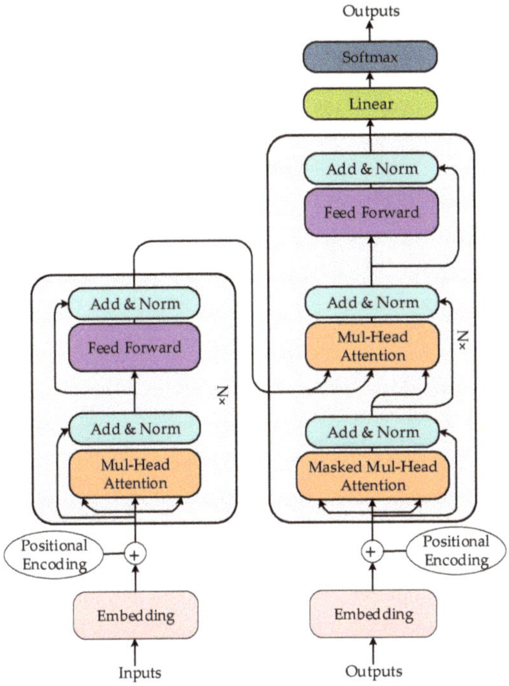

Figure 8. The Transformer Model Architecture [82].

Syam et al. [8] developed a Transformer-based deep neural network for ML time series modeling, providing fast and reliable clock bias correction predictions for single-frequency receivers with an accuracy of less than 2 ns and maximum prediction horizons up to 2 h. This model demonstrated a 50% improvement in prediction performance compared to other clock bias prediction methods, as illustrated in Figure 9. Pan et al. [91] proposed a sequence-to-sequence neural network model (Informer) based on optimized self-attention mechanisms, which effectively leverages dependencies within clock bias data to achieve deep latent feature extraction and parallel computation, exhibiting high accuracy and stability in short-term clock bias prediction.

The Transformer deep learning model utilizes parallel computation via its self-attention mechanism, enabling efficient training and superior capability in capturing global features of sequence data, particularly excelling in handling long sequences. However, it requires substantial computational resources and memory, performing optimally with large-scale datasets but exhibiting relatively poorer performance than classical RNNs, LSTMs, or convolutional neural networks when data or computational resources are limited. The Informer model, specifically designed for long-term time series prediction, offers enhanced capabilities for modeling lengthy sequences and achieves greater computational efficiency compared to conventional Transformer architectures. Nevertheless, its complex structure imposes stringent hardware and algorithmic requirements, and its performance heavily depends on abundant training data; insufficient data volume may hinder the model from reaching its full potential.

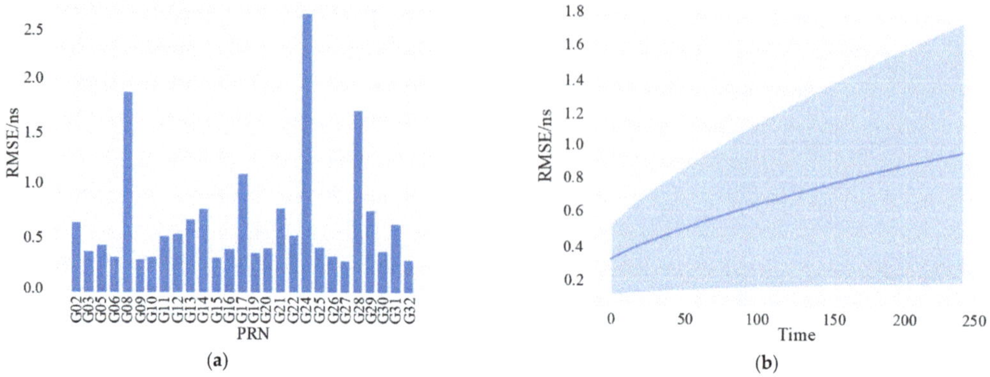

Figure 9. Clock Bias Prediction Results for two hours Based on the Transformer Model [8]. (**a**) RMSE of PRN Clock Bias Prediction. (**b**) RMSE of Clock Bias Prediction for the Next.

4.5. Summary of AI Models

Satellite atomic clocks are vulnerable to external environmental factors and possess inherently complex characteristics, causing their clock bias to exhibit pronounced nonlinear behavior that challenges classical models in accurately describing these variations [84]. In contrast, artificial neural networks exhibit strong sensitivity to nonlinear dynamics and can overcome the limitations of conventional models, rendering them particularly suitable for modeling within big data environments. By leveraging extensive historical data through deep learning and optimization techniques, these networks automatically extract salient features from the data, making them well-suited for various complex time series prediction tasks.

5. Conclusions

The clock bias prediction models introduced above each possess distinct characteristics in practical applications, all capable of achieving relatively good prediction results under certain conditions. An analysis and summary of the advantages and disadvantages of current major clock bias prediction models are presented in Table 1.

Table 1. Advantages and disadvantages of satellite clock bias prediction model.

Prediction Models		Model Features
Classical Prediction Models	QP	**Advantages:** Simple implementation, suitable for short-term prediction, and high computational efficiency. **Limitations:** Poor nonlinear adaptability, weak extrapolation performance
	GM	**Advantages:** Strong small-sample adaptability, low data requirements, fast modeling. **Limitations:** Unstable long-term prediction, limited applicability.
	Kalman	**Advantages:** Strong dynamic tracking capability, suitable for multi-variable fusion, and effective noise handling. **Limitations:** High model complexity, dependent on model assumptions
	ARIMA	**Advantages:** Captures temporal dependencies, and suitable for long-term prediction. **Limitations:** Requires data stationarity, poor adaptability to nonlinear variations
AI Prediction Models	ML-based	**Advantages:** Simple model structure with fast training speed. **Limitations:** Relies on manual feature extraction, suitable for relatively simple datasets or limited computing resources.
	MLP-based	**Advantages:** Simple structure with fast computation, ideal for small-scale high-dimensional data. **Limitations:** Unable to capture temporal dependencies in sequential data.
	RNN-based	**Advantages:** Capable of handling long-term dependencies, and suitable for longer sequence prediction. **Limitations:** Prone to vanishing/exploding gradients, slower computation speed.
	Transformer-based	**Advantages:** High computational efficiency, parallel processing capability, captures long-range dependencies, suitable for long time series. **Limitations:** High model complexity, requires large training datasets. Demands substantial computing resources with a relatively slow training process

Classical mathematical models rely on well-defined formulas and parameters, employing specific algorithms that make them suitable for scenarios characterized by strong regularity or short-term predictions. These conventional models demonstrate higher efficiency with linear and regular data patterns, exhibit better applicability for small sample sizes, and incur low computational costs, rendering them appropriate for short-term, regular predictions or resource-constrained environments.

Artificial intelligence prediction models leverage neural networks and deep learning techniques to automatically extract data features and adapt to complex nonlinear characteristics, proving particularly effective in big data contexts. While AI models excel at handling nonlinear features and complex data patterns, they require substantial data volume and computational resources, making them more appropriate for large-scale datasets and predictions involving intricate dynamics. Despite these advances, the fundamental trade-off between model complexity and interpretability remains a inherent limitation of many state-of-the-art AI systems.

Table 2 systematically reviews both classical and recent representative studies in the field, compiling key performance metrics (e.g., RMSE, MAE) reported across different research works on official datasets. This approach allows for evidence-based extraction from existing literature, objectively illustrating the relative performance of various models and providing a well-founded basis for our comparative analysis.

Table 2. The accuracy of each model from classical and recent representative studies.

Type	Model	Institution	Year	Periodical	Model Accuracy
Classical Model	polynomial model	GFZ WUM	2014 2017	GPS Solutions	real-time prediction accuracy is superior to 0.55 ns. The standard deviation of the clock bias is 0.15 ns.
	Grey Model	NTSC	2018	Acta Astronomica Sinica	24-h prediction accuracy is 1.27 ns.
	Kalman Filter	XISM	2023	Acta Geodaetica et Cartographica Sinica	6-h prediction accuracy is 8~9 ns.
		Tampere	2019	ICL-GNSS	30-days prediction accuracy is 5.22~241.79 ns
Deep-learning Model	ARIMA	IEU	2021	GPS Solutions	30-days prediction accuracy is 18.16-86.28 ns
	MLPs		2020		1-h prediction accuracy is superior to 0.3 ns.
	RNNs	XSCC	2021		3-days prediction accuracy is 1.59 ns
		SYSU	2023		3~7 days prediction accuracy is 8.6~19.2 ns.
		SSC	2024	Satellite Navigation	10-days prediction accuracy is 0.316 ns
	Transformer	GMV UK	2023	ION GNSS	2-h prediction accuracy is below 2 ns

Table 3 provides a decision-making framework for the practical application of different models in real-world scenarios. Model selection is not merely a benchmarking exercise based on performance rankings, but rather a trade-off process constrained by practical considerations. While academic research often prioritizes State-of-the-Art (SOTA) results and tends to favor complex AI models such as Transformers, real-world industrial applications are typically governed by critical factors including robustness, interpretability, computational cost, and maintenance overhead. It is considered best practice to begin with a simple model to establish a strong baseline. More complex deep learning models should only be introduced when simpler approaches prove insufficient and the anticipated benefits clearly justify the additional complexity and cost.

Table 3. Decision-making reference framework for model application.

The Dimensions Considered	Preferred Model
The amount of data is very small ($n < 100$)	GM, QP
High computational efficiency/interpretability	ML-based, QP
The data is stable and the trend is obvious.	ARIMA
Dynamic system, multi-sensor fusion	Kalman Filter
Medium-sized sequence data	LSTM/GRU
Massive data, long sequence prediction	Transformer-based
Limited computing resources	ML-based, ARIMA
Requires rapid deployment and debugging	ML-based, ARIMA

The future of satellite clock bias prediction will depend on the integrated optimization of classical mathematical and artificial intelligence models, advancements in real-time data acquisition and processing technologies, and efficient utilization of computational resources. Through multifaceted innovation and refinement, the accuracy, robustness, and adaptability of clock bias prediction will be significantly enhanced, thereby better supporting the advancement of navigation and positioning systems. Considering the critical importance of satellite clock bias prediction and the limitations of existing research, several promising research directions and their associated key technologies are proposed:

1. Multi-source data fusion and processing

Satellite clock bias data come from diverse sources, including ground observation data, measured atomic clock data, inter-satellite link data, and satellite orbit parameters, with coupled relationships existing between clock bias data and orbit parameter determination. Future research could explore multi-source data fusion methods to improve the accuracy and stability of prediction models by integrating data from different sources.

2. Integration of short-term and long-term prediction

Current satellite clock bias prediction models mostly focus on either short-term or long-term prediction while neglecting their complementary nature. Short-term prediction models typically rely on recent data and can quickly respond to short-term variations but are susceptible to sudden events and noise; long-term prediction focuses on global trends and better captures stable long-term variation patterns but responds relatively slowly to dynamic changes. Future research should emphasize the combination of short-term and long-term prediction to achieve accurate full-cycle prediction of satellite clock bias.

3. Improving model generalization capability and robustness

Although existing models perform well in specific scenarios, their generalization capability and robustness across different satellites and environments still need improvement. Future research could focus on cross-satellite and cross-constellation adaptability studies to enhance prediction capability in complex environments. Further improvements could also be made to existing models by incorporating actual physical characteristics of on-board clocks, such as considering more complex periodic and random terms, and applying more advanced algorithms. Deep learning techniques could be further utilized to improve the capture of complex nonlinear clock bias variations, while optimizing neural network architectures and parameter tuning strategies to obtain higher-precision prediction results.

4. Emphasizing real-time performance and computational efficiency

As application scenarios of satellite navigation systems become increasingly complex, the real-time requirements for clock bias prediction are also growing. Future research could focus on improving computational efficiency by applying parallel and distributed computing technologies to enhance model training and prediction efficiency, particularly for processing long time series data. Hardware acceleration devices such as GPUs and TPUs could be utilized to improve the training speed and prediction efficiency of deep learning models.

5. Innovating deep learning model architectures

Future research could also explore how to integrate these new deep learning models with existing models to form a multi-model ensemble system. By combining the advantages of different models and optimizing ensemble strategies, the robustness and accuracy of satellite clock bias prediction could be further improved. Such innovative deep learning architectures are expected to break through current bottlenecks of prediction models and provide more reliable clock bias prediction support for future satellite navigation systems.

Author Contributions: Conceptualization, Y.L. and Z.M.; Methodology, Y.L.; Software, G.W.; Validation, Y.L., M.L. and E.Y.; Formal Analysis, Y.L.; Investigation, Y.L.; Resources, Z.M.; Data Curation, G.W.; Writing—Original Draft Preparation, Y.L.; Writing—Review & Editing, Z.M. and G.W.; Visualization, M.L.; Supervision, Z.M.; Project Administration, Z.M. All authors have read and agreed to the published version of the manuscript.

Funding: This research received no external funding.

Data Availability Statement: The experimental data in the manuscript are all public data and can be downloaded from https://cddis.nasa.gov/archive (accessed on 7 September 2025).

Conflicts of Interest: The authors declare no conflict of interest.

References

1. Hadas, T.; Kazmierski, K.; Kudłacik, I.; Marut, G.; Madraszek, S. Galileo High Accuracy Service in Real-Time PNT, Geoscience and Monitoring Applications. *IEEE Geosci. Remote Sens. Lett.* **2024**, *21*, 8000905. [CrossRef]
2. Sun, R.; Yang, Y.; Chiang, K.-W.; Duong, T.-T.; Lin, K.-Y.; Tsai, G.-J. Robust IMU/GPS/VO Integration for Vehicle Navigation in GNSS Degraded Urban Areas. *IEEE Sens. J.* **2020**, *20*, 10110–10122. [CrossRef]
3. Lutwak, R. Micro-Technology for Positioning, Navigation, and Timing towards PNT Everywhere and Always. In Proceedings of the 2014 International Symposium on Inertial Sensors and Systems (INERTIAL), Laguna Beach, CA, USA, 25–26 February 2014; pp. 1–4. [CrossRef]
4. Iannucci, P.A.; Humphreys, T.E. Economical Fused LEO GNSS. In Proceedings of the 2020 IEEE/ION Position, Location and Navigation Symposium (PLANS), Portland, OR, USA, 20–23 April 2020; pp. 426–443. [CrossRef]
5. Zhai, W.; Wu, J.; Sun, C. Real Time Precision Clock Difference Determination and Accuracy Evaluation Based on iGMAS/IGS Orbit. In Proceedings of the 9th China Satellite Navigation Conference (CSNC), Harbin, China, 23–25 May 2018; pp. 5, 31–35.
6. Breakiron, L.A.; Smith, A.L.; Fonville, B.C.; Powers, E.; Matsakis, D.N. The Accuracy of Two-Way Satellite Time Transfer Calibrations. In Proceedings of the 36th Annual Precise Time and Time Interval Systems and Applications Meeting, Washington, DC, USA, 7–9 December 2004.
7. Yang, Y.; Mao, Y.; Sun, B. Basic Performance and Future Developments of BeiDou Global Navigation Satellite System. *Satell. Navig.* **2020**, *1*, 1. [CrossRef]
8. Syam, W.P.; Priyadarshi, S.; Roqué, A.A.G.; Conesa, A.P.; Buscarlet, G.; Orso, M.D. Transformer Deep Learning for Accurate Orbit Corrections in Real-Time. In Proceedings of the 36th International Technical Meeting of the Satellite Division of The Institute of Navigation (ION GNSS+ 2023), Denver, CO, USA, 11–15 September 2023; pp. 159–174. [CrossRef]
9. Chen, Q.; Yi, T. Brief analysis on global four major navigation satellite systems. *J. Navig. Position.* **2020**, *8*, 115–120. [CrossRef]
10. Montenbruck, O.; Steigenberger, P.; Prange, L.; Deng, Z.; Zhao, Q.; Perosanz, F.; Romero, I.; Noll, C.; Stürze, A.; Weber, G.; et al. The Multi-GNSS Experiment (MGEX) of the International GNSS Service (IGS)–Achievements, Prospects and Challenges. *Adv. Space Res.* **2017**, *59*, 1671–1697. [CrossRef]
11. Nie, Z.; Gao, Y.; Wang, Z.; Ji, S.; Yang, H. An Approach to GPS Clock Prediction for Real-Time PPP during Outages of RTS Stream. *GPS Solut.* **2017**, *22*, 14. [CrossRef]
12. Guo, J.; Xu, X.; Zhao, Q.; Liu, J. Precise Orbit Determination for Quad-Constellation Satellites at Wuhan University: Strategy, Result Validation, and Comparison. *J. Geod.* **2016**, *90*, 143–159. [CrossRef]
13. Yin, Q.; Lou, Y.; Yi, W. Comparison and analysis of igs real-time products. *J. Geod. Geodyn.* **2012**, *32*, 1671–5942.
14. Guo, F.; Li, X.; Zhang, X.; Wang, J. Assessment of Precise Orbit and Clock Products for Galileo, BeiDou, and QZSS from IGS Multi-GNSS Experiment (MGEX). *GPS Solut.* **2017**, *21*, 279–290. [CrossRef]

15. Gao, Y.; Chen, G.; Fu, W.; Chen, X. A Real-Time Linear Prediction Algorithm for Detecting Abnormal BDS-2/BDS-3 Satellite Clock Offsets. *Remote Sens.* **2023**, *15*, 1831. [CrossRef]
16. Han, Z.; Zhao, J.; Leung, H.; Ma, K.F.; Wang, W. A Review of Deep Learning Models for Time Series Prediction. *IEEE Sens. J.* **2021**, *21*, 7833–7848. [CrossRef]
17. Hassan, T.; Hassan, A. Employing artificial intelligence in Galileo orbital error prediction for real-time offline positioning. *GPS Solut.* **2025**, *29*, 135. [CrossRef]
18. Mei, G.; Zhao, F.; Qi, F.; Zhong, D.; An, S. Characteristics of the space-borne rubidium atomic clocks for the BeiDou III navigation satellite system. *Sci. Sin.-Phys. Mech. Astron.* **2021**, *51*, 118–124. [CrossRef]
19. Zhang, H.; JU, B.; GU, D.; Liu, Y. Precise orbit determination for TH02-02 satellites based on BDS3 and GPS observations. *Chin. J. Aeronaut.* **2023**, *36*, 475–485. [CrossRef]
20. Jiao, Y.; Kou, Y. Analysis, modeling and simulation of GPS satellite clock errors. *Sci. Sin. Phys. Mech. Astron.* **2011**, *41*, 596. [CrossRef]
21. Liu, L.; Du, L.; Zhu, L.; Han, C. Satellite clock parameter short-term prediction using piece-wise adaptive filter with state noise compensation. In *China Satellite Navigation Conference (CSNC) 2012 Proceedings, Proceedings of the CSNC2012, Guanzhou, China, 15–19 May 2012*; Lecture Notes in Electrical Engineering; Springer: Berlin/Heidelberg, Germany, 2012; Volume 160, pp. 527–538. [CrossRef]
22. Xie, X. Precise Orbit and Clock Determination for BDS-3 Satellites Using Inter-Satellite Link Observations. Ph.D. Dissertation, Wuhan University, Wuhan, China, 2020. [CrossRef]
23. Ding, S.; Yang, H.; Yang, X.; Zhang, Z. Influence of the second-order ionospheric delay on GPS PPP time transfer. *Xi'an Dianzi Keji Daxue Xuebao/J. Xidian Univ.* **2021**, *48*, 50–56, 82. [CrossRef]
24. Han, S.C.; Kwon, J.H.; Jekeli, C. Accurate Absolute GPS Positioning through Satellite Clock Error Estimation. *J. Geod.* **2001**, *75*, 33–43. [CrossRef]
25. Jiang, M.; Dong, S.; Wu, W. Research on Time Scale Algorithm Based on Hydrogen Masers. *IEEE Instrum. Meas. Mag.* **2020**, *23*, 35–40. [CrossRef]
26. Panfilo, G.; Tavella, P. Atomic Clock Prediction Based on Stochastic Differential Equations. *Metrologia* **2008**, *45*, S108–S116. [CrossRef]
27. Allan, D.W.; Levine, J. A Historical Perspective on the Development of the Allan Variances and Their Strengths and Weaknesses. *IEEE Trans. Ultrason. Ferroelect. Freq. Contr.* **2016**, *63*, 513–519. [CrossRef]
28. Lei, Y.; Cai, H.; Zhao, D. A Novel Method for Satellite Clock Bias Prediction Based on Phase Space Reconstruction and Gaussian Processes. *Acta Metrol. Sin.* **2016**, *37*, 318–322. [CrossRef]
29. Wang, Y.; Lv, Z.; Wang, N.; Li, L. Prediction of Navigation Satellite Clock Bias Considering Clock's Stochastic Variation Behavior with Robust Least Square Collocation. *Acta Geod. Cartogr. Sin.* **2016**, *45*, 646–655. [CrossRef]
30. Liao, J.; Zhang, Y. Satellite Clock Error Prediction of Improved Polynomial and Periodic Model. *GNSS World China* **2018**, *43*, 91–95. [CrossRef]
31. Liang, Y.; Ren, C.; Yang, X.; Pang, G.; Lan, L. A Grey Model Based on First Differences in the Application of Satellite Clock Bias Prediction. *Chin. Astron. Astrophys.* **2016**, *40*, 79–93. [CrossRef]
32. Mei, C.; Huang, H.; Jiang, K.; Xia, L.; Pan, X. Application of Discrete Grey Model Based on Stepwise Ratio Sequence in the Satellite Clock Offset Prediction. *Geomat. Inf. Sci. Wuhan Univ.* **2021**, *46*, 1154–1160. [CrossRef]
33. Jiang, Y. Application research of GM(1,1) model of power function transform in BDS satellite clock bias prediction. *GNSS World China* **2020**, *45*, 49–54. [CrossRef]
34. Yu, Y.; Zhang, H.; Li, X.; Xiao, B. Prediction Method of Satellite Clock Bias Based on Grey Model of First-order Difference of Vondrfik Filter. *Acta Astron. Sin.* **2018**, *59*, 75–84. [CrossRef]
35. Zheng, Z.; Lu, X.; Chen, Y. Improved Grey Model and Application in Real-Time GPS Satellite Clock Bias Prediction. In Proceedings of the 2008 Fourth International Conference on Natural Computation, Jinan, China, 18–20 October 2008; Volume 2, pp. 419–423. [CrossRef]
36. Yu, Y.; Huang, M.; Duan, T.; Wang, C.; Hu, R. Enhancing Satellite Clock Bias Prediction Accuracy in the Case of Jumps with an Improved Grey Model. *Math. Probl. Eng.* **2020**, *2020*, 81–86. [CrossRef]
37. Yu, Y.; Huang, M.; Duan, T.; Wang, C.; Hu, R. Satellite clock bias prediction based on particle swarm optimization and weighted grey regression combined model. *J. Harbin Inst. Technol.* **2020**, *52*, 144–151, 182. [CrossRef]
38. Li, C.; Chen, X.; Liu, J.; Wu, W.; Liu, Z. Predicting Satellite Clock Errors Using Grey Model Optimizedby Adaptive TS-IPSO. *Geomat. Inf. Sci. Wuhan Univ.* **2018**, *43*, 854–859. [CrossRef]
39. Yang, C.; Liu, Z.; Xu, X. Application of L-M Algorithm Optimized Gray Model in GPS Satellite Clock Error Predication. *J. Henan Polytech. Univ.* **2020**, *39*, 47–52. [CrossRef]
40. Yuan, D.; Zhang, J.; Zhang, Z.; Wei, S. BDS Clock Error Prediction Based on SAFA-FDGM(1,1) Model. *J. Geod. Geodyn.* **2021**, *41*, 672–675. [CrossRef]

41. Zhao, X.; Huang, Z.; Chen, M.; Yang, W. Application and Accuracy Analysis of Dynamic Grey Model in Satellite Clock Error Prediction. *Explor. Sci. Technol.* **2016**, *1*, 23–25, 36. [CrossRef]
42. Tan, X.; Xu, J.; Li, F.; Wu, M.; Chen, D.; Liang, Y. Improved GM (1,1) Model by Optimizing Initial Condition to Predict Satellite Clock Bias. *Math. Probl. Eng.* **2022**, *2022*, 3895884. [CrossRef]
43. Davis, J.; Greenhall, C.; Stacey, P. A Kalman filter clock algorithm for use in the presence of flicker frequency modulation noise. *Metrologia* **2005**, *42*, 1–10. [CrossRef]
44. Greenhall, C. Forming stable timescales from the Jones-Tryon Kalman filter. *Metrologia* **2003**, *40*, S335–S341. [CrossRef]
45. Guo, H. Study on the Analysis Theories and Algorithms of the Time and Frequency Characterization for Atomic Clocks of Navi-gation Satellites. Ph.D. Dissertation, PLA Information Engineering University, Zhengzhou, China, 2006.
46. Guo, H.; Yang, Y.; He, H.; Xu, T. Determination of Covariance Matrix of Kalman Filter Used for Time Prediction of Atomic Clocks of Navigation Satellites. *Acta Geod. Cartogr. Sin.* **2010**, *39*, 146–150.
47. Howe, D.; Beard, R.; Greenhall, C.; Vernotte, F.; Riley, B. A Total Estimator of the Hadamard Function Used for GPS Operations. In Proceedings of the 32nd Annual Precise Time and Time Interval (PTTI) Meeting, Reston, VA, USA, 28–30 November 2000.
48. Hutseil, T. Relating the hadamard variance to MCS kalman filter clock estimation. In Proceedings of the 27th Annual Precise Time and Time Interval Systems and Applications Meeting, San Diego, CA, USA, 1–29 November 1995.
49. Davis, J.; Bhattarai, S.; Ziebart, M. Development of a Kalman Filter Based GPS Satellite Clock Time-Offset Prediction Algorithm. In Proceedings of the 2012 European Frequency and Time Forum, Gothenburg, Sweden, 23–27 April 2012; pp. 152–156. [CrossRef]
50. Pihlajasalo, J.; Leppäkoski, H.; Kuismanen, S.; Ali-Löytty, S.; Piché, R. Methods for Long-Term GNSS Clock Offset Prediction. In Proceedings of the 2019 International Conference on Localization and GNSS (ICL-GNSS), Nuremberg, Germany, 4–6 June 2019; pp. 1–6. [CrossRef]
51. Pratt, J.; Axelrad, P.; Larson, K.M.; Lesage, B.; Gerren, R.; DiOrio, N. Satellite Clock Bias Estimation for iGPS. *GPS Solut.* **2013**, *17*, 381–389. [CrossRef]
52. Zhang, J.; Tang, L. Improved Kalman filter for the rubidium atomic clock error prediction. *J. Xi'an Univ. Posts Telecommun.* **2019**, *24*, 1–5. [CrossRef]
53. Wang, J.; Hu, Y.; He, Z.; Yang, H. Clock Bias Prediction Based on Linear Weighted Combination Kalman Filter. *Acta Astron. Sin.* **2012**, *53*, 213–221. [CrossRef]
54. Qafisheh, M.; Martín, A.; Capilla, R.M.; Anquela, A.B. SVR and ARIMA Models as Machine Learning Solutions for Solving the Latency Problem in Real-Time Clock Corrections. *GPS Solut.* **2022**, *26*, 85. [CrossRef]
55. Zhou, W.; Huang, C.; Song, S.; Chen, Q.; Liu, Z. Characteristic Analysis and Short-Term Prediction of GPS/BDS Satellite Clock Correction. In *China Satellite Navigation Conference (CSNC) 2016 Proceedings: Volume III, Proceedings of the CSNC2016, Changsha, China, 18–20 May 2016*; Springer Nature: Berlin/Heidelberg, Germany, 2016; Volume 390, pp. 187–200. [CrossRef]
56. Jiang, S.; Li, B. Application of ARIMA Model in Short-Term Satellite Clock Error Prediction. *J. Navig. Position.* **2019**, *7*, 118–124. [CrossRef]
57. Yan, K.; Chang, G.; Zhou, T.; Zhao, Z. Sparse auto-regressive modeling for satellite clock offsets prediction. *Hydrogr. Surv. Charting* **2021**, *41*, 28–32, 37. [CrossRef]
58. González, S.G. Advanced ARIMA Model for Clock Bias Prediction. Master's Thesis, Technical University of Madrid, Madrid, Spain, 2018.
59. Kim, M.; Kim, J. Predicting IGS RTS Corrections Using ARMA Neural Networks. *Math. Probl. Eng.* **2015**, *2015*, 851761. [CrossRef]
60. Han, S.; Zhang, G.; Zhang, N.; Zhu, J. New algorithm for detecting AO outliers in AR model and its application in the prediction of GPS satellite clock errors. *Acta Geod. Cartogr. Sin.* **2019**, *48*, 1225–1235. [CrossRef]
61. Ma, Z.; Li, G.; Zhang, Q.; Li, X. Likelihood Ratio Method for Outlier Detection and Estimation in BeiDou Satellite Clock Offset. *J. Geomat. Sci. Technol.* **2019**, *36*, 221–226. [CrossRef]
62. Zhang, G.; Han, S.; Ye, J.; Hao, R.; Zhang, J.; Li, X.; Jia, K. A Method for Precisely Predicting Satellite Clock Bias Based on Robust Fitting of ARMA Models. *GPS Solut.* **2022**, *26*, 3. [CrossRef]
63. Xiao, Y.; Tang, S. Short-term prediction of satellite clock bias based on PSO-SVM refined QP model. *J. Guilin Univ. Technol.* **2019**, *39*, 893–898. [CrossRef]
64. He, L.; Zhou, H.; Liu, Z.; Wen, Y.; He, X. Improving Clock Prediction Algorithm for BDS-2/3 Satellites Based on LS-SVM Method. *Remote Sens.* **2019**, *11*, 2554. [CrossRef]
65. Zhu, J.; Wang, X.; Gao, Y.; Zhang, J.; Wang, S. Hydrogen Atomic Clock Difference Prediction Based on the LSSVM. *J. Eng.* **2019**, *2019*, 9017–9021. [CrossRef]
66. He, L.; Zhou, H.; Zhu, S.; Zeng, P. An Improved QZSS Satellite Clock Offsets Prediction Based on the Extreme Learning Machine Method. *IEEE Access* **2020**, *8*, 156557–156568. [CrossRef]
67. Lei, Y.; Guo, J. A network structure design method for ELM and its application to prediction of satellite clock bias. *J. Time Freq.* **2015**, *38*, 209–215. [CrossRef]

68. Ya, S.; Zhao, X.; Liu, C.; Chen, J.; Liu, C. Improved BDS-2/3 Satellite Ultra-Fast Clock Bias Prediction Based with the SSA-ELM Model. *Sensors* **2023**, *23*, 2453. [CrossRef]
69. Sun, W.; Zou, Y. Short Term Load Forecasting Based on BP Neural Network Trained by PSO. In Proceedings of the 2007 International Conference on Machine Learning and Cybernetics, Hong Kong, China, 19–22 August 2007; Volume 5, pp. 2863–2868. [CrossRef]
70. Du, L.; Chen, H.; Yuan, Y.; Song, L.; Meng, X. Global Navigation Satellite System Receiver Positioning in Harsh Environments via Clock Bias Prediction by Empirical Mode Decomposition and Back Propagation Neural Network Method. *Sensors* **2024**, *24*, 2342. [CrossRef]
71. Sun, P.; Wei, D.; Sun, B. Genetic Algorithm Optimization in the Prediction of Satellite Clock Bias by BP Neural Network. *Acta Astron. Sin.* **2020**, *61*, 92–104. [CrossRef]
72. Lv, D.; Ou, J.; Yu, S. Prediction of the satellite clock bias based on MEA-BP neural network. *Acta Geod. Cartogr. Sin.* **2020**, *49*, 993–1003. [CrossRef]
73. Bai, H.; Cao, Q.; An, S. Mind Evolutionary Algorithm Optimization in the Prediction of Satellite Clock Bias Using the Back Propagation Neural Network. *Sci. Rep.* **2023**, *13*, 2095. [CrossRef] [PubMed]
74. Meng, C.; Wu, D.; Lei, Y. Neural Network Satellite Clock Bias Prediction Based on the Whale Optimization Algorithm. *Adv. Nat. Comput. Fuzzy Syst. Knowl. Discov.* **2022**, *89*, 1152–1160. [CrossRef]
75. Ya, S.; Zhao, X.; Liu, C.; Chen, J.; Liu, C.; Hu, H. Enhancing Short-Term Prediction of BDS-3 Satellite Clock Bias Based with BSO Optimized BP Neural Network. *Int. J. Aerosp. Eng.* **2022**, *2022*, 1–18. [CrossRef]
76. Lv, D.; Liu, G.; Ou, J.; Wang, S.; Gao, M. Prediction of GPS Satellite Clock Offset Based on an Improved Particle Swarm Algorithm Optimized BP Neural Network. *Remote Sens.* **2022**, *14*, 2407. [CrossRef]
77. Wang, Y.; Lu, Z.; Qu, Y.; Li, L.; Wang, N. Improving Prediction Performance of GPS Satellite Clock Bias Based on Wavelet Neural Network. *GPS Solut.* **2017**, *21*, 523–534. [CrossRef]
78. Wang, X.; Chai, H.; Wang, C. A High-Precision Short-Term Prediction Method with Stable Performance for Satellite Clock Bias. *GPS Solut.* **2020**, *24*, 105. [CrossRef]
79. Ai, Q.; Xu, T.; Li, J.; Xiong, H. The Short-Term Forecast of BeiDou Satellite Clock Bias Based on Wavelet Neural Network. In *China Satellite Navigation Conference (CSNC) 2016 Proceedings: Volume I, Proceedings of the CSNC2016, Changsha, China, 18–20 May 2016*; Sun, J., Liu, J., Fan, S., Wang, F., Eds.; Lecture Notes in Electrical Engineering; Springer: Singapore, 2016; Volume 388, pp. 145–154. [CrossRef]
80. Zhang, J. Research of Satellites Clock Error Prediction Based on Neural Networks. *Comput. Eng. Des.* **2014**, *35*, 3254–3257.
81. Wang, R.; Cai, H.; Pan, Z. Satellite Clock Bias Prediction Algorithm with Multi System Based on RBF Neural Network. *J. Ocean Technol.* **2019**, *38*, 56–61. [CrossRef]
82. Gnyś, P.; Przestrzelski, P. Application of long short term memory neural networks for GPS satellite clock bias prediction. *TASK Q.* **2021**, *25*, 381–395. [CrossRef]
83. Huang, B.; Ji, Z.; Zhai, R.; Xiao, C.; Yang, F.; Yang, B.; Wang, Y. Clock Bias Prediction Algorithm for Navigation Satellites Based on a Supervised Learning Long Short-Term Memory Neural Network. *GPS Solut.* **2021**, *25*, 80. [CrossRef]
84. He, S.; Liu, J.; Zhu, X.; Dai, Z.; Li, D. Research on Modeling and Predicting of BDS-3 Satellite Clock Bias Using the LSTM Neural Network Model. *GPS Solut.* **2023**, *27*, 108. [CrossRef]
85. Cai, C.; Liu, M.; Li, P.; Li, Z.; Lv, K. Enhancing Satellite Clock Bias Prediction in BDS with LSTM-Attention Model. *GPS Solut.* **2024**, *28*, 92. [CrossRef]
86. Yang, S.; Yi, X.; Dong, R.; Wu, Y.; Shuai, T.; Zhang, J.; Ren, Q.; Gong, W. Long-Term Autonomous Time-Keeping of Navigation Constellations Based on Sparse Sampling LSTM Algorithm. *Satell. Navig.* **2024**, *5*, 15. [CrossRef]
87. Liu, H.; Liu, F.; Kong, Y.; Yang, C. Improved SSA-Based GRU Neural Network for BDS-3 Satellite Clock Bias Forecasting. *Sensors* **2024**, *24*, 1178. [CrossRef]
88. Liang, Y.F.; Xu, J.N.; Li, F.N.; Jiang, P.F. Nonlinear Autoregressive Model With Exogenous Input Recurrent Neural Network to Predict Satellites' Clock Bias. *IEEE Access* **2021**, *9*, 24416–24424. [CrossRef]
89. Liang, Y.; Xu, J.; Wu, M. Elman Neural Network Based on Particle Swarm Optimization for Prediction of GPS Rapid Clock Bias. In *China Satellite Navigation Conference (CSNC 2022) Proceedings, Proceedings of the CSNC 2022, Beijing, China, 25–27 May 2022*; Springer: Berlin/Heidelberg, Germany, 2022; Volume 910, pp. 361–371. [CrossRef]
90. Vaswani, A.; Shazeer, N.; Parmar, N.; Uszkoreit, J.; Jones, L.; Gomez, A.N.; Kaiser, Ł.; Polosukhin, I. Attention Is All You Need. In Proceedings of the 31st International Conference on Neural Information Processing Systems, Long Beach, CA, USA, 4–9 December 2017.
91. Pan, X.; Zhao, W.; Huang, W.; Zhang, S.; Jin, L. Short-term prediction of satellite clock bias based on improved self-attention model. *J. Chin. Inert. Technol.* **2023**, *31*, 1092–1101. [CrossRef]

MDPI AG
Grosspeteranlage 5
4052 Basel
Switzerland
Tel.: +41 61 683 77 34

Remote Sensing Editorial Office
E-mail: remotesensing@mdpi.com
www.mdpi.com/journal/remotesensing

www.ingramcontent.com/pod-product-compliance
Lightning Source LLC
Chambersburg PA
CBHW061443210926
51850CB00008B/1599
* 9 7 8 3 7 2 5 8 6 3 9 0 7 *